Advances in All-optical Communication

Online at: https://doi.org/10.1088/978-0-7503-5623-7

Advances in All-optical Communication

Edited by

Shanmuga Sundar Dhanabalan

School of Computing, Engineering and Mathematical Sciences, La Trobe University, Melbourne, Victoria 3086, Australia

Arun Thirumurugan

Sede Vallenar, Universidad de ATACAMA, Costanera #105, Vallenar 1612178, Chile

Sridarshini Thirumaran

Department of Electronics and Communication Engineering, College of Engineering Guindy, Anna University, Chennai, Tamil nadu, India

IOP Publishing, Bristol, UK

ISBN   978-0-7503-5623-7 (ebook)
ISBN   978-0-7503-5621-3 (print)
ISBN   978-0-7503-5624-4 (myPrint)
ISBN   978-0-7503-5622-0 (mobi)

DOI   10.1088/978-0-7503-5623-7

Version: 20241101

IOP ebooks

British Library Cataloguing-in-Publication Data: A catalogue record for this book is available from the British Library.

Published by IOP Publishing, wholly owned by The Institute of Physics, London

IOP Publishing, No.2 The Distillery, Glassfields, Avon Street, Bristol, BS2 0GR, UK

US Office: IOP Publishing, Inc., 190 North Independence Mall West, Suite 601, Philadelphia, PA 19106, USA

This book is dedicated to the pioneers of innovation, the visionaries of tomorrow, and the champions of flexibility in thought and deed. It is a tribute to those who tirelessly explore the frontiers of materials, fabrication, and applications in the world of optics and photonics.

Contents

Preface   xiii

Acknowledgments   xv

Editor biographies   xvi

List of contributors   xviii

**1   Highly efficient materials for photonic crystal-based optical   1-1
       components**
M Soroosh, F Parandin, F Haddadan, M J Maleki and F Bagheri

1.1   Introduction   1-1
      1.1.1 Light as a carrier for data transmission   1-1
      1.1.2 Optical processing   1-2
      1.1.3 Optical communication   1-3
      1.1.4 Photonic crystal structure   1-4
1.2   Photonic crystal and light propagation   1-5
1.3   Photonic crystal-based devices   1-6
      1.3.1 Demultiplexer/multiplexer   1-8
      1.3.2 Flip-flop   1-12
      1.3.3 Comparator   1-13
1.4   Materials used in photonic crystal structures   1-18
      References   1-19

**2   Unidirectional bulk growth of 1,3,5-triphenylbenzene single   2-1
       crystal and doping effect on its optical properties**
Narayanan Durairaj, Sakthivel Raja, Sridharan Moorthi Babu,
Sivaperuman Kalainathan and Jayaraman Jayapriya

2.1   Introduction   2-1
2.2   Experiment   2-2
      2.2.1 Conventional crystal growth   2-2
      2.2.2 Bulk growth of stilbene-doped 3PB   2-4
      2.2.3 Characterization techniques   2-6
2.3   Result and discussion   2-7
      2.3.1 X-ray diffraction analysis   2-7
      2.3.2 UV–visible NIR spectra analysis   2-8
      2.3.3 Fluorescence emission spectrum analysis   2-8
      2.3.4 Lifetime measurement   2-12

	2.3.5 FTIR spectral studies	2-13
	2.3.6 NMR spectra analysis	2-13
2.4	Conclusions	2-16
	References	2-16

3 Performance analysis of SOA-based all-optical logic gates over FSO channel   3-1

Margarat Michael, Elizabeth Caroline Britto, J Jeyarani and K Bhuvaneshwari

3.1	Introduction	3-1
3.2	Applications, advantages, challenges, and models of FSO	3-3
3.3	Related works	3-5
3.4	Basics of semiconductor optical amplifiers	3-6
	3.4.1 Types of SOA	3-8
	3.4.2 Principal operation of SOA and design of XOR gate	3-12
3.5	Design of OR gate using RSOA	3-14
3.6	All-optical device design using SOA and RSOA	3-15
3.7	Simulation setup of encoded inputs over wired and wireless optical channel	3-18
3.8	Conclusion	3-21
	References	3-22

4 Switching characteristics of optical solitons through inelastic interactions   4-1

M S Mani Rajan and P Nithya

4.1	Introduction	4-1
	4.1.1 Telecommunication windows	4-2
	4.1.2 Optical soliton formation in an optical fiber	4-3
	4.1.3 Nonlinearity	4-4
	4.1.4 Group velocity dispersion	4-6
	4.1.5 Attenuation	4-8
4.2	Governing theoretical model	4-9
4.3	Lax pair for the system (4.10)	4-10
4.4	Two soliton solutions through Darboux method	4-12
4.5	Discussion on switching characteristics of femtosecond solitons	4-13
4.6	Conclusions	4-15
	References	4-16

5	**Silicon photonic modulators for high-speed applications—a review**	**5-1**

R G Jesuwanth Sugesh, V R Balaji, A Sivasubramanian, Gopalkrishna Hedge, M A Ibrar Jahan and Richards Joe Stanislaus

5.1	Introduction	5-1
5.2	Phase shifters	5-3
	5.2.1 Silicon-based phase shifter	5-3
	5.2.2 Hybrid phase shifters	5-7
5.3	Mach–Zehnder modulator (MZM)	5-8
5.4	Ring modulator	5-10
5.5	Modulator performance metrics	5-12
5.6	Data centre requirements	5-13
5.7	Conclusion	5-15
	References	5-15

6	**MIMO-FSO system for various weather conditions**	**6-1**

C Palaniappan and S Robinson

6.1	Introduction to free-space optical communication	6-1
	6.1.1 Free-space optical communication	6-3
6.2	FSO communication principles	6-3
	6.2.1 FSO transceiver	6-4
	6.2.2 FSO classifications	6-6
	6.2.3 FSO applications	6-7
6.3	Performance evaluation of the FSO system	6-8
	6.3.1 Link budget	6-8
	6.3.2 FSO received optical power	6-10
	6.3.3 Data rate	6-10
	6.3.4 Signal-to-noise ratio and BER	6-11
	6.3.5 Channel models	6-12
6.4	Introduction to MIMO system	6-13
	6.4.1 From SISO to MIMO to mMIMO	6-13
	6.4.2 mMIMO	6-17
	6.4.3 Benefits of mMIMO	6-18
	6.4.4 mMIMO FSO system	6-19
6.5	Result analysis	6-20
6.6	Conclusions	6-23
	References	6-23

7 AI in optics and photonics   **7-1**
Vineeth Palliyembil and E G Anagha

7.1 Introduction   7-2
     7.1.1 Introduction to optics and photonics   7-2
     7.1.2 Introduction to AI   7-3
7.2 Intersection of AI/ML in optics   7-4
     7.2.1 Major applications   7-4
     7.2.2 Major challenges   7-10
7.3 Intersection of AI in photonics   7-10
     7.3.1 Major applications   7-11
     7.3.2 Major challenges   7-18
7.4 Conclusion and future scope of AI/ML in optics and photonics   7-19
     References   7-20

8 Blood components detection in octagonal-cored photonic   **8-1**
crystal fiber biosensor for healthcare applications
Abdul Mu'iz Maidi, Nianyu Zou and Feroza Begum

8.1 Introduction   8-1
8.2 Design   8-3
8.3 Methodology   8-5
8.4 Results and discussion   8-6
8.5 Conclusion   8-14
     References   8-14

9 Photonic biosensors for healthcare applications   **9-1**
Seemesh Bhaskar, Kalathur Mohan Ganesh and Sai Sathish Ramamurthy

9.1 Introduction   9-2
9.2 Biosensors: plasmonic and photonic platforms   9-4
9.3 Overview of biosensing technologies and plasmonics for healthcare   9-5
     applications
9.4 Overview of nanophotonics and plasmonics for healthcare applications: 9-7
     introducing SPCE and PCCE technology
9.5 CryoSoret nano-engineering techniques for SPCE-based biosensing   9-10
     applications
9.6 Surface plasmon-coupled emission (SPCE): applications in early   9-14
     diagnostics
     9.6.1 Utility of chromaticity plot for tyrosine and spermidine sensing   9-14
     9.6.2 Utility of luminosity plot for the mercury ion sensing   9-16

9.7   Ferroplasmon-on-Mirror (FPoM): applications in early diagnostics   9-17

9.8   Photonic crystal-coupled emission: applications in early diagnostics   9-20

9.9   Futuristic scope and opportunities   9-23

9.10 Conclusions   9-25

9.11 Exercises   9-26

     References   9-26

10   Integrated photonic devices for cancer detection   **10-1**

M A Ibrar Jahan, V R Balaji, N R G Sreevani, D Sasikala,
R G Jesuwanth Sugesh, C Jenila and Gopalkrishna Hegde

10.1 Introduction   10-1

10.2 Surface plasmon resonance (SPR)-based biosensors   10-3

    10.2.1 Principle of SPR   10-3

10.3 Grating-based biosensors   10-6

    10.3.1 Bragg principle   10-7

10.4 2D photonic crystal-based biosensors   10-9

    10.4.1 Basics of 2D PhC   10-9

    10.4.2 Types of PhCs   10-10

    10.4.3 Numerical methods   10-11

    10.4.4 Performance parameters of sensor   10-12

    10.4.5 Schematic representation of 2D PhC-based biosensing   10-12

    10.4.6 Inference of advance technologies   10-14

10.5 Conclusion   10-15

    References   10-15

11   Absorbers as biosensors: leveraging absorption phenomena for   **11-1**
enhanced biosensing

Madurakavi Karthikeyan, J Pradeep, M Harikrishnan, R Sitharthan,
M Rajesh, Avinash Chandra and Rajkishor Kumar

11.1 Introduction   11-2

11.2 Factors influencing sensing performance   11-5

    11.2.1 Metamaterial design's composition and geometry   11-5

    11.2.2 Frequency range   11-5

    11.2.3 Electromagnetic properties   11-5

    11.2.4 Sensitivity and selectivity   11-6

    11.2.5 Integration with other materials/devices   11-6

    11.2.6 Fabrication techniques   11-6

    11.2.7 Environmental factors   11-6

11.2.8 Signal processing   11-6

11.2.9 Application-specific considerations   11-6

11.2.10 Power usage   11-6

11.3 Materials employed for designing biosensor absorbers   11-6

    11.3.1 Importance of metal layer and various metals used in metamaterial-based biosensor   11-7

11.4 Popular designs of absorbers for biosensing   11-8

11.5 Fabrication techniques   11-10

    11.5.1 Substrate preparation   11-11

    11.5.2 Catalyst deposition   11-11

    11.5.3 Graphene growth   11-11

    11.5.4 Transfer process   11-11

    11.5.5 Evaluation and optimization   11-12

11.6 Conclusion   11-13

    References   11-14

Preface

In an era marked by rapid technological advancements, all-optical communication has become a cornerstone of modern telecommunication systems, offering unmatched speed and efficiency. The ongoing development of optical materials, devices, and systems has not only transformed data transmission but also opened doors to innovative applications in healthcare, environmental monitoring, and artificial intelligence.

This book, *Advances in All-Optical Communication*, is a comprehensive collection of the latest research and developments in the field. It features contributions from leading experts, presenting cutting-edge advancements and exploring their practical applications. The chapters cover a broad spectrum of topics, reflecting the multi-disciplinary nature of all-optical communication technologies.

Chapter 1 discusses highly efficient materials for photonic crystal-based optical components, providing insights into the foundational elements driving modern optical systems' performance. Chapter 2 examines the unidirectional bulk growth of 1,3,5-triphenylbenzene single crystal and the impact of doping on its optical properties, highlighting material science advancements.

Chapter 3 offers a performance analysis of semiconductor optical amplifier-based all-optical XOR gates over free-space optical (FSO) channels, showcasing the potential of optical logic gates in free-space optical communication. Chapter 4 investigates the switching characteristics of optical solitons through inelastic interactions, illustrating the intricate dynamics of soliton-based systems.

Chapter 5 reviews silicon photonic modulators for high-speed applications, emphasizing the importance of silicon photonics in modern communication networks. Chapter 6 evaluates the performance of optimal massive multiple-input and multiple-output (MIMO) FSO systems under various weather conditions, addressing the challenges and solutions in FSO communication.

Chapter 7 delves into the applications and challenges of artificial intelligence in optics and photonics, highlighting AI's transformative impact on the field. Chapter 8 focuses on detecting blood components using octagonal-cored photonic crystal fiber biosensors, showcasing the intersection of photonics and healthcare.

Chapter 9 further explores photonic biosensors for healthcare applications, emphasizing their potential in medical diagnostics and monitoring. Chapter 10 discusses integrated photonic devices for cancer detection, presenting innovative approaches to cancer diagnostics. Finally, chapter 11 examines absorbers as biosensors, utilizing absorption phenomena for enhanced biosensing.

This book aims to be a valuable resource for researchers, engineers, and students in optical communication. By presenting a diverse range of topics, it provides a comprehensive overview of current trends and future directions. We hope the insights and knowledge shared in these chapters will inspire further advancements and innovations in all-optical communication.

We extend our heartfelt gratitude to all the contributors for their exceptional work and dedication. Their expertise and vision have made this compilation possible. We also thank the readers for their interest in this book and trust it will be a significant addition to their professional libraries.

Acknowledgments

We would like to express our sincere gratitude to all the contributors whose expertise and dedication have made this book, *Advances in All-Optical Communication*, possible. Their valuable insights and research have greatly enriched the content of this compilation. We extend our appreciation to the Institute of Physics for their support and guidance throughout the publication process. Their commitment to advancing scientific knowledge and education has been instrumental in bringing this project to fruition. Our heartfelt thanks go to our reviewers for their meticulous evaluation and constructive feedback, which have significantly enhanced the quality of this book.

Shanmuga Sundar Dhanabalan would like to express his sincere thanks to Professor Sharath Sriram, Professor Madhu Bhaskaran (Functional Materials and Microsystem Research Group, RMIT University, Australia), Professor Wenny Rahayu, Professor Prakash Veeraraghavan, Professor Aniruddha Desai (School of Computing, Engineering and Mathematical Sciences, La Trobe University, Australia), and Professor Sivanantha Raja Avaninathan (Alagappa Chettiar Government College of Engineering and Technology, Karaikudi, Tamil Nadu, India) for their continuous support, guidance, and encouragement. He would also like to extend his gratitude to Mrs Preethi Chidambaram and his family for support. Additionally, he acknowledges the personnel and project funding from the Australian Research Council Industry Fellowship Scheme.

Arun Thirumurugan extends his heartfelt gratitude to Dr Justin Joseyphus (National Institute of Technology, Tiruchirappalli, India) for his invaluable guidance and mentorship and to Professor P V Satyam (Institute of Physics, Bhubaneswar, India) for his steadfast support and encouragement. He is also thankful to Dr Ali Akbari-Fakhrabadi (Facultad de Ciencias Físicas y Matemáticas, University of Chile, Chile) for his guidance and collaboration and to Professor R V Mangala Raja (University of Adolfo Ibanez, Santiago, Chile) for his mentorship and support. Special thanks are due to Dr R Udaya Bhaskar and Mauricio J Morel (University of Atacama, Chile) for their support and collaboration, as well as to Carolina Venegas and Juan Campos (Sede Vallenar, Universidad de Atacama, Chile) for their assistance and cooperation. Additionally, he acknowledges the financial support provided by the Agencia Nacional de Investigación y Desarrollo de Chile (ANID) through the SA 77210070 project and the continuous support of the Universidad of Atacama.

Sridarshini Thirumaran would like to express her heartfelt gratitude to all the editors, the authors of each technical chapter, different teams of the publication unit, and the Magnificent Universe to make this book a successful one. Lastly, we are deeply grateful to our families and colleagues for their unwavering support and encouragement during the development of this book. Their patience and under-standing have been invaluable.

Editor biographies

Shanmuga Sundar Dhanabalan

Dr Shanmuga Sundar Dhanabalan is working as a senior lecturer in the School of Computing, Engineering and Mathematical Sciences, La Trobe University, Melbourne, Australia. He is an accomplished researcher with a proven track record in designing, developing, and translating micro- and nano-scale devices. His primary objective is to create next-generation smart products that enhance quality of life and well-being, making a significant contribution to society. He has 10 years of research and 6 years of teaching experience in diverse countries such as Australia, Chile, and India. Dr Dhanabalan brings a wealth of knowledge in soft electronics, wearable sensors, optics, and photonics. He has been listed among the top 2% of scientists globally, according to the World's Top 2% of Scientists list released by Stanford University, USA, and Elsevier.

He has been awarded a prestigious and competitive ARC Industry Fellowship 2024 and listed in the top 5% of successful applications. He also secured a competitive fellowship grant from the Chilean government as a Principal Investigator. He is the chief investigator of two CRC-P projects. In the last three years, he has generated more than $4.1 million through industry and university collaborations, securing funding from Cooperative Research Centres Projects (CRC-P) and the ARC. Throughout his career, he has demonstrated strong research leadership as Principal/Chief Investigator on multiple funded projects.

Dr Dhanabalan's leadership has driven multidisciplinary teams to achieve significant milestones, resulting in groundbreaking publications, products, and patents. In the past five years, Dr Dhanabalan has 57 peer-reviewed publications (career total of 77), 15 presentations (11 invited and 4 keynote presentations), with a Field Weighted Citation Index (FWCI) of 1.90. He holds three patents (1 India, 2 Australia). He serves as a topical editor for IEEE, Elsevier, and Springer journals. He is a member of professional societies such as IEEE (Senior Member), and Royal Society of Victoria. In addition, he has served as a session chair and technical committee member at various international conferences.

Arun Thirumurugan

Dr Arun Thirumurugan is an assistant professor at the University of Atacama, Sede Vallenar, Vallenar, Chile. He completed his PhD at the National Institute of Technology (NIT), Tiruchirappalli, India. He has worked as a postdoctoral fellow at the Institute of Physics, Bhubaneswar, and the University of Chile, Santiago, Chile. His research interests include the synthesis and surface modification of nanomaterials for energy and environmental applications. He has reviewed over 160 articles for various publishers and has served in editorial roles for several journals, including the *Journal of Energy Chemistry, International Journal of Energy Research, Journal of Nanomaterials, Social Impacts, Journal of Physics: Condensed*

Matter, Optical and Quantum Electronics, and *Micromachines.* He has published more than 110 papers and edited six books.

T Sridarshini

Dr T Sridarshini is working as Assistant Professor in Department of Electronics and Communication Engineering, College of Engineering Guindy, Anna University, Chennai, India. She has completed her doctoral degree in the field of Photonics from Anna University, Chennai. Her areas of research are optical communication and networks and photonics. She has contributed about 20 papers in reputed international/national journals and four book chapters.

List of contributors

E G Anagha
Department of ECE, Adi Shankara Institute of Engineering and Technology, Mattoor, Kalady, Kerala 683574, India

F Bagheri
Department of Electrical Engineering, Shahid Chamran University of Ahvaz, Ahvaz, Iran

V R Balaji
School of Electronics Engineering, Vellore Institute of Technology, Chennai Campus, Tamil Nadu 600127, India

Feroza Begum
Faculty of Integrated Technologies, Universiti Brunei Darussalam, Jalan Tungku Link, Gadong, Bandar Seri Begawan BE1410, Brunei Darussalam

Seemesh Bhaskar
STAR Laboratory, Central Research Instruments Facility (CRIF), Department of Chemistry, Sri Sathya Sai Institute of Higher Learning, Prasanthi Nilayam, Puttaparthi, Anantapur, Andhra Pradesh 515134, India

K Bhuvaneshwari
Department of Electronics and Communication Engineering, IFET College of Engineering, Villupuram, Tamil Nadu, India

Elizabeth Caroline Britto
Department of Electronics and Communication Engineering, IFET College of Engineering, Villupuram, Tamil Nadu, India

Avinash Chandra
School of Electronics Engineering, Vellore Institute of Technology, Vellore, Tamil Nadu, India

Narayanan Durairaj
Crystal Growth Centre, Anna University, Chennai, Tamil Nadu 600025, India

Kalathur Mohan Ganesh
Department of Electrical and Computer Engineering, Holonyak Micro and Nanotechnology Laboratory, Carl R. Woese Institute for Genomic Biology, University of Illinois at Urbana–Champaign, Urbana, IL, 61801, USA

F Haddadan
Department of Electrical Engineering, Shahid Chamran University of Ahvaz, Ahvaz, Iran

M Harikrishnan
Department of Electronics and Communication Engineering, Sri Manakula Vinayagar Engineering College, Madagadipet, Puducherry, India

Gopalkrishna Hedge
Centre for Nanoscience and Engineering, Indian Institute of Science, Bangalore, India

M A Ibrar Jahan
RNS Institute of Technology, Visvesvaraya Technological University, Bengaluru, India

C Jenila
Department of Electronics and Communication Engineering, Kalasalingam Academy of Research and Education, Virudhunagar, Tamil Nadu, India

R G Jesuwanth Sugesh
Department of Electronics and Communication Engineering, School of Engineering and Technology, CHRIST (Deemed to be University), Bangalore, India

J Jeyarani
Department of Electronics and Communication Engineering, National Institute Technology, Trichy, Tamil Nadu, India

Madurakavi Karthikeyan
School of Electronics Engineering, Vellore Institute of Technology, Vellore, Tamil Nadu, India

Rajkishor Kumar
School of Electronics Engineering, Vellore Institute of Technology, Vellore, Tamil Nadu, India

Abdul Mu'iz Maidi
Faculty of Integrated Technologies, Universiti Brunei Darussalam, Jalan Tungku Link, Gadong, Bandar Seri Begawan BE1410, Brunei Darussalam

M J Maleki
Department of Electrical Engineering, Shahid Chamran University of Ahvaz, Ahvaz, Iran

M S Mani Rajan
Department of Physics, University College of Engineering, Anna University, Ramanathapuram 623513, India

Margarat Michael
Department of Electronics and Communication Engineering, IFET College of Engineering, Villupuram, Tamil Nadu, India

P Nithya
Department of EEE, University College of Engineering, Anna University, Ramanathapuram 623513, India

C Palaniappan
Department of Electronics and Communication Engineering, Mount Zion College of Engineering and Technology, Pudukkottai, Tamil Nadu, India

Vineeth Palliyembil
Department of ECE, Indian Institute of Information Technology (IIIT), Kottayam, Kerala 686635, India

F Parandin
Department of Electrical Engineering, Kermanshah Branch, Islamic Azad University, Kermanshah, Iran

J Pradeep
Department of Electronics and Communication Engineering, Sri Manakula Vinayagar Engineering College, Madagadipet, Puducherry, India

Sakthivel Raja
Crystal Growth Centre, Anna University, Chennai, Tamil Nadu 600025, India

M Rajesh
Department of Computer Engineering, Sanjivani College of Engineering, Kopargaon, Ahmednagar, Maharashtra, India

Sai Sathish Ramamurthy
Department of Electrical and Computer Engineering, Holonyak Micro and Nanotechnology Laboratory, Carl R. Woese Institute for Genomic Biology, University of Illinois at Urbana–Champaign, Urbana, IL, 61801, USA

S Robinson
Department of Electronics and Communication Engineering, Mount Zion College of Engineering and Technology, Pudukkottai, Tamil Nadu, India

D Sasikala
Department of Electronics and Communication Engineering, Mookambigai College of Engineering, Pudukkottai 622502, Tamil Nadu, India

R Sitharthan
School of Electrical Engineering, Vellore Institute of Technology, Chennai Campus, Tamil Nadu, India

Kalainathan Sivaperuman
Centre for Nanotechnology Research, VIT Vellore, Tamil Nadu 632014, India

A Sivasubramanian
School of Electronics Engineering, Vellore Institute of Technology, Chennai, India

M Soroosh
Department of Electrical Engineering, Shahid Chamran University of Ahvaz, Ahvaz, Iran

N R G Sreevani
Department of Computer Science and Engineering, S A Engineering College, Chennai 600077, India

Moorthi Babu Sridharan
Crystal Growth Centre, Anna University, Chennai, Tamil Nadu 600025, India

Richards Joe Stanislaus
School of Electronics Engineering, Vellore Institute of Technology, Chennai, India

Nianyu Zou
Research Institute of Photonics, Dalian Polytechnic University, Dalian 116024, China

IOP Publishing

Advances in All-optical Communication

Shanmuga Sundar Dhanabalan, Arun Thirumurugan and Sridarshini Thirumaran

Chapter 1

Highly efficient materials for photonic crystal-based optical components

M Soroosh, F Parandin, F Haddadan, M J Maleki and F Bagheri

Increasing demands for high-bandwidth and high-speed processing have attracted researchers' attention to designing and extending all-optical devices. Photons are known as suitable carriers for transferring and processing information. The wave nature of photons causes light propagation in all directions in response to the wave vector components. Photonic crystals are periodic arrays of dielectric materials and present a photonic bandgap corresponding to the lattice constant and the radius of rods (or holes). By removing some rows and columns of rods (or holes), designing the waveguides for desired wavelengths is possible. Manipulating and transferring optical waves through the photonic crystal-based waveguides and lines is simply obtained. Using the ring resonators, one can obtain a transmission efficiency higher than 90% so approaching a high contrast ratio is possible for digital applications. Recently, different materials such as doped glasses, polymers, chalcogenide, and graphene have been used in photonic crystal structures for enhancement in performance. Accordingly, some optical components such as multiplexers, demultiplexers, comparators, and flip-flops have been considered.

1.1 Introduction

1.1.1 Light as a carrier for data transmission

Light is widely used as a carrier for data transmission due to its numerous advantages over other forms of signal transmission, such as electrical signals. Some key points about light as a high carrier for data transmission are as follows:

Speed: Light travels at an extremely fast speed in a vacuum, approximately 299 792 km s^{-1}. This high speed allows for rapid data transmission, making it ideal for applications that require quick communication, such as optical processors, optical systems, and high-speed internet.

doi:10.1088/978-0-7503-5623-7ch1

Bandwidth: Light has a wide spectrum of frequencies, enabling the transmission of large amounts of data simultaneously. This vast bandwidth can support high data rates, facilitating the transfer of multimedia content, large files, and real-time video streaming.

Low attenuation: Light signals experience less attenuation (weakening) over long distances compared to electrical signals. This characteristic enables light to transmit data over long-haul fiber-optic cables without significant loss, making it suitable for long-distance communication networks.

Immunity to electromagnetic interference: Unlike electrical signals, light signals are not affected by electromagnetic interference, which can degrade the quality of the transmitted data. This immunity makes light-based data transmission more reliable, especially in environments with high levels of electromagnetic noise.

Security: Light-based data transmission provides improved security compared to other transmission methods. Fiber-optic cables used to transmit light signals do not emit detectable electromagnetic radiation, reducing the risk of unauthorized interception or eavesdropping.

1.1.2 Optical processing

Optical processing refers to the use of light and optics to manipulate and process information. It holds significant importance in various fields due to its unique advantages over traditional electronic processing methods. The main reasons why optical processing is important are considered in the following.

Speed: Optical processing enables extremely high-speed data processing. Optical signals can travel at the speed of light, which is significantly faster than electronic signals. This makes optical processing ideal for applications that require real-time data analysis, such as telecommunications, image processing, and scientific research.

Bandwidth: Light signals have a much broader bandwidth compared to electrical signals. This means that optical processing can handle a larger volume of data simultaneously, enabling more efficient and faster communication and computation. It is particularly crucial in industries that deal with massive amounts of data, such as data centers, cloud computing, and high-performance computing.

Energy efficiency: Optical processing offers improved energy efficiency compared to electronic processing. Optical components consume less power and generate less heat, resulting in reduced energy consumption and cooling requirements. This is beneficial for applications where energy efficiency is critical, such as portable devices, data centers, and environmentally conscious technologies.

Parallelism: Another advantage of optical processing is its inherent parallelism. With optics, multiple light beams can be processed simultaneously, allowing for parallel computations. This parallelism can greatly accelerate complex computations and data-intensive tasks like matrix operations, pattern recognition, and machine learning.

Security: Optical processing provides enhanced security for data transmission and processing. Optical signals are less susceptible to interception and interference since they do not radiate electromagnetic waves like electronic signals. Additionally, optical encryption techniques can be utilized for secure communication, making it valuable for applications requiring high-level data protection, such as military communications and financial transactions.

Miniaturization: Optics offers the potential for miniaturization and integration into compact devices. Optical components can be made smaller and lighter compared to their electronic counterparts. This is advantageous in applications that demand portability and space efficiency, including medical devices, aerospace systems, and wearable technology.

Overall, the importance of optical processing lies in its ability to provide high-speed, high-bandwidth, energy-efficient, secure, and parallel processing capabilities. These advantages make it a promising technology for various industries, enabling advancements in communication systems, computing power, data analysis, and scientific research.

1.1.3 Optical communication

Optical communication is known as the transmission of information using light signals through optical fibers or free space. It has become a crucial technology in modern telecommunications and networking systems due to several important reasons:

High bandwidth: Optical communication offers significantly higher bandwidth compared to traditional copper-based communication systems. Light signals can carry a large amount of information, allowing for faster data transfer rates and greater capacity to handle increasing data demands.

Long-distance transmission: Optical fibers enable long-distance transmission of data without significant loss or degradation. Light signals can travel over hundreds of kilometers with minimal signal loss, making optical communication ideal for global communication networks.

Speed and reliability: Light travels at a very high speed, allowing for rapid data transmission. Optical communication provides low latency, reducing delays and ensuring real-time or near-real-time communication. Additionally, fiber-optic cables are not susceptible to electromagnetic interference, providing a highly reliable and secure communication medium.

Immunity to interference: Optical communication is not affected by electromagnetic interference from other electronic devices or power lines. This immunity makes it an excellent choice for environments with high levels of electrical noise, such as industrial settings or urban areas.

Scalability: Optical communication systems can be easily scaled up to meet growing data demands. By utilizing wavelength-division multiplexing techniques, multiple channels of data can be transmitted simultaneously over a single fiber, maximizing the available capacity.

Security: Optical communication offers enhanced security features. Since light signals are confined within the fiber, it is difficult to intercept or tap into the transmission without physical access to the cable. This makes optical communication inherently more secure than other forms of communication.

Energy efficiency: Compared to traditional copper-based communication systems, optical communication is more energy-efficient. Fiber-optic cables have lower attenuation, which reduces the need for signal amplification, resulting in lower power consumption.

Broad applications: Optical communication is used in various fields, including telecommunications, internet connectivity, data centers, cable television, and scientific research. It provides the backbone for high-speed internet connections, enables global networking, facilitates cloud computing, and supports high-definition video streaming.

1.1.4 Photonic crystal structure

Photonic crystals play a crucial role in the development of optical devices by manipulating the behavior of light at the nanoscale. They are periodic structures that exhibit a unique property known as photonic bandgap, which is analogous to the electronic bandgap in semiconductors. This bandgap prohibits the propagation of certain wavelengths of light within specific frequency ranges, resulting in the manipulation and control of light propagation. The key roles of photonic crystals in optical devices are investigated in the following [1].

Light control: Photonic crystals can control how light interacts with materials. By creating a periodic structure with alternating regions of high and low refractive index, photonic crystals can bend, reflect, or scatter light in specific directions. This enables the design of optical components such as lenses, waveguides, and filters with precise control over light propagation.

Wavelength selectivity: The photonic bandgap property of photonic crystals allows them to act as wavelength-selective filters. They can selectively transmit or block certain wavelengths of light, enabling the creation of optical devices like narrow-band filters and color displays.

Enhanced light-matter interaction: Photonic crystals can enhance the interaction between light and matter by confining and concentrating light within small volumes. This property is utilized in sensors, detectors, and solar cells to increase light absorption and improve device efficiency.

Slow light and dispersion engineering: Photonic crystals can slow down the speed of light by manipulating their group velocity. This phenomenon is exploited in devices such as delay lines, signal processing elements, and optical buffers. Additionally, photonic crystals can engineer the dispersion properties of light, allowing for the control of chromatic dispersion and the generation of ultrafast pulses.

Nonlinear optics: Photonic crystals can enhance nonlinear optical effects, where the optical response of a material depends on the intensity of light. By tailoring the band structure and mode confinement, photonic crystals enable efficient nonlinear frequency conversions, such as second-harmonic generation and parametric amplification.

In conclusion, photonic crystals provide a versatile platform for manipulating light at the nanoscale, enabling the development of advanced optical devices with improved performance, miniaturization, and functionality.

1.2 Photonic crystal and light propagation

Photonic crystals are an array of dielectrics that are arranged in a spatial period. This arrangement makes attractive features for guiding and manipulating optical waves. The difference between the permittivity of dielectrics in different directions results in three categories of photonic crystal structures including one-dimensional (1D), two-dimensional (2D), and three-dimensional (3D) devices [1].

The photonic crystal structures are provided by forming some columns of dielectric rods with air gaps and are called rod-type photonic crystals. Also, with the air holes in a dielectric background, a hole-type structure is formed. Each type of photonic crystal has some merits. In a rod-type photonic crystal, light can have greater coupling with less coupling loss in comparison with a hole-type one, while there is more confinement in the hole-type photonic crystal [2].

There are two important parameters that can determine the behavior of light in each photonic crystal: lattice constant and radius of rods or holes. The lattice constant is known as the distance between the center of two dielectric rods or air holes located in the vicinity of each other. With regard to the form of the periodic array of dielectrics, there are two structures of photonic crystal lattice: square-lattice and triangular-lattice [1].

The periodic form of the photonic crystal structure creates a forbidden range of frequencies, which is named the photonic bandgap (PBG). To simulate the optical waves with different powers and wavelengths in photonic crystal structures, the PBG of the structure should be calculated. The PBG is attributed to the concept of Bragg's condition. The fundamental structure of the 1D-photonic crystal consists of glass rods with radius r, effective linear and nonlinear indices along with air gaps a. The size of the designed structure is the main parameter of photonic crystal-based devices that should be considered. An optical signal with a particular wavelength in PBG can be propagated through a waveguide. To reduce power leakage in the waveguide, one should use Bragg's condition. This condition is considered as follows [3]:

$$n_a(2r) + n_b(a - 2r) = \lambda/2$$

where n_a and n_b are the refractive indices of dielectric and air and $2r$ and $(a-2r)$ are the thickness of the rod and the distance between rods, respectively.

So, according to the propagation of incoming light at different angles, Bragg's condition is different from one angle to another. So, there is a group of wavelengths that have the same properties and create a bandgap in a photonic crystal. This bandgap is used for light propagation in photonic crystals because incoming light which is placed in the limitation of the bandgap does not disperse.

Having said that, there is a wide range of wavelengths that are placed in the bandgap. Therefore, Maxwell's equations need to be solved to calculate the components of magnetic and electric fields. By calculating the components, the propagation of the waves through the photonic crystal structure in all directions is simulated. With no source in the structure, Maxwell's equations are defined as follows [3]:

$$\nabla \times E = -\partial B/\partial t, \nabla \cdot B = 0$$

$$\nabla \times H = \partial D/\partial t, \nabla \cdot D = 0$$

where E and H are the electric and magnetic components of the optical waves, respectively, and D and B represent the displacement and magnetic induction fields as $D = \varepsilon_0 \varepsilon_r$ and $B = \mu_0 \mu_r H$. ε_0 and μ_0 refer to the permittivity and permeability of free space, respectively, and ε_r is the relative permittivity of the dielectric medium. Generally, the relative permeability (μ_r) is assumed to be unit. Based on Bloch's theorem, Maxwell's equations have a solution in the form of the following [4]:

$$E(r) = \phi(r)e^{-ikr}$$

where k is the wave number and $\phi(r)$ is defined as a function that changes periodically in the space. By using the mentioned form, Maxwell's equations are written as follows [4]:

$$\nabla^2 E(r) + \varepsilon_r(r)\omega^2 c^2 E(r) = 0$$

Solving the above equation provides the electric field distribution along the eigenvalues for bandgap. Based on the direction of the electric field and magnetic field in a photonic crystal, there are two types of bandgap: TE-bandgap when an electric field is transversal and TM-bad gap for the transverse magnetic field. In the design of photonic crystal-based devices, the TE- or TM-bandgap is selected, considering the incoming wavelength. Also, in some cases, one of the bandgaps is too narrow, which is likely to be improper. However, having a wide bandgap is unsuitable, because it can cause cross-talk (see figure 1.1). For having a wider bandgap, the difference between the refractive index dielectric and background must be larger [1].

With the help of plane-wave expansion method, the dispersion of the structure in the reduced Brillouin zone and then the band structure is calculated as depicted in figure 1.1. It shows that this structure has three PBG regions, but the researchers chose the one with the largest width, which is at $0.28 < a/\lambda < 0.41$ in TM mode. It is equal to 1243 nm $< \lambda <$ 1821 nm [5].

After the calculation of photonic bandgap, waveguides are formed by removing some rods or holes. The main features of photonic crystal structures are simplicity of fabrication and high transmission factor in a small area.

1.3 Photonic crystal-based devices

Designing optical devices based on photonic crystals involves utilizing the unique properties and characteristics of photonic crystals to manipulate and control the behavior of light. Photonic crystals are periodic structures composed of materials with varying refractive indices, which create a bandgap that prohibits the propagation of certain wavelengths of light. Here are the key steps involved in designing optical devices based on photonic crystals [6–11]:

Understanding photonic crystals: Firstly, you need to have a thorough understanding of photonic crystals, their properties, and how they interact with light. This includes knowledge of concepts such as bandgaps, dispersion relations, and modes.

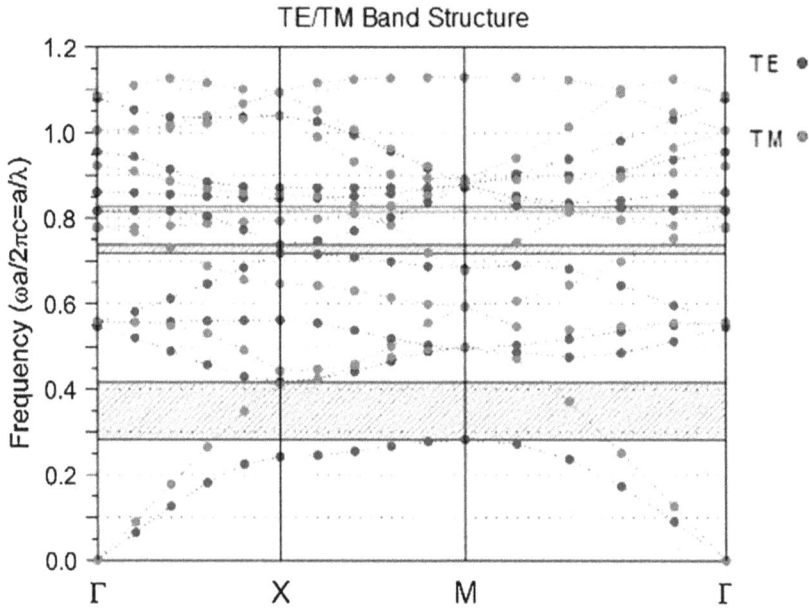

Figure 1.1. Dispersion of the encoder for the reduced Brillouin zone with the index of 3.46 and the lattice constant of 510 nm [5]. (2018) (Copyright © 2018, Springer Science Business Media, LLC, part of Springer Nature). With permission of Springer.

Device objectives: Identify the specific objectives of the optical device you want to design. For example, it could be a filter, waveguide, resonator, or sensor. Determine the desired wavelength range, bandwidth, efficiency, and other relevant parameters.

Material selection: Choose appropriate materials for constructing the photonic crystal structure. The refractive index contrast between different materials is crucial in creating the necessary bandgap and manipulating light propagation.

Structural design: Determine the geometric configuration and dimensions of the photonic crystal structure. This includes the arrangement of constituent materials, lattice type (e.g., square, hexagonal), and unit cell size. The structural parameters significantly affect the bandgap properties and functionality of the device.

Simulation and modeling: Utilize computational methods, such as finite-difference time-domain or plane-wave expansion algorithms, to simulate the behavior of light within the designed photonic crystal structure. This helps evaluate the device performance, optimize the design, and predict the effects of various parameters.

Fabrication techniques: Based on the simulation results and design specifications, select suitable fabrication techniques to realize the physical photonic crystal structure. Common methods include electron beam lithography, nanoimprint lithography, and self-assembly techniques.

Characterization and testing: Once fabricated, characterize the optical device to validate its performance. Use experimental techniques like optical microscopy, spectroscopy, or near-field scanning to measure properties such as transmission, reflection, dispersion, and efficiency.

Optimization and iteration: Based on the characterization results, iterate the design process to refine and optimize the device's performance. This may involve adjusting structural parameters, material choices, or fabrication techniques.

Integration and application: Finally, integrate the photonic crystal-based device into a larger system or application, if applicable. This could involve connecting it with other optical components or incorporating it into a specific platform or device.

Designing optical devices based on photonic crystals requires a multidisciplinary approach, combining knowledge from fields such as optics, materials science, electromagnetism, and fabrication techniques. It often involves a combination of theoretical modeling, numerical simulations, experimental characterization, and iterative design optimization to achieve the desired functionality and performance.

1.3.1 Demultiplexer/multiplexer

A multiplexer is a combinational circuit with 2^n inputs, n select lines, and one output. The inputs are routed to the output based on the selection line value. In other words, the values of the select lines determine which input will be transferred to the output. A demultiplexer performs the reverse of multiplexing. It has an input and several outputs, which are transferred to one of the outputs according to the value of the input select lines. The wavelengths can be separated in the demultiplexer, and each wavelength can be emitted in the desired output. One of the new structures for designing optical logic circuits is photonic crystals. Various circuits for multiplexers and demultiplexers have been designed based on photonic crystals.

In the design of a multiplexer, its size should be as small as possible. Also, the optical power should be small in logic 0 mode and higher in logic 1 to reduce the detection error. These parameters should also be considered in the design of a demultiplexer. In addition, the output spectral width should be reduced to increase the quality factor. The quality factor is obtained by dividing the central wavelength by the output spectral width.

Talebzadeh *et al* introduced a 4-channel demultiplexer for dropping at wavelengths 1549, 1549.7, 1550.3, and 1550.9 nm (as shown in figure 1.2) [12]. The lattice is composed of chalcogenide rods with a square arrangement. The structure has four resonant rings including some defects and one core for improving the coupling of the incoming waves to output waveguides. The main advantage of the proposed device is the low channel spacing needed for high bit rate transmission systems.

Talebzadeh *et al* used some defects in the form of two circles at the center of a ring to improve the coupling efficiency (as shown in figure 1.3) [13]. They succeeded in increasing the efficiency up to 97% at the second optical communication window. The channels were adjusted at the wavelengths of 1307.7, 1309, 1311, and 1313 nm through the different values for the defect's rods. They demonstrated that the core of the rings plays an important role in the dropping operation. The solid core assists in reducing the channel spacing for resonant rings while defects at the core increase the coupling efficiency.

Mehdizadeh *et al* used the resonant cavity to achieve the dropping operation for designing a 4-channel demultiplexer in a small area of 201 μm^2 as illustrated in

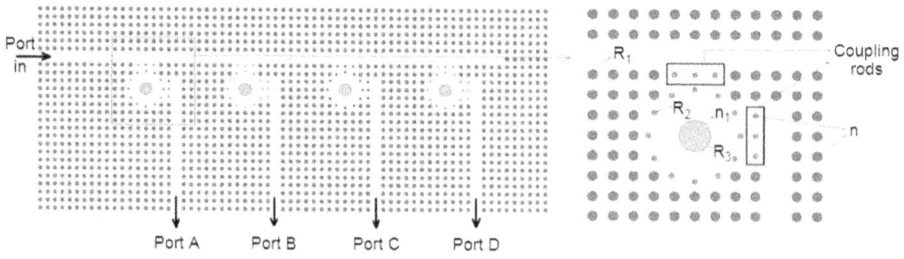

Figure 1.2. A low channel space demultiplexer including defects in the resonant rings. Reprinted with permission from [12].

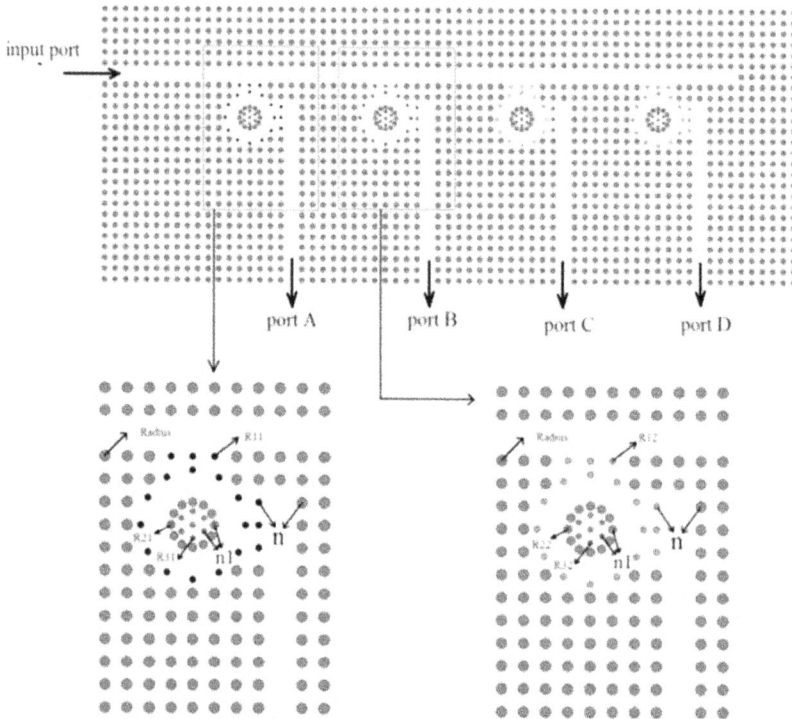

Figure 1.3. A 4-channel demultiplexer including some defects at the core for improvement of the transmission efficiency. Reproduced with permission from [13].

figure 1.4 [14]. The platform was made of silicon rods with air gaps in the form of a hexagonal lattice. According to Bragg's theory, three defects reflect the waves in the cavity that make an instructive interference at the resonant wavelength. They showed the transmission efficiency of the resonant cavity is lower than the resonant ring; however, the resonant cavity results in a narrow bandwidth.

Zahedi *et al* designed a 4-output demultiplexer based on a 2D photonic crystal [15]. Four ring resonators made of doped glass rods are used in this structure. This demultiplexer can separate four wavelengths between 1548 and 1559 nm. These

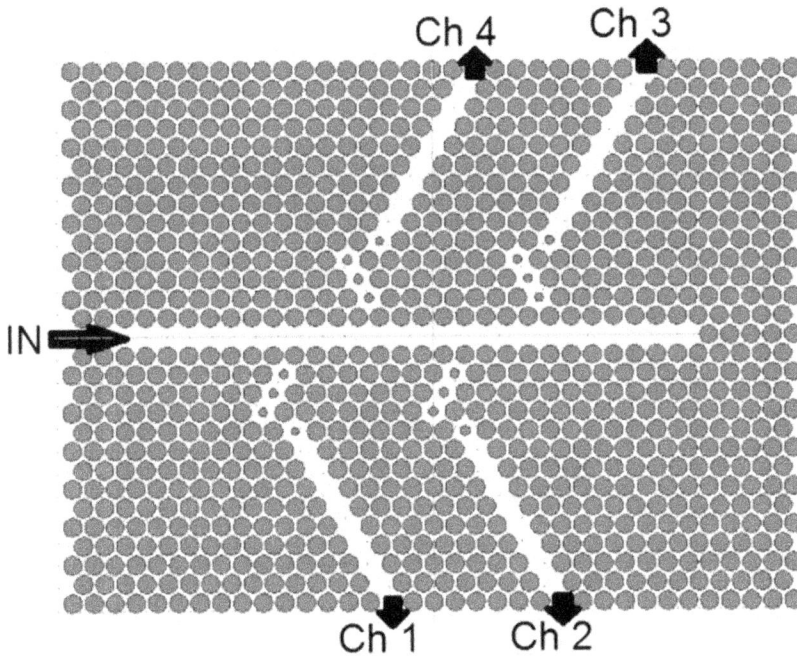

Figure 1.4. The cavity-based demultiplexer for 4-channel dropping. Reproduced with permission from [14].

cases' minimum and maximum quality factors are 1641 and 3225, respectively. The quality factor is the ratio of the central wavelength to the full width at half maximum. Figure 1.5 shows the designed demultiplexer structure and the normalized output powers regarding wavelength. The rods used in this design have a refractive index of 3.37. The lattice constant is also considered to be 530 nm. The lowest value of normalized power in the outputs is about 0.95.

Mohammadi *et al* designed a structure that can be used as an 8-channel demultiplexer [16]. One of the features of this design is that no ring resonator is used, and wavelengths are selected only with linear and point InP defects. The designed structure and simulation results are shown in figure 1.6. This structure can distinguish wavelengths from 1545 to 1571 nm without overlap. The minimum value of the quality factor for the proposed structure is 1030, and the maximum value is 1741. The normalized power transmitted to the output in all modes is 0.98, which is a high value.

A structure for the multiplexer was designed by Parandin *et al* [17] and silicon rods in the air were used in the structure. The structure includes two inputs, one selection line, and one output. The proposed design for this circuit and the output power in different states are shown in figure 1.7. The highest normalized power for logic 0 equals 0.2, and the lowest power value in logic 1 is equal to 0.52. The structure was also redesigned to a smaller size than the previous structure. As a result, the maximum power in logic 0 becomes about 0.25, and the minimum power in logic 1 changes to 0.55.

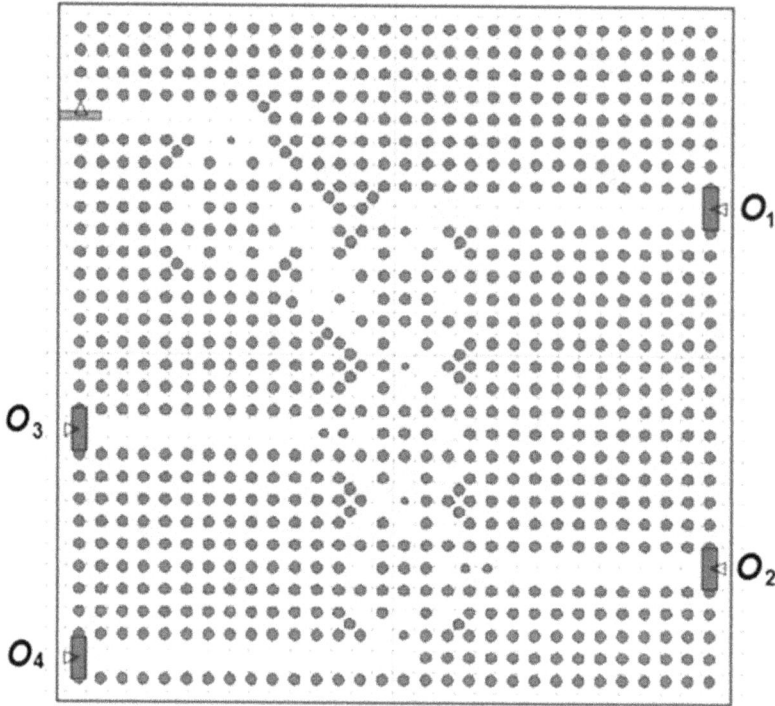

Figure 1.5. The proposed 4-output demultiplexer Reproduced with permission from [15]. © 2019 Walter de Gruyter GmbH, Berlin/Boston.

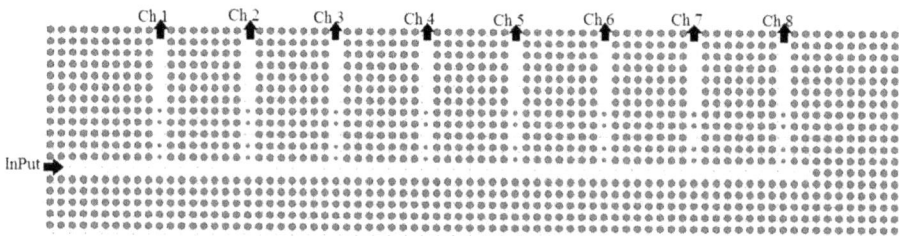

Figure 1.6. The 8-channel demultiplexer with resonant cavities [16]. (2019) (Copyright © 2019, Springer Science Business Media, LLC, part of Springer Nature). With permission of Springer.

Another structure was recently presented by Maleki *et al* for a two-input multiplexer [18]. This structure includes silicon rods with a refractive index of 3.1, and the lattice constant equals 513 nm (see figure 1.8). Two ring resonators are used to drop the waves toward the output port. The power in logic 0 is very low, and that in logic 1 equals 0.85 in a normalized form. In the designed structures, one of the important parameters is the output power in logic 0 and 1, and increasing the difference between these two values is one of the design goals. Also, a high-quality factor indicates a low spectral width and, as a result, the ability to separate very close

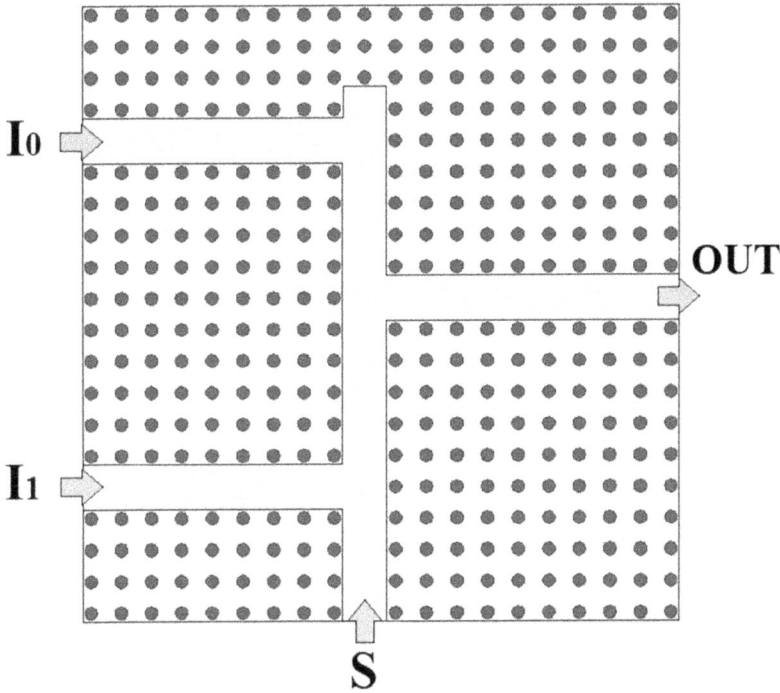

Figure 1.7. The multiplexer composed of two inputs and a selector. Reprinted from [17]. Copyright (2022), with permission from Elsevier.

wavelengths in the demultiplexer. Also, fewer ring resonators reduce the structure's complexity and delay time. Finally, the small size of the designed structure is also suitable for use in optical integrated circuits.

Maleki *et al* replaced the resonant rings with the resonant cavities to achieve a compact multiplexer (as depicted in figure 1.9) [19]. The resonant cavities $C1$ and $C2$ correspond to the guiding waves toward the output port O. According to the selector S, different intensities reach the cavities for working states. A rise time of 180 fs in response to the input pulse shows the fast response of the structure. Moreover, the area of the device is as small as 80 μm^2.

1.3.2 Flip-flop

An all-optical SR flip-flop based on doped glass photonic crystals was designed using the nonlinear optical Kerr effect [20]. This structure consists of a main part and two optical keys as depicted in figure 1.10. The main part consists of two connected resonant cavities, whose resonant modes are at 1586 and 1620 nm. The cavities are designed in such a way that the resonance of one cavity prevents signal coupling to another cavity. To design the key section, a bias port is used to store data when the input is zero. The refractive index of dielectric rods, the lattice constant (a), and the footprint of this structure are 3.5, 575 nm, and 361 μm^2, respectively. The radii of rods for switch 1 and switch 2 are $0.2227a$ and $0.237a$, respectively.

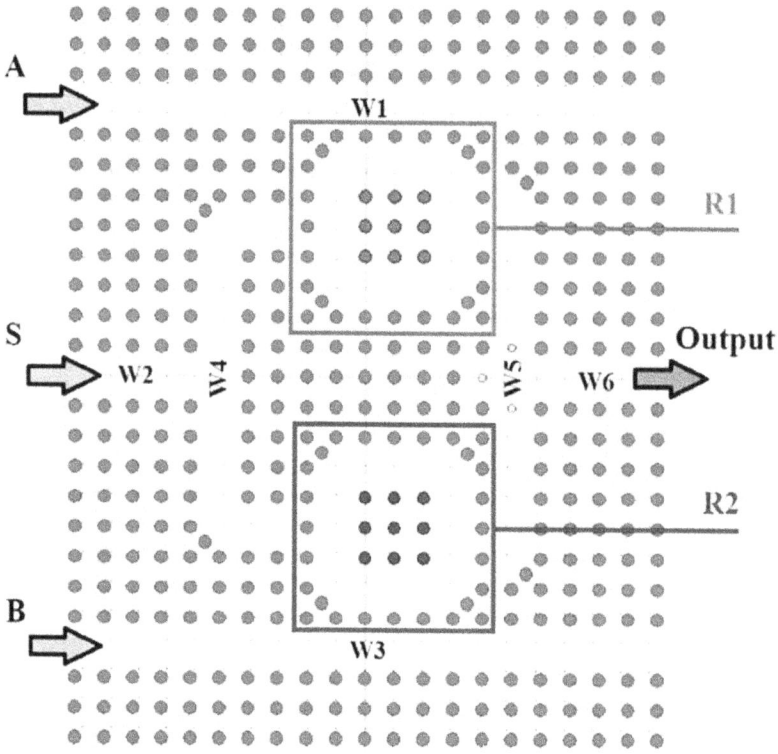

Figure 1.8. The resonator-based 2-to-1 multiplexer with two inputs *A* and *B* and selector *S* [18]. (2022) (Copyright © 2022, Springer Science Business Media, LLC, part of Springer Nature). With permission of Springer.

Parandin *et al* designed a compact structure for the silicon-based D flip-flop using the cross-connected waveguides (as depicted in figure 1.11) [21]. The structure has the capability to remain feature for input D. The interference between signals D and CP makes an appropriate signal at two output ports. The fast response of 30 fs and the small area of 33 μm^2 are the main advantages of the presented device.

1.3.3 Comparator

Recently, a photonic crystal-based structure was proposed for the all-optical 1-bit comparator [22]. To design this structure, three inputs (X, Y, REF), two outputs(O_1, O_2), and seven waveguides (OWG_1, OWG_2, OWG_3, OWG_4, OWG_5, OWG_6, OWG_7) have been used to guide the signals (see figure 1.12). This structure consists of a 31 × 33 cubic lattice of silicon rods immersed in air. Tthe refractive index, the lattice constant, and the radius of the rods at operating wavelength of 1550 nm are about 3.46, 590 nm, and 118 nm, respectively. To guide the incoming waves toward port O_1, three defect points with a radius of 59 nm are placed at the junction of the waveguides OWG_1 and OWG_6.

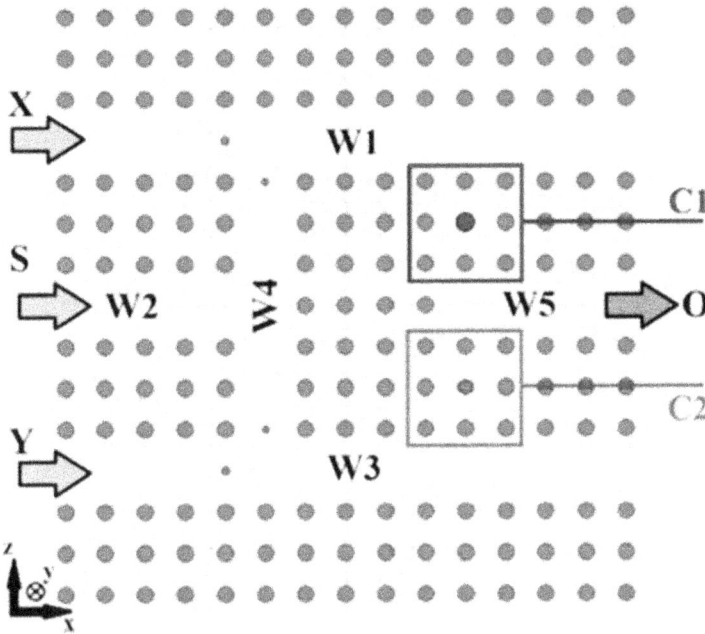

Figure 1.9. The cavity-based structure including nonlinear rods for 2-to-1 multiplexer [19]. (2022) (Copyright © 2022, Springer Science Business Media, LLC, part of Springer Nature). With permission of Springer.

The two input bits X and Y are compared according to table 1.1. According to this table, the input REF is always equal to one and is used to correctly compare two input bits X and Y.

A two-dimensional crystal photonic structure consisting of 24×19 silicon rods with a circular cross-section of radius 0.2 was designed with an air gap [23]. The lattice constant and footprint of the designed structure are equal to 0.6 μm and 149 μm² respectively. The refractive index of the rods is 3.46, which was designed for approximately 1.55 μm wavelength. In this structure, linear defects were used to obtain the comparator and create the input and output paths. This alternating structure led to the creation two the photonic bandgap. As depicted in figure 1.13, this structure has three inputs, A, B, and Ref, and three outputs F_1 ($A < B$), F_2 ($A = B$), and F_3 ($A > B$). The Ref input is always on and is used when both the A and B inputs are off and the F_2 output must be equal to one. The contrast ratio for F_1, F_2, and F_3 outputs is equal to 5.2, 5.3, and 5.2 dB, respectively.

A one-bit comparator with two inputs (A, B) and two outputs (F_1, F_2) is proposed in figure 1.14 [24]. This structure consists of GaAs rods with a refractive index of 3.37 on a background of air. The lattice structure (a), the radius of the rods, and the working wavelength are equal to 600 nm, 0.18a, and 1550 nm, respectively. In this structure, an X-shaped path and a small number of point defects are used to guide the input signals toward the output. The operation of the structure is such that when only one of the inputs is active, one of the outputs will be in the state of logical one and the other one will be in the state of logical zero. When both inputs are active, the

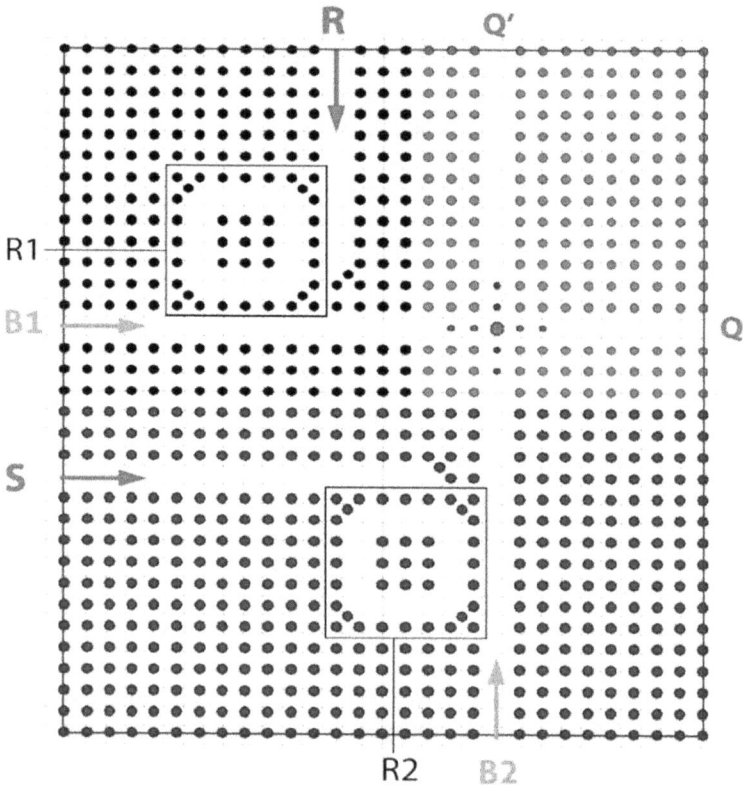

Figure 1.10. A cross-connection with defect rods to interfere among the entered waves in flip-flop operation [20]. (2018) (Copyright © 2018, Springer Science Business Media, LLC, part of Springer Nature). With permission of Springer.

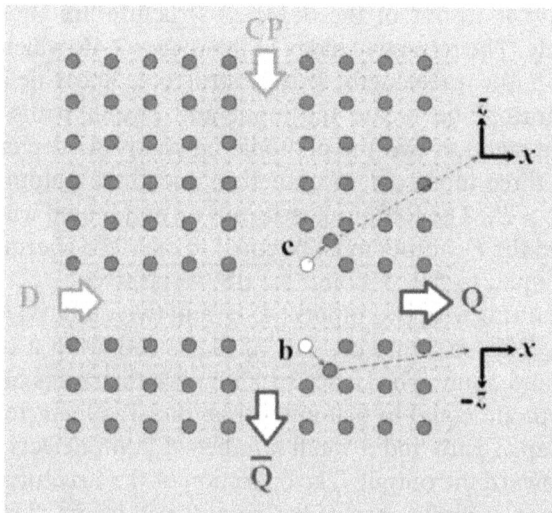

Figure 1.11. The cross-connected structure and defects for designing a flip-flop [21]. (2024) John Wiley & Sons. (© Wiley Periodicals, LLC.)

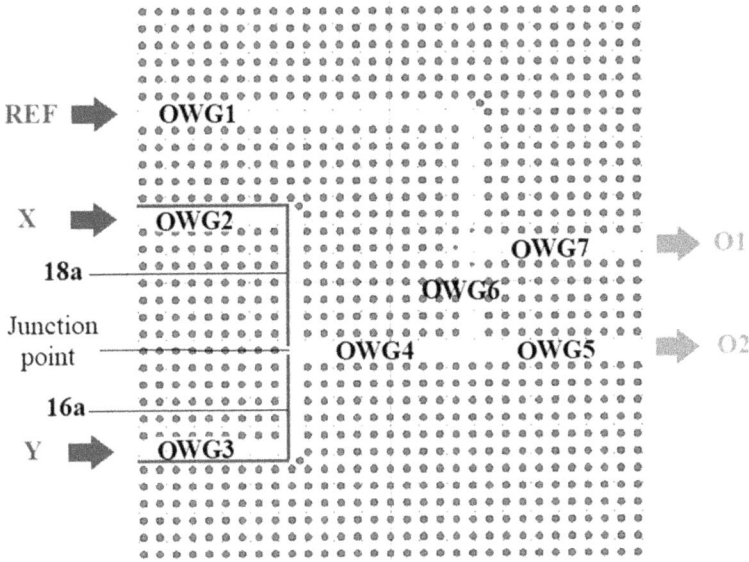

Figure 1.12. Proposed structure for a one-bit comparator including three inputs and two outputs [22]. (2023) (Copyright © 2023, Springer Science Business Media, LLC, part of Springer Nature). With permission of Springer.

Table 1.1. Accuracy table for the proposed structure in reference [22].

Ref. signal number	X	Y	O_1	O_2	Status
1	0	0	0	1	$X = Y$
	0	1	1	0	$X < Y$
	1	0	1	1	$X > Y$
	1	1	0	1	$X = Y$

signal sent from the inputs will interfere with each other in the X-shaped path and a very weak signal will reach the output. In this case, the outputs will be assumed to be in a logical zero state.

A one-bit ultrafast comparator was designed using ring resonators based on nonlinear doped glass rods [25]. As illustrated in figure 1.15, the structure has three inputs (B, X, Y), three outputs (Q_1, Q_2, Q_3), and the comparison operation is performed using four resonant rings (R_1, R_2, R_3, R_4) and seven waveguides (w_1, w_2, w_3, w_4, w_5, w_6, w_7). They reported an index of 3.46, a spatial period of 595 nm, and a radius of 0.2 a, respectively. A doped glass with an index of 1.4 and an optical Kerr effect of 10^{-14} m^2 W^{-1} is used for nonlinear rods. To confine the waves in the resonator and amplify the amplitude, 31 rods were used at the center and 4 rods at the corners. Also, some nonlinear rods (with green color) are inserted to provide the sensitivity and index dependence to incoming optical power as shown in figure 1.15.

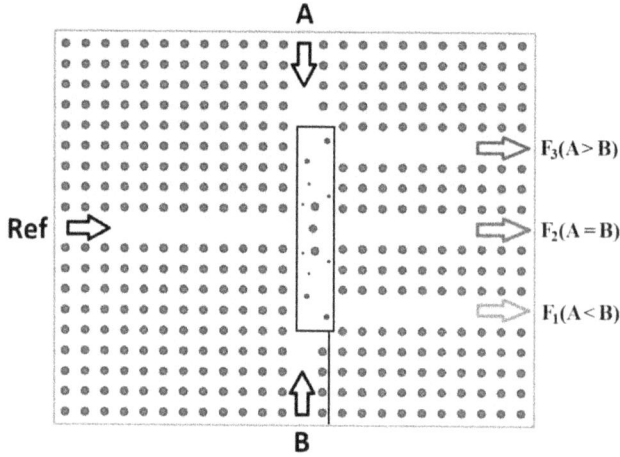

Figure 1.13. A one-bit comparator with some defects in the main waveguide. Reprinted with permission from [23]. © 2022 by the authors.

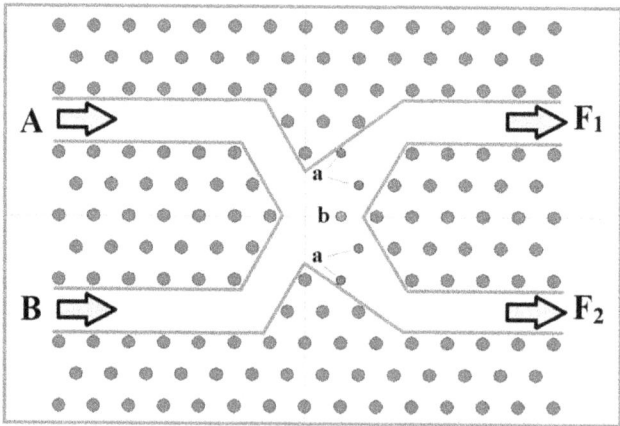

Figure 1.14. Designing an all-optical logic comparator using cross-connected waveguides including five defects. Reprinted from [24]. Copyright (2021), with permission from Elsevier.

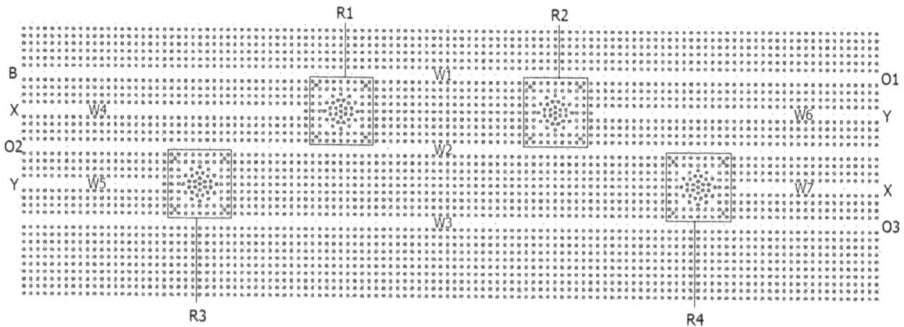

Figure 1.15. The resonance ring-based comparator including some nonlinear rods for achieving the sensitivity to entered optical power into rings. Reproduced from [25]. CC BY 4.0.

Table 1.2. Comparison of the important parameters of the all-optical comparators.

Work	Contrast ratio (dB)	Footprint (μm^2)	Structure
[22]	6.87	351	Defect-base
[23]	5.2	149.04	Defect-base
[24]	6.43	60	Ring & defect-base
[25]	3.68	1525	Ring-base

Table 1.2 compares the important parameters of comparators. According to this table, the structure proposed in reference [24] has the smallest footprint, and the structure proposed in reference [22] has the highest contrast ratio.

1.4 Materials used in photonic crystal structures

Photonic crystals are typically composed of dielectric materials with a periodic variation in refractive index. The choice of materials depends on the desired properties and applications of the photonic crystal. Some commonly used materials for photonic crystals include:

Silicon (Si): Silicon is a widely used material due to its compatibility with complementary metal-oxide-semiconductor (CMOS) fabrication processes, which allows for integration with electronic devices. It has a high refractive index and can be engineered to have a photonic bandgap.

Gallium arsenide (GaAs): GaAs is a semiconductor material that offers excellent optical properties and is commonly used in optoelectronic devices. It has a high refractive index and can be used to create photonic crystals for various applications.

Indium phosphide (InP): InP is another semiconductor material often employed in photonic crystal structures due to its favorable optical properties, such as a direct bandgap and high refractive index. It is frequently used in the fabrication of lasers and photodetectors.

Polymer-based materials: Polymers can also be used to create photonic crystals due to their flexibility, low-cost fabrication, and tunable optical properties. Polymer-based photonic crystals are commonly used in applications such as sensors and integrated photonics.

Titania/silica (TiO_2/SiO_2): These materials are frequently used in combination to create photonic crystals with a large refractive index contrast. They can be fabricated using techniques like sol-gel processing or atomic layer deposition.

Chalcogenides: Chalcogenide glasses, such as As_2S_3 or GeSbS, are suitable for creating photonic crystals in the infrared wavelength range. These materials offer a high refractive index and low losses in the infrared region.

Graphene: It has several potential applications in photonic crystals due to its unique optical and electrical properties. Here are a few notable applications:

(a) Tunable photonic devices: Graphene's conductivity can be dynamically controlled by applying an external electric field or changing the Fermi level. This property allows for the development of tunable photonic devices such as modulators, switches, and filters based on graphene-integrated photonic crystal structures.

(b) Enhanced light-matter interaction: Graphene's two-dimensional structure and zero bandgap enable strong light-matter interactions. When integrated into photonic crystal structures, graphene can enhance light absorption, emission, and scattering processes, leading to improved efficiency in various optoelectronic devices like solar cells, LEDs, and sensors.

(c) Ultrafast optical devices: Graphene exhibits ultrafast carrier dynamics, with electron mobility reaching up to several tens of thousands of $cm^2 \, V^{-1} \, s^{-1}$. Photonic crystals incorporating graphene can be used to design ultrafast optical devices, including ultrafast lasers, photodetectors, and optical modulators operating at high frequencies.

(d) Surface plasmon resonance (SPR) sensing: Photonic crystal structures enhanced with graphene layers can be employed for highly sensitive SPR sensing. The combination of graphene's excellent electrical conductivity and the localized surface plasmons in the photonic crystal structure enables precise detection of changes in refractive index, making it useful for biosensing applications and environmental monitoring.

(e) Terahertz (THz) applications: Graphene's unique electronic properties make it suitable for terahertz applications. By integrating graphene within photonic crystals designed for the terahertz regime, it is possible to develop efficient THz waveguides, modulators, polarizers, and other devices.

It's important to note that the choice of material relies on the specific requirements of the photonic crystal, such as the desired bandgap characteristics, operating wavelength range, and fabrication techniques available. Different materials may be suitable for different applications within the field of photonics.

References

[1] Soroosh M, Farmani A, Maleki M J, Haddadan F and Mansouri M 2023 Highly efficient graphene-based optical components for networking applications *Photonic Crystal and Its Applications for Next Generation Systems* 1st edn (Singapore: Springer Nature) pp 15–35

[2] Chen J, Mehdizadeh F, Soroosh M and Alipour-Banaei H 2021 *Opt. Quantum Electron.* **53** 510

[3] Yasumoto K 2006 *Electromagnetic Theory and Applications for Photonic Crystals* (London: Taylor and Francis) pp 32–66

[4] Sukhoivanov I A and Guryev I V 2009 *Photonic Crystals: Physics and Practical Modeling* (Berlin: Springer) pp 23–54

[5] Haddadan F and Soroosh M 2019 *Photonic Netw. Commun.* **37** 83–9

[6] Geerthana S, Sridarshini T, Syedakbar S, Nithya S, Balaji V R, Thirumurugan A and Dhanabalan S S 2023 *Phys. Scr* **98** 105975

[7] Geerthana S, Sridarshini T, Balaji V R, Sitharthan R, Madurakavi K, Thirumurugan A and Dhanabalan S S 2023 *Opt. Quantum Electron.* **55** 778

[8] Sridarshini T, Geerthana S, Balaji V R, Thirumurugan A, Sitharthan R and Dhanabalan S S 2023 *Laser Phys.* **33** 076207

[9] Sridarshini T, Chidambaram P, Geerthana S, Balaji V R, Thirumurugan A, Madurakavi K and Dhanabalan S S 2022 *Sustain. Comput. Inform. Syst.* **36** 100815

[10] Dhanabalan S S, Thirumurugan A, Raju R, Kamaraj S K and Thirumaran S 2023 *Photonic Crystal and Its Applications for Next Generation Systems* (London: Springer Nature) pp 23–45

[11] Sridarshini T, Dhanabalan S S, Balaji V R, Manjula A, Indira Gandhi S and Sivanantha R A 2023 Photonic crystal based routers for all optical communication networks *Modeling and Optimization of Optical Communication Networks* (New York: Wiley) pp 137–62

[12] Talebzadeh R, Soroosh M and Mehdizadeh F 2016 *Opt. Appl.* **46** 553–64

[13] Talebzadeh R and Soroosh M 2015 *Optoelectron. Adv. Mater. Rapid Commun* **9** 5–9

[14] Mehdizadeh F and Soroosh M 2015 *Optoelectron. Adv. Mater. Commun* **9** 324

[15] Zahedi A, Parandin F, Karkhanehchi M M, Habibi-Shams H and Rajamand S 2019 *J. Opt. Commun.* **40** 17–20

[16] Mohammadi B, Soroosh M and Kovsarian A 2019 *Photonic Netw. Commun.* **38** 115–20

[17] Parandin F and Sheykhian A 2022 *Opt. Laser Technol.* **151** 108021

[18] Maleki M J and Soroosh M 2022 *Opt. Quantum Electron.* **54** 397

[19] Maleki M J and Soroosh M 2022 *Opt. Quantum Electron.* **54** 818

[20] Zamanian-Dehkordi S S, Soroosh M and Akbarizadeh G 2018 *Opt. Rev.* **25** 523–31

[21] Parandin F, Sheykhian A and Askarian A 2024 *Microw. Opt. Technol. Lett.* **66** e34006

[22] Askarian A and Parandin F 2023 *J. Comput. Electron.* **22** 288–95

[23] Parandin F, Olyaee S, Kamarian R and Jomour M 2022 *Photonics* **25** 459

[24] Parandin F 2021 *Opt. Laser Technol.* **144** 107399

[25] Jalali S M H, Soroosh M and Akbarizadeh G 2019 *J. Optoelectron. Nanostruct* **4** 59–72

IOP Publishing

Advances in All-optical Communication

Shanmuga Sundar Dhanabalan, Arun Thirumurugan and Sridarshini Thirumaran

Chapter 2

Unidirectional bulk growth of 1,3,5-triphenylbenzene single crystal and doping effect on its optical properties

Narayanan Durairaj, Sakthivel Raja, Sridharan Moorthi Babu, Sivaperuman Kalainathan and Jayaraman Jayapriya

Fast neutron detection in gamma and other mixed radiation backgrounds is likely due to the dual emission of organic molecules. Improving the emission and decay time of organic materials is more important for enabling the detection of large events of neutron-gamma/mixed other particle discrimination. The present work investigates the scintillator crystal 1,3,5-triphenylbenzene (3PB) doped with 2,5-diphenyloxyzole (PPO) and trans-stilbene (TSB), 1,4-diphenyl-1,3-butadiene (DPB) with different mole percentage ratios. An optimized crystal growth procedure was demonstrated and achieved different diameters of 3PB single crystal cylinders for the device fabrications. The optical studies relevant to the scintillator crystals were systematically studied, which exposes the importance of growing bulk 3PB scintillator crystals. The structural and optical characterizations confirmed the dopant influences and found the optimized dopant ratio in the host (3PB) compound.

2.1 Introduction

Organic scintillator crystals are vital in fast neutron detection in a mixed radiation field. Based on the molecular nature of luminescence and decay time, it discriminates the invisible radiations in nuclear spectroscopy, particle counting, and radiation detection applications [1–4]. 3PB is a promising organic scintillator crystal that provides improved pulse shape discrimination (PSD) for neutron and gamma particles [5]. High light yield output and short decay time could be expected for the scintillator applications because they cover large radiation detection events. But 3PB has a lower light output value and longer decay time than the existing material. However, the light output can be improved by adding a small bright

doi:10.1088/978-0-7503-5623-7ch2

fluorophore as a dopant in a host compound [5–9]. The optimal ratio of the dopant plays a crucial role in modulating the luminescence intensity of a core molecule.

An attempt was made by Balamurugan *et al* to enhance the fluorescence intensity of naphthalene crystal by doping the PPO fluorophore. Also, they demonstrated the limitation of dopants in host molecules. The four concentrations, 10, 30, 50, and 70 mmol%, of PPO were doped in naphthalene ($C_{10}H_8$) and the light intensity and decay time were analyzed. The emission intensity increased at 10 and 30 mmol% among the four concentrations. The maximum intensity was observed at 30 mmol%. Further increasing the dopant concentration (above 30 mmol%) the emission intensity was decreased due to the concentration-quenching effect [7].

Gulunov *et al* described the dopant DPB distribution in *p*-terphenyl. They analyzed the DPB distribution from 0.1 to 0.7 mass% in *p*-terphenyl. The maximum light output values were observed in the range of 0.1–0.3 mass%. The quenching effect was observed above 0.3 mass% concentration [8]. For 3PB, Zaitseva *et al* suggested some of the organic fluorophores and tried PPO dopants to improve the 3PB luminescence intensity. However, they did not describe the limitation of dopants in the 3PB [5]. Durairaj *et al* performed the different concentration of DPB, PPO in 3PB and optimized dopant concentration in 3PB [10]. The authors also reported a bulk growth of DPB, PPO-doped 3PB crystals. The work has grown unidirectional 3PB single crystal and reported preliminary studies for scintillator device fabrications [11, 12]. They utilized two-ring heaters for the bulk growth of a unidirectional crystal cylinder. However, the authors utilized the constant temperature bath (CTB) instead of ring heaters in the present work. To improve the luminescence intensity of 3PB, they used trans-stilbene, PPO, and DPB as dopant. The distribution of different dopant ratios was verified. Further, x-ray diffraction (XRD), ultraviolet (UV), Fourier transform infrared (FTIR), and nuclear magnetic resonance (NMR) studies were extended to analyze the dopant incorporation in 3PB.

2.2 Experiment

2.2.1 Conventional crystal growth

As per the temperature-dependent solubility information [11], 22.54 g of the 3PB (solute) is dissolved in 100 ml tetrahydrofuran solvent at room temperature. The saturated solutions were carefully filtered and then transferred into a crystal growth vessel using Whatman filter paper. The growth vessel containing the saturated solution was securely covered with aluminium foil, allowing for slow evaporation of the solvent through a few strategically placed holes. Due to the continuous solvent evaporation, the solution reaches supersaturation. When the solution attains supersaturation, nucleation will be initiated, and the continuous accumulation of nucleation forms the clusters. When the solution maintains the supersaturation then the accumulation cluster becomes a crystal. After 15 days, the optically good-quality crystal was harvested for further characterization. The grown crystal is shown in figure 2.1.

Further, the obtained seed crystal was used to grow bulk crystals (figure 2.2) using the conventional seed hanging method to prepare the seed crystal for the unidirectional method.

The brief crystal growth procedure and schematic representation of seed preparation along the molecule orientation for the unidirectional growth technique are given below as a flowchart.

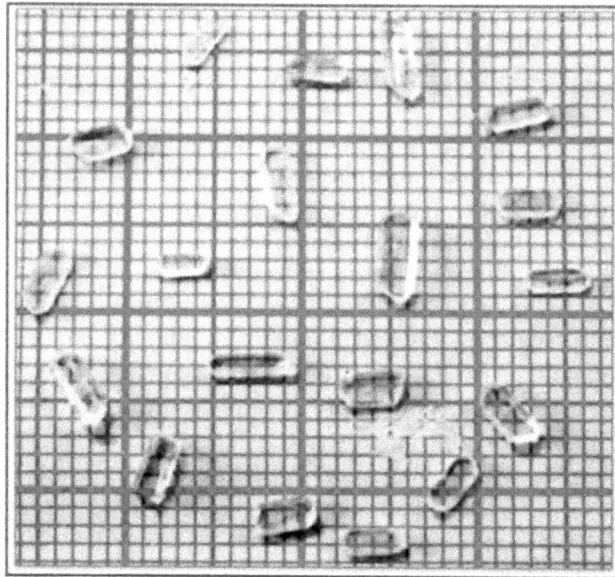

Figure 2.1. Grown 3PB seed crystals.

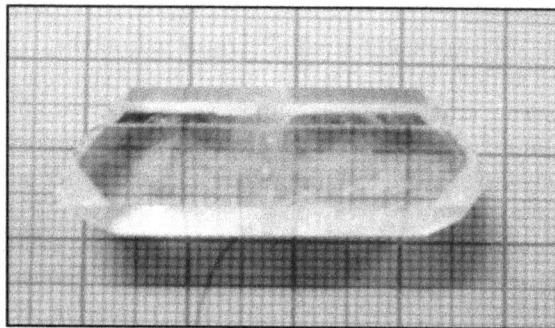

Figure 2.2. 3PB crystal grown by top seed hanging method.

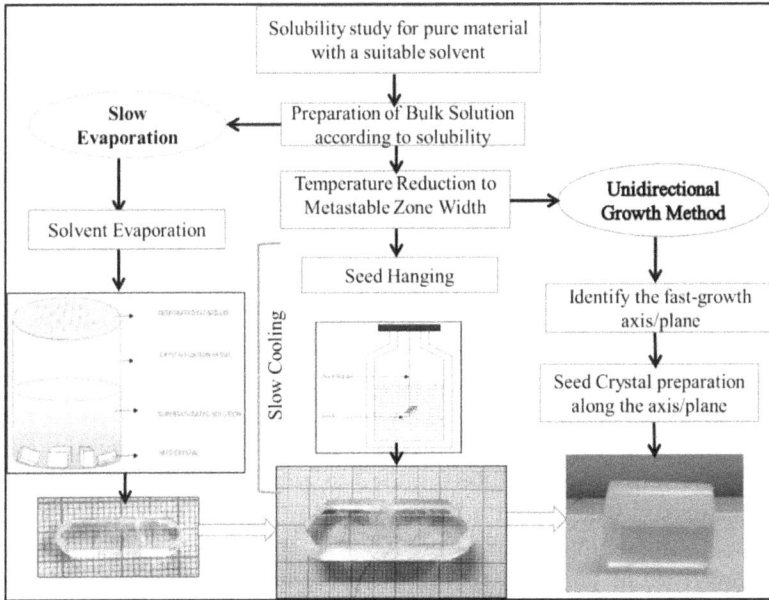

The flowchart for the unidirectional crystal growth by solution growth method.

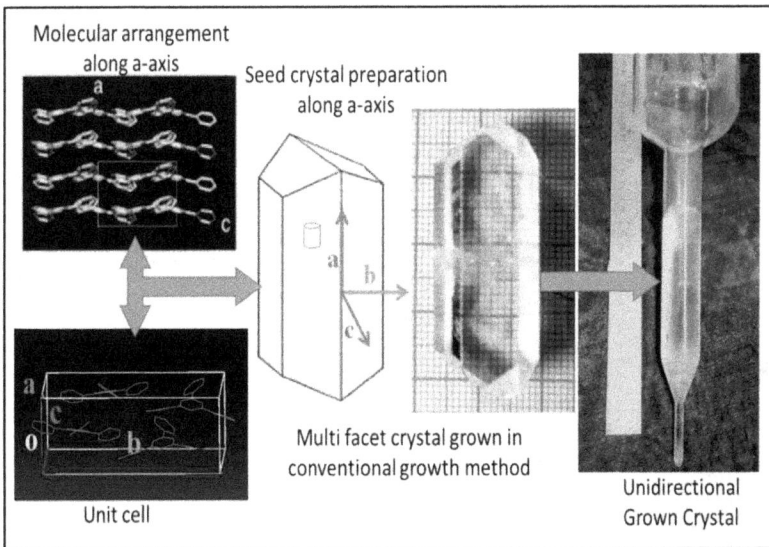

The schematic representation of seed preparation along molecule orientation for unidirectional growth method.

2.2.2 Bulk growth of stilbene-doped 3PB

The PPO, DPB, and stilbene dopant concentration ratios of 10, 30, 50, and 70 mmol % were added in 3PB. The saturated solutions of various concentrations were carefully filtered and then transferred into a crystal growth vessel using Whatman filter paper. Pristine and PPO, DPB, and stilbene-doped 3PB single crystals were harvested by the conventional solvent evaporation technique.

A bulk (300 ml) saturated solution was prepared in tetrahydrofuran solvent above all concentrations at 40 °C. A suitable seed crystal, grown from a slow solvent evaporation technique, was prepared along the *a*-axis and mounted in the bottom of a borosilicate glass ampoule.

A schematic diagram (figure 2.3) of the constant temperature water bath (CTB) and glass ampoule design is shown in figures 2.3(a) and (b). Figure 2.4 shows the CTB with a cryostat system. The prepared saturated solution was filled in a glass ampoule and tightly covered with aluminium foil to avoid uncontrollable solvent evaporation. The saturated solution-filled glass ampoule was kept in a CTB at 40 °C with an accuracy of ±0.01. The bath temperature was reduced to a nucleation temperature of 35 °C at 0.25 °C for every 6 h. After reaching the nucleation temperature, the seed crystal will be initiated to grow within two days. Further, the growth of the crystal will be monitored day by day. When the growth is reduced or stopped, the fresh solution will be added to get a bulk crystal. To get 100% crystallization a small pinhole was made on the aluminium foil to evaporate the solvent. Due to the continuous evaporation of the solvent, the supersaturation was maintained over the growth period. After 37 days, the bulk size (20 mm diameter) S-3PB crystal was harvested. Figure 2.5 shows the pure and stilbene-doped 3PB crystals. The multifaced S-3PB is due to a lower concentration of supersaturation. Figure 2.6 shows the DPB-doped 3PB (D-3PB) and PPO-doped 3PB (P-3PB) crystals.

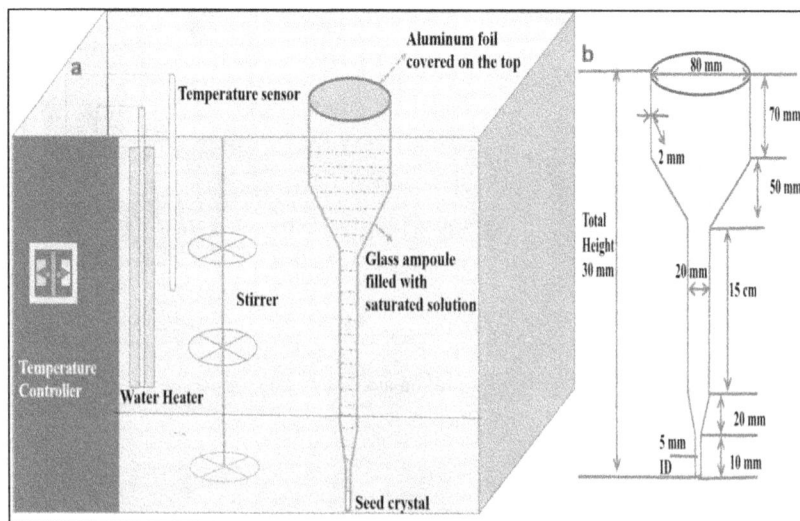

Figure 2.3. Schematic diagram of (a) constant temperature water bath and (b) ampoule design.

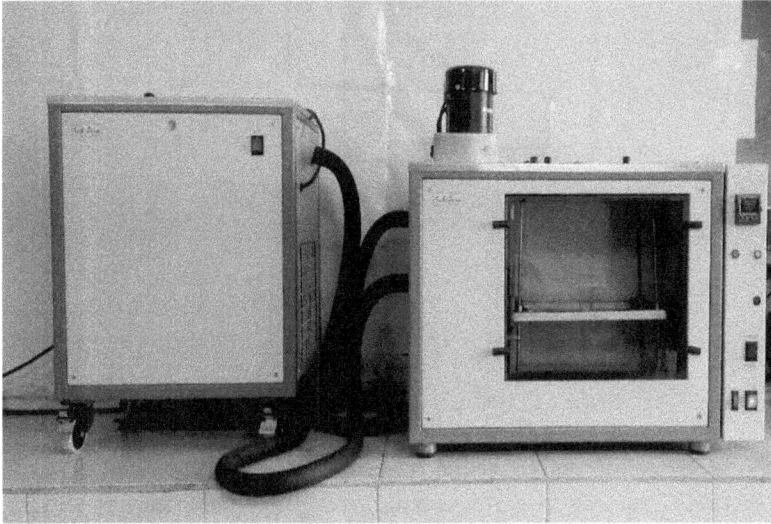

Figure 2.4. Constant temperature water bath (CTB) with cryostat unit.

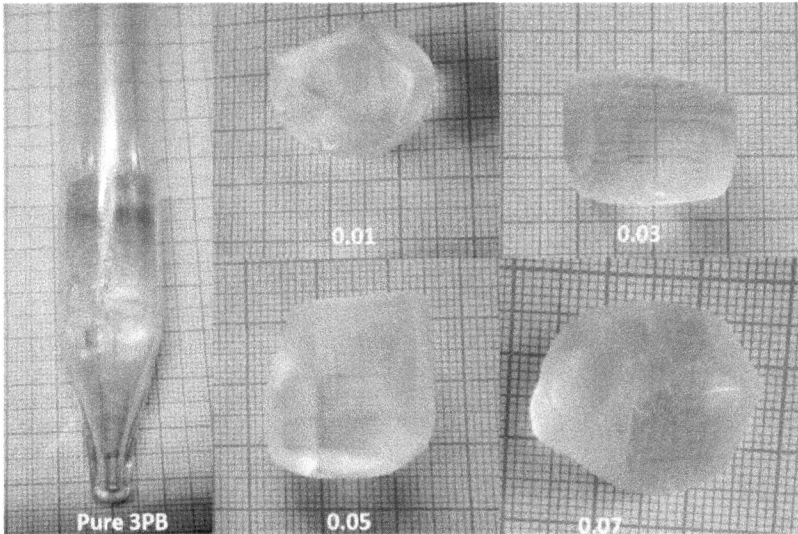

Figure 2.5. Pure and stilbene-doped 3PB crystals.

2.2.3 Characterization techniques

Pure and doped 3PB crystal cell parameters were measured at ambient temperature using Bruker Kappa APEX-II with Mo K_α radiation ($\lambda = 0.7107$ Å). The 3PB fine powder was used for phase purity and crystallinity. For the powder x-ray diffraction analysis, Bruker x-ray diffractometer with source Cu K_α radiation ($\lambda = 1.5406$ Å)

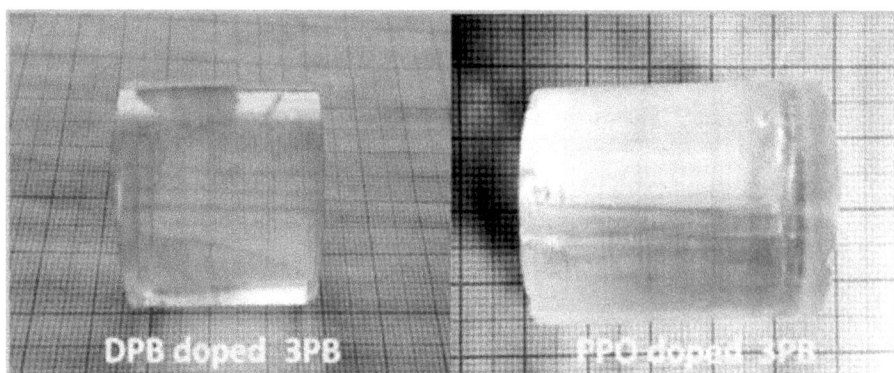

Figure 2.6. DPB- and PPO-doped 3PB crystals [11].

scanning angles ranging from 5 to 60° scan rate of $0.02°$ s^{-1} were used. The x-ray diffractions of different plans at the specific 2θ angles were identified and marked by Powder X software. Well-polished pure and doped 3PB crystals were used for the optical absorption studies using the ELICO double beam UV–Vis–NIR spectrometer (SL-218) in the 200–900 nm region. Pure and doped crystals were used for the optical absorption studies. Similar transparent and defectless crystal size was used for emission and lifetime measurement. The emission spectra were measured by an Edinburgh FLS 980 spectrometer with an excitation of 300 nm at room temperature. The proton and carbon bonding nature of all crystals was measured and verified with BRUKER-400 MHz FT-NMR and for the analysis crystals were dissolved in $CDCl_3$ solvent. The FTIR spectra of all 3PB crystals was recorded in the range of 4000–400 cm^{-1} by the SHIMADZU IRAFFINITY spectrometer to verify the functional groups.

2.3 Result and discussion

2.3.1 X-ray diffraction analysis

The unit cell parameters of pure and doped 3PB crystals are listed in table 2.1. The calculated values agree with International Centre for Diffraction Data file 33–1943 [11]. The values confirm that the doped crystals also belong to the orthorhombic crystal structure. Figures 2.7 (a) and (b) show the pure doped 3PB crystal powder XRD pattern. In powder XRD patterns, there are no additional peaks and phase changes were observed in the spectra. The dopant molecules are scattered randomly within the empty spaces or imperfections of crystal lattices [14]. The amount of doping is not high enough to cause any significant disruption to the structure of the crystal domains. The variations in peak intensity are due to differences in grain orientation and size. The broadening of the peaks can be attributed to lattice strain resulting from the presence of dopants, providing evidence of the inclusion of dopant molecules within the crystal lattices. The slight variation in the lattice parameter values confirms the strain due to the increasing dopant concentrations.

Table 2.1. Unit cell parameters of pure and doped 3PB crystals.

Lattice parameters	3PB [11]	D-3PB [10]	S-3PB [present work]	P-3PB [11]
a (Å)	7.603	7.634	7.636	7.645
b (Å)	19.782	19.84	19.90	19.94
C (Å)	11.296	11.308	11.316	11.347
Angles	$\alpha = \beta = \gamma = 90°$	$\alpha = \beta = \gamma = 90°$	$\alpha = \beta = \gamma = 90°$	$\alpha = \beta = \gamma = 90°$
Space group	$Pna2_1$	Pna21	$Pna2_1$	Pna21

2.3.2 UV–visible NIR spectra analysis

Figures 2.8 (a) and (b) display the UV–Vis absorption spectra of both pure and doped 3PB crystals. The absorption edges of all dopants were found to be identical, with no additional absorption peaks observed. The UV absorption of the 3PB crystal can be attributed to the aromatic transition of π–π^* excitation within individual molecules. Furthermore, the cut-off wavelength of the doped crystals closely matched that of the pure 3PB, indicating good agreement. The same behavior was observed in pure and doped crystals over the visible and NIR regions. However, the transmittance of doped crystals decreased when the dopant concentration increased. The distortion in absorption spectra of doped crystals reveals the incorporation of dopants in 3PB.

2.3.3 Fluorescence emission spectrum analysis

The purpose of dopant in 3PB is used to increase the light efficiency or the number of photons emitted. When a dopant is added to a core compound, it increases the density of the triplet energy level. This, in turn, enhances delayed fluorescence and reduces the decay time. In organic compounds, triplets are known to be mobile, and their energy migrates until two triplets collide and combine to form one singlet excited state. The lifetime of the delayed emission depends on the lifetime of the triplet states and the rate of triplet collisional interactions. Prompt fluorescence occurs when the singlet energy level transitions directly, while delayed emission is the result of collisional interactions between pairs of molecules in the lowest excited triplet state. The collisional interaction of the triplet energy level is particularly significant in the context of neutron detection in an organic scintillator [11, 15]. However, the high concentration of an impurity may suppress the light efficiency due to non-radioactive transitions from the triplet state. Hence, the different concentrations (10, 30, 50, and 70 mmol%) of dopants were performed and the limits of dopant concentration in 3PB. The emission spectrum was analyzed for both pure crystals and crystals doped with various concentrations of DPB and PPO. The results, depicted in figures 2.9(a) and (b), revealed that the highest emission intensity occurred at 10 mmol% of DPB and 50 mmol% of PPO-doped crystals. This increase in intensity is attributed to the enhanced exchange coupling of triplet mobility in the

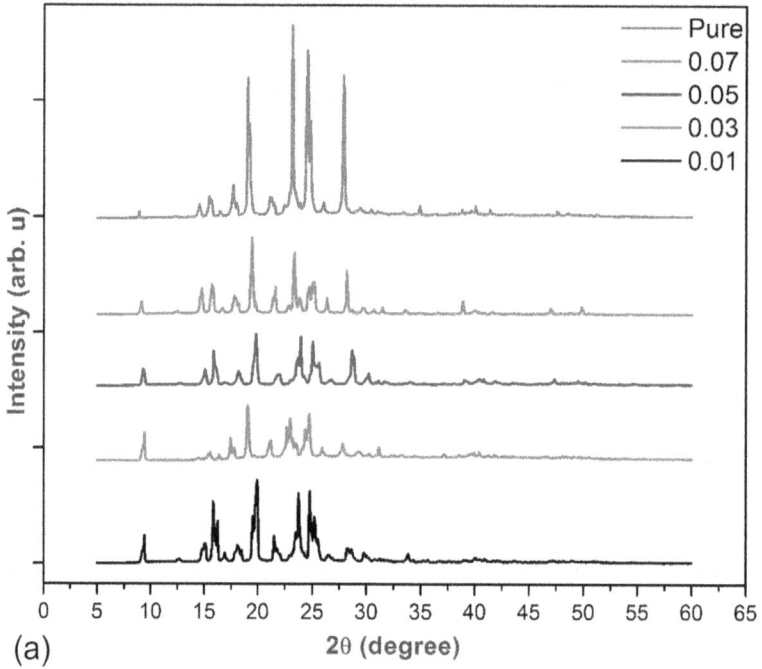

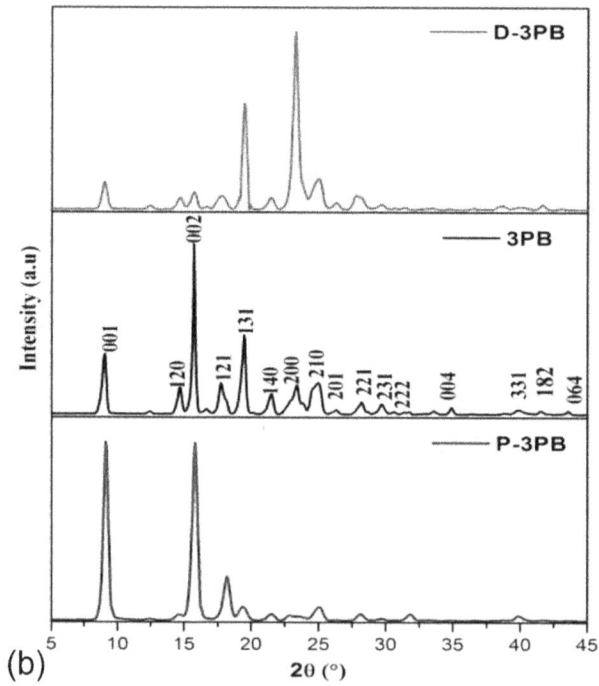

Figure 2.7. (a) XRD spectra of pure and S-3PB, (b) XRD spectrum of Pure 3PB, D-3PB, and P-3PB [11].

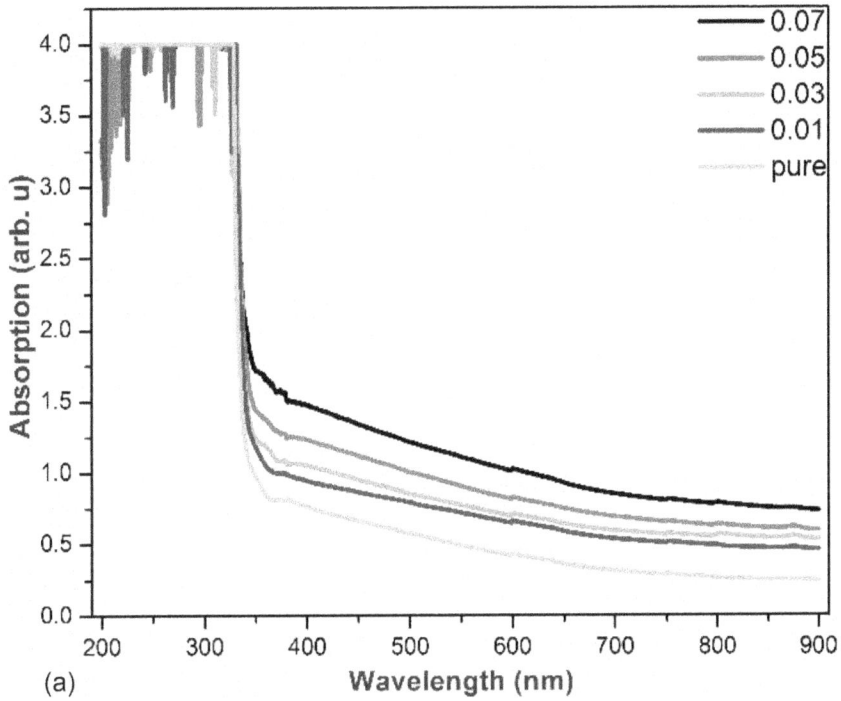

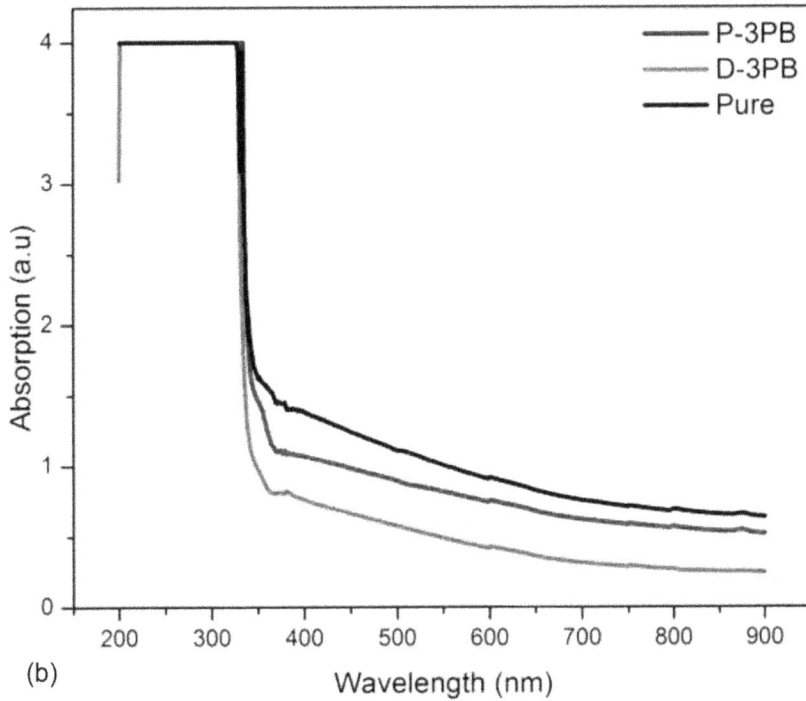

Figure 2.8. (a) UV–visible–NIR spectra of pure and S-3PB, (b) UV–Vis absorption of Spectrum pure, PPO-, and DPB-doped 3PB crystals [11].

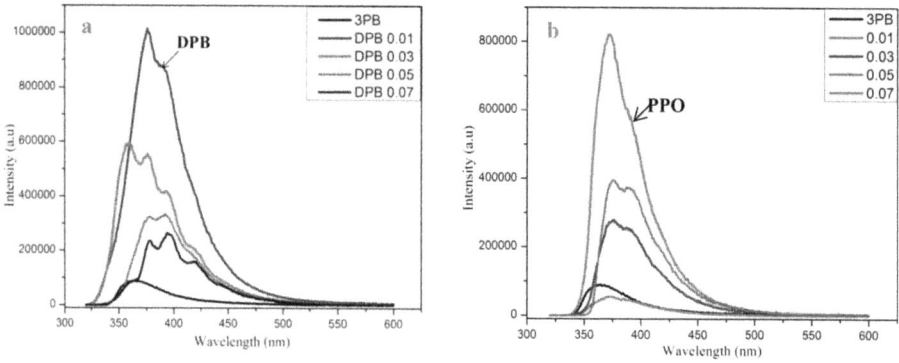

Figure 2.9. Fluorescence spectra of (a) D-3PB and (b) P-3PB crystal excited at 300 nm.

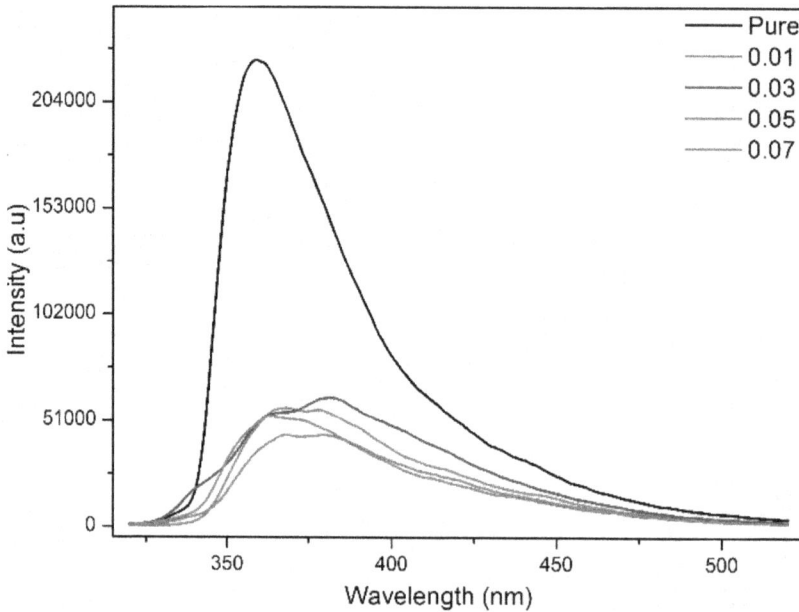

Figure 2.10. Fluorescence emission spectra of pure and stilbene-doped.

doped crystals [15]. However, concentrations exceeding these levels led to a decrease in fluorescence intensity due to the concentration-quenching effect. Additionally, the fluorescence intensity of pure 3PB was lower than that of doped crystals, indicating its less populated excited states. The fluorescence spectra provide compelling evidence of dopant incorporation into the 3PB crystals.

The fluorescence spectrum depicted in figure 2.10 illustrates the characteristics of both pure 3PB and stilbene-doped 3PB (TSB-3PB). It is widely recognized that the fluorescence of organic crystals primarily originates from the behavior of individual molecules [16]. Both the pure and doped crystal spectra span from 315 to 510 nm, with notable peaks around 360 nm. These spectra serve as evidence that the fluorescence intensity of stilbene-doped 3PB is comparatively lower than that of pure 3PB crystal. The aim of dopants in the host molecule is to increase the luminescence intensity. When stilbene was added, a noticeable decrease in emission intensity was observed. This observation may indicate that the doped molecules serve as entrapped impurities within the crystal lattices, affecting the overall emission properties of the material.

Incorporating a dopant into an organic scintillator serves to elevate the density of the triplet energy level, thereby enhancing delayed fluorescence and reducing the decay time. Given that triplets are recognized for their mobility in organic substances, the energy travels until two triplets collide and combine to create a singlet excited state. The duration of the delayed emission hinges on both the lifespan of the triplet energy level and the frequency of collisions among them [5, 17]. In a dual decay process, prompt fluorescence corresponds to the direct de-excitation of the singlet energy level transition, while delayed emission arises from the collisional interaction of pairs of the lowest excited triplet state of the molecules. The significance of the collisional interaction of the triplet energy level lies in its role in identifying neutron detection in an organic scintillator. However, in the present scenario, the rapid de-excitation of impurity molecules, both triplet and singlet excitons, has subdued the emission intensity, elucidating why the excited states of trapped impurities are not fully discernible by host molecules.

2.3.4 Lifetime measurement

A distinguishing feature of the organic scintillator as a detector was assessed using the time-correlated single photon counting method (TCSPC) [13]. In figures 2.11 (a) and (b), the decay time spectrum of the crystals doped with 10 mmol% of DPB and 10 mmol% of PPO was visually presented. The analysis of the decay time measurement involved fitting it with two exponential decay components to provide a more comprehensive understanding of the decay behavior.

$$F(x) = A_1 e^{\frac{-t}{\tau_1}} + A_2 \; e^{\frac{-t}{\tau_2}} \qquad (2.1)$$

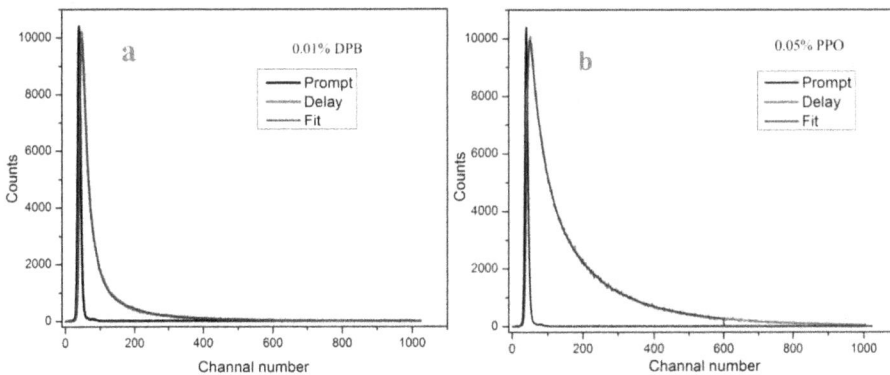

Figure 2.11. Fluorescence decay curve (a) 10% DPB (b) 50% PPO-doped 3PB crystal [10].

Table 2.2. Comparison of scintillation decay time with existing materials.

S. No.	Compound	Prompt	Delayed	Reference
1	Stilbene	4.5 ns	—	[5]
2	Anthracene	30 ns	—	[6]
3	p-Terphenyl	3.3 ns	—	[8]
4	Pure Naphthalene	14.4 ns	78.6 ns	[7]
5	PPO-doped Naphthalene	7.5 ns	50.2 ns	[7]
6	Pure 3PB	12 ns	30 ns	[11]
7	50 % PPO-doped 3PB	3.8 ns	17.8 ns	[10]
8	10 % DPB doped 3PB	2.08 ns	11.37 ns	[10]

The prompt and delayed emissions are denoted as A_1 and A_2, each with lifetimes τ_1 and τ_2. Typically, organic molecules exhibit fast scintillation decay times ranging from 2 to 30 ns. In table 2.2, the calculated fluorescence lifetime of the pure 3PB crystal and the 3PB crystal doped with 10% DPB and 50% PPO is provided, alongside data for existing organic scintillators. In comparison to previous findings, the 3PB crystal doped with 10% DPB and 50% PPO demonstrates a notably short decay time. Prompt fluorescence arises from the direct de-excitation of singlet energy level transition, while delayed emission is a result of collisional interaction between pairs of molecules in the lowest excited triplet state.

2.3.5 FTIR spectral studies

Figures 2.12(a) and (b) depict the FTIR transmission spectra of pure and stilbene-doped 3PB crystals. The absorption bands observed at 3055 and 3030 cm^{-1} can be attributed to the (C–H) stretching vibration, while the stretching vibrations at 1591 and 1492 cm^{-1} are assigned to the (C=C) group of the phenyl ring. Furthermore, the absorption bands at 1074 and 1026 cm^{-1} represent the phenyl ring C–H stretching in-plane bending vibration. The phenyl ring C–H out-plane bending vibration band is assigned at 746 and 690 cm^{-1} in the spectrum. Additionally, the absorption band at 869 cm^{-1} is due to the 1,3-meta position substitution of the phenyl groups. In comparison, as illustrated in figure 2.6, the FTIR spectrum of the doped crystal closely resembles that of 3PB. The absence of observable changes in the spectral features from the pure 3PB to the doped crystal suggests that the corresponding functional groups are sensitive to low-level impurity concentrations. Therefore, the lack of observed changes suggests that the very low-level concentration of the dopant does not significantly influence the parameters of the vibrational spectrum [18].

2.3.6 NMR spectra analysis

Figures 2.13 and 2.14 show the 1H NMR spectra of pure and doped 3PB. The singlet of three protons from the central phenyl ring was assigned at 7.78 ppm. The doublet of six protons present in the ortho position of the outer phenyl ring was observed at 7.71 ppm. Similarly, two triplets of six and three protons in the outer

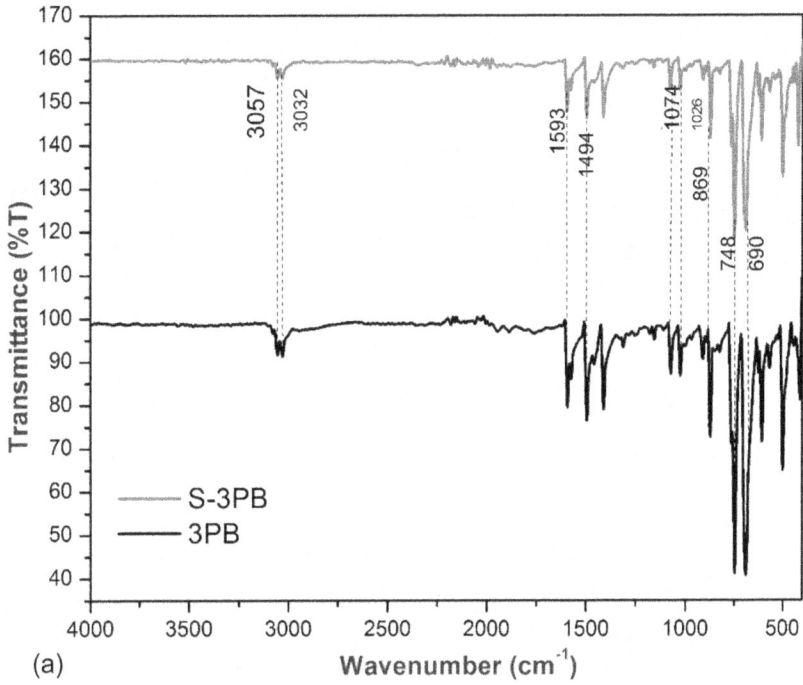

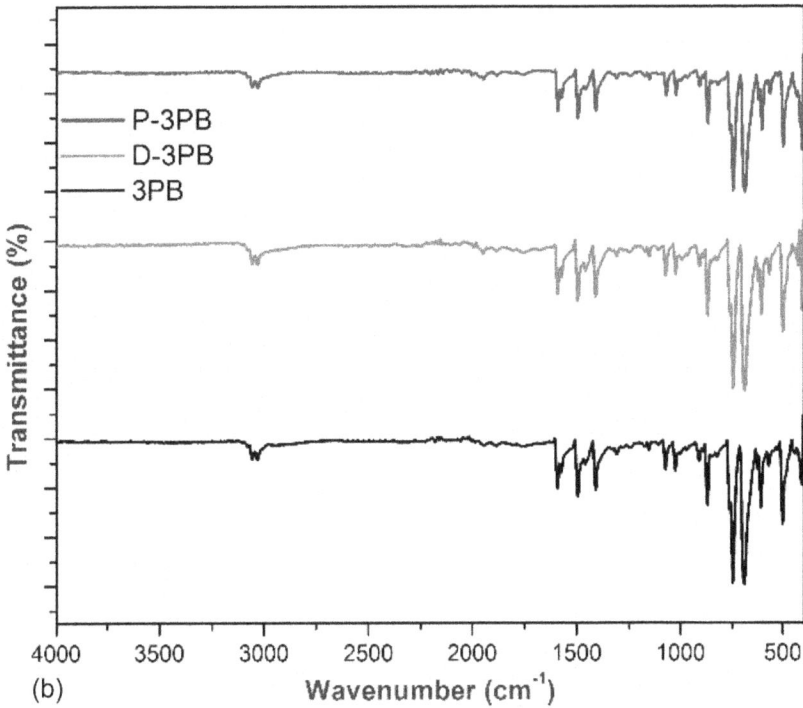

Figure 2.12. (a) Comparison of pure and stilbene-doped 3PB FTIR Spectrum, (b) FTIR spectra of pure 3PB crystal and D-3PB and P-3PB crystals.

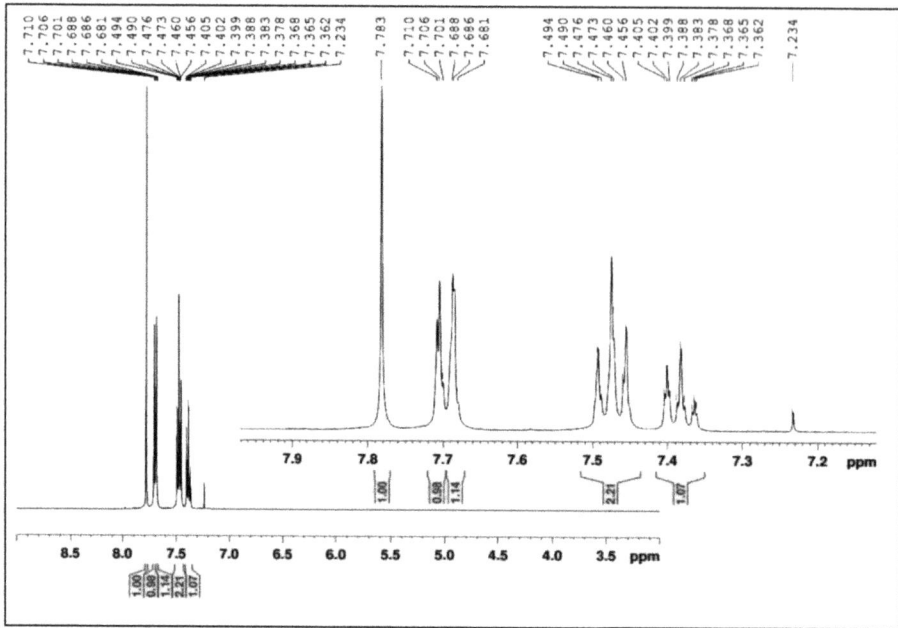

Figure 2.13. The ^{1}H NMR spectrum of pure 3PB insert picture extends the spectrum and shows all the positions of protons noticed in the molecular structure.

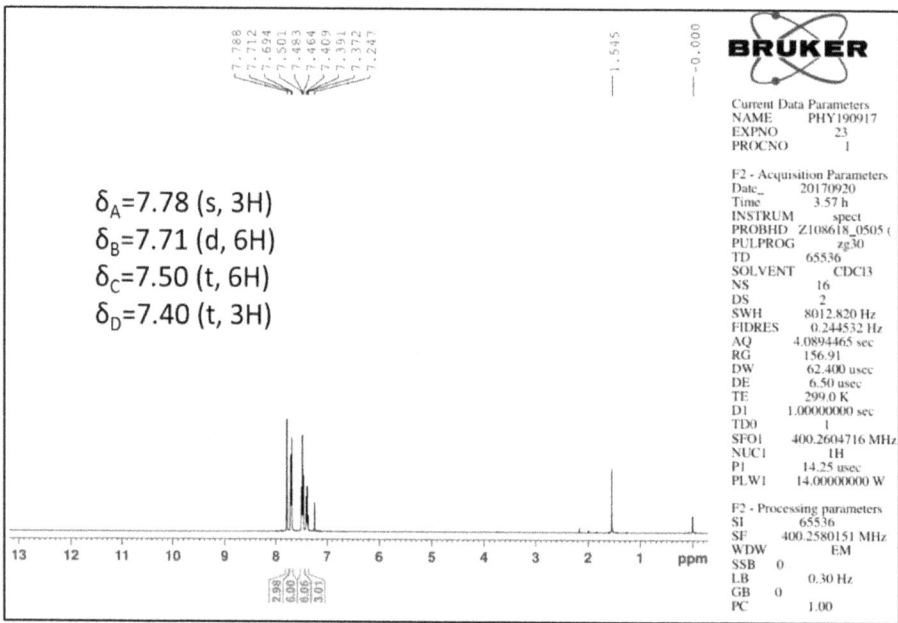

δ_A=7.78 (s, 3H)

δ_B=7.71 (d, 6H)

δ_C=7.50 (t, 6H)

δ_D=7.40 (t, 3H)

Figure 2.14. ^{1}H NMR spectrum of stilbene-doped 3PB.

phenyl ring's meta and para positions were observed at 7.49 and 7.39 ppm, respectively. The NMR spectrum of stilbene-doped 3PB is almost similar in figure 2.12. There are no other peaks observed in the spectrum except for 3PB. The difference is that each peak's chemical shift (δ) is slightly different due to a small addition of stilbene-doped in 3PB.

2.4 Conclusions

Unidirectional stilbene-doped 3PB crystal was grown by the unidirectional growth method. Different dopant low-level concentrations were verified to improve the fluorescence intensity of 3PB. Among the four different concentrations tested, it was found that 10 mmol% of DPB and 50 mmol% of PPO resulted in the highest emission intensity. The fluorescence spectra clearly indicate the successful incorporation of the dopants into the 3PB crystals. Further analysis involved measuring the fluorescence decay time, which showed a significant reduction compared to pure 3PB and other organic scintillators. The study demonstrated that the 10 mmol% DPB-doped crystal exhibited the maximum emission intensity and a short decay time of 2.08 ns, which is a notable improvement when compared to pure and other scintillator crystals. This enhancement can be attributed to the exchange coupling and mobility of triplets in the guest and host molecules.

The DPB, PPO-doped unidirectional bulk size (25 mm diameter) 3PB crystals were grown from the solution. The unit cell parameters of pure 3PB crystals were calculated and compared to those of 3PB crystals doped with DPB and PPO at dopant ratios of 0.01 and 0.05 mol%. The calculations revealed slight changes in the lattice parameters as the dopant ratio increased. However, the XRD pattern has no new phase formation or phase changes. The difference in peak intensity peak broadening reveals the lattice strain due to the addition of dopants. In the optical analysis, the same absorption edges were observed, but distortion in transmission spectra confirmed the dopants attributed to the pure 3PB crystal. The FTIR spectra for doped crystals concerning the corresponding spectral feature of pure 3PB. Hence, XRD and FTIR spectra show that toping such a small amount does not lead to noticeable changes in intermolecular interactions. The distortion in the UV absorption spectrum may indicate dopant incorporation in 3PB. The decrease in fluorescence intensity suggests that the dopant molecules appear as an impurity in crystal lattices, bringing about reduced fluorescence emission intensity of doped crystals. In the further results of FTIR and NMR spectra of doped crystals, no changes were observed, revealing that the small amount of dopant does not lead to noticeable changes in intermolecular interaction energy.

References

[1] Angelone M *et al* 2013 Properties of para-terphenyl as detector for α, β and γ radiation (arXiv:1305.0442v1) [physics. ins-det]
[2] Durairaj N, Kalainathan S and Babu S M 2023 Bulk growth of unidirectional 1,3,5-triphenylbenzene (TPB) single crystal for fast neutron detection by time of flight technique *Phys. Open* **17** 100170

[3] Yanagida T, Watanabe K and Fujimoto Y 2015 Comparative study of neutron and gamma-ray pulse shape discrimination of anthracene, stilbene, and p-terphenyl *Nucl. Instrum. Methods Phys. Res.* A **784** 111–4

[4] Bourne M M, Clarke S D, Adamowicz N, Pozzi S A, Zaitseva N and Carman L 2016 Neutron detection in a high-gamma field using solution-grown stilbene *Nucl. Instrum. Methods Phys. Res.* A **806** 348–55

[5] Zaitseva N, Carman L, Glenn A, Newby J, Faust M, Hamel S, Cherepy N and Payne S 2011 Application of solution techniques for rapid growth of organic crystals *J. Cryst. Growth* **314** 163–70

[6] Knoll G F 2000 *Radiation Detection and Measurement* 3rd edn (New York: Wiley)

[7] Balamurugana N, Arulchakkaravarthi A and Ramasamy P 2008 Growth of 2,5-diphenyloxazole-doped naphthalene crystal by Bridgman method and its fluorescence studies *J. Cryst. Growth* **310** 2115–9

[8] Galunov N Z, Lazarev I V, Martynenko E V, Vashchenko V V and Vashchenko E V 2015 Distribution coefficient of 1,4-diphenyl-1,3-butadiene in p-terphenyl single crystal and its influence on scintillation crystal light output *Mol. Cryst. Liq. Cryst.* **616** 176–86

[9] Ai Q, Chen P, Feng Y and Xu Y 2017 Growth of pentacene-doped p-terphenyl crystals by vertical Bridgman technique and doping effect on their characterization *Cryst. Growth Des.* **17** 2473–7

[10] Durairaj N, Kalainathan S and Kumar R 2017 Fluorescence emission and decay time studies on doped 1,3,5-triphenylbenzene scintillator crystal grown by solution growth technique *Mech. Mater. Sci. Eng. MMSE J. Open Access* **9**

[11] Durairaj N, Kalainathan S and Krishnaiah M V 2016 Investigation on unidirectional growth of 1,3,5-triphenylbenzene by Sankaranarayanan–Ramasamy method and its characterization of life time, thermal analysis, hardness and etching studies *Mater. Chem. Phys.* **181** 529–37

[12] Kalainathan S, Durairaj N and Kumar R 2018 Bulk growth, structural and optical properties of pure and doped 1,3,5-triphenylbenzene crystal by Sankaranarayanan–Ramasamy method for scintillator applications. *Int. J. Soc. Mater. Eng. Resour.* **23** 64–7

[13] Milledge H *et al* 1982 University College London, England, ICDD, Grant in-Aid

[14] Mullin J W 2001 *Crystallization* 4th edn (Oxford: Butterworth-Heinemann) pp 205–84

[15] Durairaj N and Kalainathan S 2018 Growth and characterization of stilbene doped bibenzyl scintillator crystal by solution growth technique *J. Mater. Sci.: Mater. Electron.* **29** 10480–6

[16] Wolf H C 1968 Energy transfer in organic molecular crystals: a survey of experiments *Adv. At. Mol. Phys.* **3** 119–42

[17] Zaitseva N, Glenn A, Carman L, Hatarik R, Hamel S, Faust M, Schabes B, Cherepy N and Payne S 2011 Pulse shape discrimination in impure and mixed single-crystal organic scintillators *IEEE Trans. Nucl. Sci.* **58** 3411–20

[18] Amorim da Costa A M and Amado A M Doping effects in p-terphenyl molecular crystals: a study by Raman spectroscopy *Solid State Ionics* **125** 263–9

IOP Publishing

Advances in All-optical Communication

Shanmuga Sundar Dhanabalan, Arun Thirumurugan and Sridarshini Thirumaran

Chapter 3

Performance analysis of SOA-based all-optical logic gates over FSO channel

Margarat Michael, Elizabeth Caroline Britto, J Jeyarani and K Bhuvaneshwari

Currently, the semiconductor optical amplifier (SOA) stands as a rapidly advancing technology in the realm of optical amplifiers, catering to a wide array of applications in the domain of all-optical signal processing. Optical amplifiers play a pivotal role in elevating optical powers, particularly for ultrashort pulses. With the escalating demands of high data rates in digital communication, optical amplifiers find extensive utility in enabling swift and efficient communication. Among the various SOA types, the reflective SOA (RSOA) emerges as a high-performance solution, frequently employed in tandem with cost-effective transceivers for modulation and data amplification in radio-over-fiber systems. Thanks to its dual functionality, RSOA exhibits superior gain and switching performance when compared to conventional SOAs. Within the scope of this chapter, we present the design of an all-optical XOR gate, configured to secure free-space optical (FSO) channels at 10 Gbps for binary inputs. This holds significant implications for FSO systems, which are poised to support secure and high-speed future-generation applications through the integration of innovative all-optical gate technologies. The results derived from the Optisystem 7 software simulations affirm that the designed SOA-based logic gate designs offer enhanced performance metrics in FSO system applications, encompassing output power of 10.9 dBm, negligible bit error rate (BER), and quality factor (QF) of around 60.

3.1 Introduction

Future generation networks and their data handling require the entire transmission channel to be all-optical in nature. Prominent technology benefactors such as IBM, INTEL, LUXTERA, and many more have already incorporated photonics in the form of photonic-integrated circuits. The basic building block of any photonic-integrated unit is the design of all-optical logic gates. Already researchers are experimenting on multiple methodologies including SOA, optical fibers, photonic crystals, etc., based all-optical logic gates to efficiently support such a huge data-handling requirement. The main requirement of all-optical logic devices is to be compatible with the transmission channel to ensure error-free transmission.

An all-optical communication network is used in optical telecommunication applications for increasing the productivity at lower operating cost. All-optical networks use digital optical elements compared to electronic devices and gates. All-optical signal processing fare in high-capacity core networks to keep away from inadequate of optoelectronic conversion. It requires optical add drop multiplexers and optical cross connects to carry out the functions data encoding encryption, addressing, etc. Optical communication technology plays a vital part in many industries today, as most applications today [1] demand more secure, faster, and more extensive systems for their communication network operations [2, 3]. While the idea of fully optical computers and processors is no longer a central focus, optical signal processing continues to be explored as a potential solution to many of the challenges faced by modern technology. Without doubt the increasing complication of mathematical and scientific problems and the requirement for novel ultrafast and mass-producible intermediate components for sensing and telecommunication applications require all-optical gate-based components capable of manipulating data signals at superfast speed levels above 1 Tbps. All-optical logic gates are the critical element [4] to support all-optical signal processing in the intermediate data processing devices that are used in data encoding, pattern recognition, code conversion, signal regeneration, addressing, optical computing, etc.

In electronic devices, logic levels 0 and 1 are represented by current or voltage. In optical devices, these levels are indicated by the intensity, polarization, or phase of light, with a specific threshold used to differentiate between them. In electronic processors, digital data can be transmitted through an optical fiber with light speed, but its switching speed is limited to maximum of 50 ps for the mean power of 0.5 mW and the energy value around 25 fJ for each switching [5]. Generally, in the semiconductor-based logic gates, the switching speed is limited by p-n junction capacitance. Simultaneously, the optical logic gate switching speed is in fs range [6] and is limited only by the propagating light velocity in it. This means comparatively it is a thousand times faster than equivalent electronic devices.

In an electronic device, input data are processed serially, but in optical devices photons with different wavelengths travel simultaneously in the same fiber without any cross or interference and process the data in parallel. Optical processing of data is free from electrical short circuits and is highly immune to electromagnetic interference.

In photonic transmission, free-space optics (FSO) delivers the vital combination of required qualities to bring the communicated traffic to the backbone of the optical fiber network. This work considers an all-optical logic gate such as XOR and OR using SOA. Its performance is verified over FSO link and also with fiber channel. In order to compare the performance of the FSO system, encoded binary data with an all-optical XOR logic gate [7, 8] using Semiconductor Optical Amplifier (SOA) over Mechzender Interferometric Structure (MZI) is transmitted both in wired and wireless modes of communication.

3.2 Applications, advantages, challenges, and models of FSO

FSO, also known as optical wireless communications, is a type of communication that sends data over the air to more than one transceiver. FSO communication is highly secure, directional, license-free, wide bandwidth interference-free communication with reduced cost and deployment time [9]. FSO systems operate at 1.5 Gbps and also have the potential to operate at high speeds. The most often utilized optical windows of FSO communication systems are 850 and 1600 nm [10]. 1550 nm is the most appealing wavelength among these since it has the low absorption properties of air. Based on the applications, the FSO can be classed as indoor or outdoor.

FSO is a line-of-sight (LOS) technology that sends and receives data, voice, and video using light as a carrier. FSO systems make bandwidth scalability, portability, rapid deployment, and redeployment possible [11]. This optical link does not necessitate the purchase of costly fiber-optic cables or the acquisition of spectrum licenses. Light is required for FSO technology, which can be provided by Light Emitting Diodes (LEDs) or Light Amplification by Stimulated Emission of Radiation (LASER) [12]. FSO utilizes LASER input analogous to fiber-optic communication, except with air as the propagation medium.

FSO, which was created by NASA and the military, has been utilized to offer fast communication links in remote regions for more than 30 years. Fiber optics and FSO technology are comparable and allow for the transmission of large amounts of data. Wavelength division multiplexing (WDM) technologies can now operate in free space with the usage of similar optical transmitters and receivers. FSO is a straightforward technique that connects optical transceiver devices. By combining a transmitter and a receiver, each transceiver unit delivers complete duplex transmission.

Each FSO unit consists of an optical transmitter with a laser optical source, as well as a telescope or lens for transmitting light in the atmosphere and an optical data receiver. The receiving telescope or lens is then connected by means of optical fiber to a highly sensitive receiver. By using relays in between source and destination nodes, FSO systems are operated across several kilometers even if there is no direct LOS path. Photons, which make up the light emitted by lasers, travel faster than electrons in wired telecommunication systems and can carry a large amount of data [13]. FSO is superior to RF communication because it has significantly bigger dimensions, involves no spectrum licenses, uses less power, has a more portable transmitter and receiver construction, is more secure, is less expensive, and provides superior interference protection [14]. Satellite and deep space communications make use of FSO [15].

FSO is an upgradeable technology because it is supported by a range of suppliers [16]. This functionality assists businesses and service providers in safeguarding their investments on embedded telecommunication networks. FSO does not require any security software and is extremely resistant to saturation and interference. FSO is easy to set up; it can be installed behind windows, removing the need for expensive rooftop rights. FSO combines fiber-like speed with wireless flexibility. FSO offers a solution to last-mile bottleneck issues. It connects consumers to the fiber-optic backbone, bridging the bandwidth gap. The service providers of FSO Applications include DS3 Services, Access and Web Hosting–Metro Network Extension, Access and Backhaul, Gigabit Ethernet, Service Access/Provider–Network Restoration, Ring Closure—SONET and FSO—The Metro Ring Access.

FSO encompasses various aspects of telecommunication networks, including applications for wireless communication service providers. It offers access to consolidated, block-level data storage via a storage area network, and provides connectivity in the last-mile and LAN segment interconnects to link remote locations. FSO can also serve as a backup for fiber optic link failures and assist with network extensions. The FSO system is quick to deploy, and new networks and core equipment can also be connected quickly. FSO makes it simple to implement SONET rings, Backhaul: FSO can serve as a backhaul network, transporting high-speed and high-data-rate cellular phone traffic from antenna towers to the PSTN with the increased transmission speed, Service acceleration: FSO may provide immediate service during the fiber infrastructure installation process, WAN Access bridging: FSO provides high-speed data in the WAN and serves as the satellite terminal backbone, FSO allows for communication across point-to-point linkages such as airplane to ground, satellite to ground, and short and long-range communication, Military access: FSO system is a highly secured because an intruder cannot intercept FSO and is used in many military applications.

The short wavelengths of the optical system, which includes air attenuation in the form of scattering and absorption, as well as optical scintillation, affect the overall performance of FSO devices [17]. Gas particles such as water vapor, CO_2, and CH_4 and particles such as haze, fog, smoke, and clouds suspended in the atmosphere produce attenuation in FSO systems. The FSO attenuation causes reduction in the optical power received and scintillation leads to random fluctuations in both phase and amplitude of the detected optical signal. In general, the performance of FSO systems is limited by pointing errors caused by factors such as weak earthquakes, building sway, strong winds, cooling, and thermal expansion. These issues lead to transceiver misalignment, which in turn degrades the overall performance in the FSO system.

The significant challenges in FSO communication systems include absorption loss and scattering loss [18], and the fact that the input laser beams need to interact with gas particles and molecules in the atmosphere while it is propagating through the atmosphere of the earth. The primary causes of loss in an atmospheric channel are absorption and scattering. Absorption loss in wireless optical communication depends on its wavelength and is affected by the amount of water vapor in the atmosphere. Scattering occurs as a result of atoms interacting with the transmitted optical signal. During fog, the water particle become denser and diffracts the optical signal transmitted [19]. Absorption as well as scattering contribute to the attenuation. High attenuation of 350 dB km^{-1} has been observed during dense fog conditions. The wavelength utilized for FSO transmission is comparable to the fog size, causing the transmitted optical signal to be attenuated. Fog attenuation is inversely proportional to transmitted light wavelength. Rain does not have the same effect as fog [20]. Raindrops are larger than the wavelength utilized in an FSO transmission system. Attenuation would be 6 dB km^{-1} for a rainfall rate of around

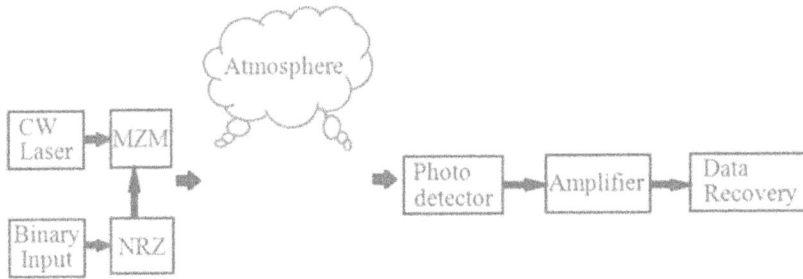

Figure 3.1. The fundamental FSO system.

2.5 cm h^{-1} and would increase with increased rainfall rate. Snow particles are between fog and rain particles in size. Snow has greater attenuation than rain but lower than fog particles [21]. The fundamental FSO system [22] is made up of three basic blocks–a transmitter, transmission channel, and a receiver–as shown in figure 3.1.

A coder, modulator, driving circuit, and optical lens make up the transmitter block in any FSO system. The input source data is modulated in the optical carrier, and through atmospheric propagation, the modified optical signal reaches the receiver. In the literature, various modulation schemes are suggested. External and internal modulations are the two forms of modulation employed in FSO. The later modulation techniques have good quality with increased bit rate. On the other hand, the earlier modulation techniques have advantages like compactness, simplicity, and cost effectiveness. Intensity modulation (IM), in which optical source intensity is changed by an RF or mm-wave signal, is the most preferable type of modulation. Demodulation is accomplished over direct optical carrier detection and conversion with photo detector. Numerous researchers have described optical carriers that are modulated in terms of frequency and phase. The information bearing system additionally includes state of polarization as a parameter. FSO is a line of sight (LOS) technology with no obstructions along the propagation track. In order to prevent aiming errors, the transmitter and receiver should be properly positioned. The information signal propagates through free space to reach the intended receiver. The optical signal is gathered by the receiver and transformed to an electrical signal by photo detectors. Following detection, the original data is retrieved using demodulators.

In this chapter, all-optical logic XOR and OR gates using SOA-based non-linearity are presented and the performance of the designed system verified over a FSO channel.

3.3 Related works

The SOA can be classified into three categories: Travelling Wave SOA (TWSOA), Fabry–Perot SOA (FPSOA), and Reflective SOA (RSOA). The TWSOA is good at detecting and amplifying the optical signals. All-optical gates such as OR, XOR, NOR, and NAND are designed with the SOA-MZI, which exhibits high operational speed and high extinction ratio [23]. The anti-reflecting-based SOA was observed to be operating well in storing high-speed optical data, which is in the range of 80–100 Gb s^{-1}, for long-term [24]. As an advancement in the design of all-optical devices, the SOA-MZI based half-adder and subtractor are designed, which are constructed from the basic logic gates [25]. The design of all-optical shift register using the D-flip flop as the basic building block is proposed, having the features of compact size, high speed and low-power consumption [26]. The ripple carry adder is a very promising logic circuit in the field of digital signal processing and microprocessors, and it is designed with the basic logic gates using the SOA-MZI [27]. After this, the parity generator and checker circuits are designed for error-free data transmission possessing high data bit rate and extinction ratio [28]. Recent advancements in all-optical logic gate computing for the application of parallel information processing are presented in [29]. Fabry–Perot SOA-based clock and data recovery circuits have been demonstrated, incorporating a logic AND operation. This operation occurs between signals to ensure proper data synchronization [30]. Both these categories of SOA have the disadvantages of nonlinearity and polarization, and in order to overcome these issues, the reflective SOA (RSOA) is used. It works on the mechanism of reflection of the system, giving the features of modulation linearity, high gain, and low injection current. An high-speed all-optical XOR gate was designed and simulated. Presented in [31], it was observed to have better switching performance than the previously discussed SOAs. A WDM is used for the design of bidirectional triple-play services, since it is economical and provides better signal-to-noise ratio, high receiver sensitivity, and an eye diagram [32]. SOA direct modulation was studied along with the micro-ring resonator notch filter to remove the narrowband frequency and to repair the electrical bandwidth of SOA [33]. The all-optical parity generator and checker were simulated in MATLAB based on the TSOA using the soliton pulses. Its performance was analyzed with parameters like extinction ratio and quality factor [34]. Enhancement in the performance of the all-optical XOR gate results in high data bit rate and its usage in high-speed signal processing applications [35]. The role of reversible gates in the optical field is very promising in computing and ultrafast data processing with high gain and low noise. Recently a TSOA-based all-optical logic device design with a latch circuit made up of two RSOA-based switches was proposed. It operates at 200 Gb s^{-1} with high contrast ratio (CR), extinction ratio (ER), and quality factor (Q) values [36]. A recently designed all-optical parity generator and checker using RSOA-based optical XOR logic gates achieves quality factors up to 57.15 and extinction ratios up to 37 dB, validated through OptiSystem simulations [37]. Thus, the RSOA has proven to be superior to existing SOAs by exhibiting polarization and wavelength dependency, as well as a low noise figure.

3.4 Basics of semiconductor optical amplifiers

An SOA is a compact nonlinear optical amplifier used to design various all-optical logic gates in optical communication systems, particularly for applications requiring high-speed data processing. The optical devices designed with an SOA exhibit low-power consumption, increased switching speed, reduced latency, and outstanding rejection to electromagnetic interference. The nonlinear effects of SOAs extend their application in all-optical logic gate design as promising modules [38].

Fundamentally, an SOA is an optoelectronic unit characterized by its accessing direction. Its basic structural design differs slightly from that of conventional laser diode and includes the formation of the population inversion as a result of input electric current injection, non-radiative recombination, stimulated and spontaneous emission. Unlike semiconductor lasers, extremity mirrors are replaced with an antireflection coating, and the light reflections are reduced by using a window or angled facet arrangement in the structure. General SOA manufacturing is done with various combinations of III–V group alloys such as indium phosphide (InP) and gallium arsenide (GaAs), depending on the bandgap and crystal lattice characteristics requirement [39]. Specifically for operation around 1550 nm, InP is typically used as the substrate, while InGaAsP serves as the active layer.

The primary purpose of any amplifier is to enhance the power of the incoming signal. In an SOA, the light or input signal is amplified in a specific area known as the active region. The basic functional structure of an SOA, illustrated in figure 3.2, consists of a front and rear facet, along with a waveguide that facilitates the propagation of light to amplify the incoming signal power. The reflectivity of the structure depends on the front and rear facet characteristics, and active medium gain values are very important in SOA design applications. Facet reflectivity is the main parameter in SOAs to create optical cavities, travelling wave devices, and for light propagation in a specific direction. Propagating mode selection also depends on the reflectivity losses. Hence, it is controlled either by increasing or decreasing the reflectivity of the facets based on the application requirement. The semiconductor material gain is the result of current injection over the device structure. The input power amplification is materialized in the active region under an external current input supply.

In general, SOA-based techniques in optical system applications consume less power but speed limitations exist due to slow gain and phase recovery dynamics of the element. The various nonlinear effects of this element can be applied in several areas, including format conversion, wavelength conversion, and switching [40, 41]. The SOA nonlinear functionality is utilized for the design of all-optical logic gates where the input probe and control pump light signals are coupled into the SOA under phase control mechanism. The SOA nonlinearities are broadly classified into

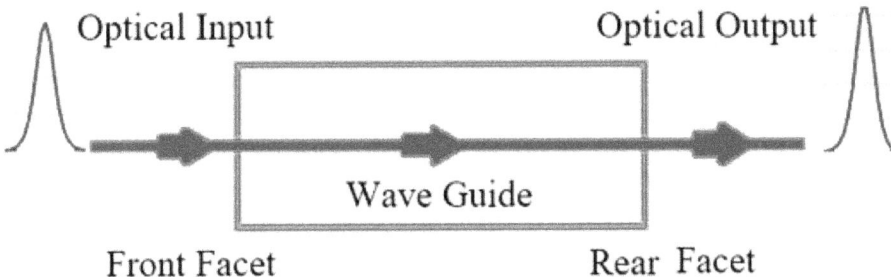

Figure 3.2. SOA basic illustration.

Table 3.1. Simulation parameters of SOA.

S. No.	Parameters	Value 10 Gbps
1	Power of probe signal	1 mW (XOR) 1.05 mW (OR)
2	Power of pump signal	0.3 mW
3	Wavelength of probe beam	1552.5 nm
4	Wavelength of pump beam	1558 nm
5	Injection current	0.1 A
6	Pulse width	0.3 Bit
7	Length	500 μm
8	Height	0.08 μm
9	Width	3 μm
10	Differential gain	$2.78e^{-020}$ m^2
11	Optical confinement factor (OCF)	0.3
12	Line width enhancement factor (LWF)	±5
13	Carrier density at transparency	1.4e + 024 m^3
14	Recombination coefficient A	143e + 008 1/s
15	Recombination coefficient B	1e−016 m^3 s^{-1}
16	Recombination coefficient C	3e−041 m^6 s^{-1}
17	Initial carrier density	3e + 024 m^{-3}

two categories. One is due to alteration of basic properties of the input signals like phase, amplitude, and polarization and the other is based on the generated new frequency components inside the device active medium. They are classified in detail as self-gain modulation (SGM), self-phase modulation (SPM), self-induced non-linear polarization rotation, cross-phase modulation (XPM), cross-gain modulation (XGM), cross-polarization modulation (XPolM,) and four-wave mixing (FWM) [42]. The standard simulation parameters for the design of SOA-based logic devices using OptiSystem software are given in table 3.1.

3.4.1 Types of SOA

The optical gain and gain medium of an SOA are based on the stimulated emission process with recombination of electron-hole pair [43]. The incoming photons are absorbed by the SOA when there is no electrical injection. However, at the same time the linear and nonlinear noise is present at the output. SOA devices operate in any one of three arrangements [44]:

(a) Traveling wave SOA (TWSOA)

(b) Fabry–Perot SOA (FPSOA)

(c) Reflective SOA (RSOA)

3.4.1.1 TWSOA

In the TWSOA the input signal is passed only once to get amplified along the active region without using any reflecting facets as shown in figure 3.3. Reflectivity for the SOA must be very small ($<0.1\%$) to function as TW amplifier. The active gain is increased through length extension of an active medium and keeping the reflectance value as zero. The advantages of TWSOA include low polarization sensitivity and high optical bandwidth. The three different approaches used in the structure to decrease the reflectivity in the mirrors are:

(a) An antireflection coating at the mirrors

(b) The active region tilting in the direction of facets

(c) Introducing buffer material between the facets and active region

The carrier density and electric field of an SOA is described with rate equation estimation [18] in the following equations (3.1–3.9). Material gain g_m is related to carrier density $N(t)$ of an SOA by

$$g_m(t) = A_g(N(t) - N_0) \tag{3.1}$$

The relationship among material gain g_m and net gain coefficient g is given by

$$g(t) = \Gamma g_m(t) - \alpha \tag{3.2}$$

where N_0 is the carrier density at transparency point; A_g and α are differential gain and effective loss coefficients, which contain absorption and scattering losses; and Γ represents confinement factor of optical signal.

The polarization state is assumed to be constant during the XPM process [45] as few active devices including SOA use low-power input in the order of 0 to 10 dBm. Hence, gain G over a propagating distance z is defined as

$$G(t, z) = e^{(g(t)z)} \tag{3.3}$$

The carrier conservation of the active medium is defied by carrier density rate equation as stated in equation (3.4). It accounts for carrier generation rate, current density, and averaged recombination inside the active layer.

$$\frac{N(t)}{\tau_S} = R_A N(t) + R_B N^2(t) + R_C N^3(t) \tag{3.4}$$

where τ_S is lifetime of the carrier and R_A, R_B, and R_C are non-radiative, radiative spontaneous recombination, and auger recombination coefficients, respectively.

Carrier density $N(t)$ is calculated by neglecting shot noise, carrier diffusion, and amplified spontaneous emission noise of the active device using the following equation:

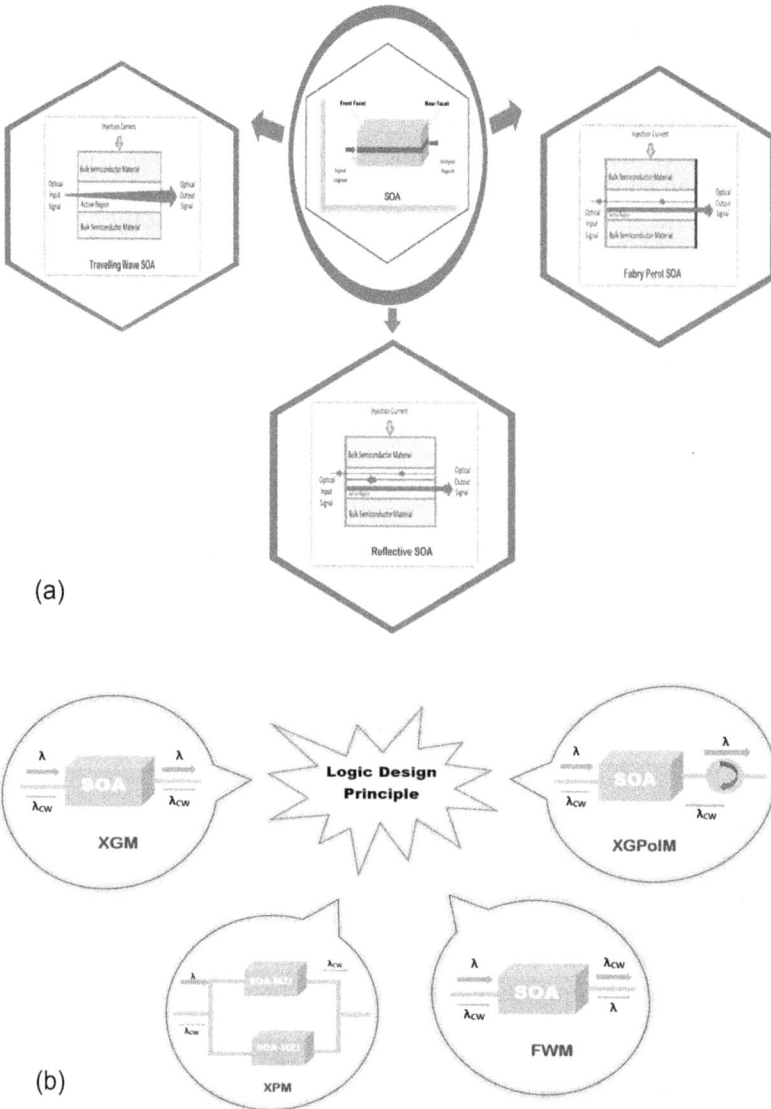

(a)

(b)

Figure 3.3. SOA types and their nonlinearity.

$$\frac{dN}{dt} = \frac{J}{qd} - \frac{N}{\tau_s} A_g (N - N_0) \frac{I}{hf} \qquad (3.5)$$

where I is light intensity, J is injection current density, q is the electron charge, h is Planck's constant, f is light frequency, t is time, and d is the active layer thickness.

$$\frac{dN}{dt} = \frac{I_P}{qV} - \frac{N}{\tau_s} \Gamma A_g (N(t) - N_0) \frac{P(N, t)L}{Vhf} \qquad (3.6)$$

where I_P is the input injection current, $V = L \times w \times d$ is volume of the active region, and w and L are the amplifier's width and length, respectively.

Average power over the SOA length is given by

$$P(N, t) = \int_0^L \frac{P(N, z)}{L} dz = \int_0^L \frac{P_{in} G(t, z)}{L} dz = P_{in} \frac{e^{(g(t)L)} - 1}{g(t)} \qquad (3.7)$$

The subsequent output optical field is

$$E_{out}(t) = E_{in}(t) e^{\frac{((1+j\delta)g(t)L)}{2}} \qquad (3.8)$$

$$P_{out} = P_{in} e^{(g(t)L)} \qquad (3.9)$$

3.4.1.2 FPSOA

The FPSOA is identical to the Fabry–Perot laser. The two surfaces at the end of lasers are cleaved in order to make them function as mirrors as shown in figure 3.3. The input light propagating inside the active region gets reflected from cleaved facets many times and is amplified. The back-and-forth reflection at the mirror surfaces of FPSOA amplifies the input signals with higher intensity. The amplified output is sensitive to optical input frequency and temperature. FPSOA offers good optical feedback and enhances the SOA gain with narrow bandwidth. The free spectral range ΔV_L of the Fabry–Perot cavity is used to calculate the amplification factor of the FPSOA as in equation (3.10):

$$\Delta V_A = \frac{2\Delta V_L}{\pi} \sin^{-1} \frac{1 - G\sqrt{R_1 R_2}}{(4G\sqrt{R_1 R_2})^{1/2}} \qquad (3.10)$$

The FPSOA amplifier bandwidth is only a small fraction of the Fabry–Perot cavity's free spectral range and is typically $\Delta V_A < 10$ GHz and $\Delta V_L \sim 100$ GHz. This feature makes Fabry–Perot amplifiers an inappropriate choice for most optical system applications

3.4.1.3 RSOA

This is a special form of SOA made using an anti-reflective and high reflective coating at the front and rear facet, respectively. A mirror is placed in the front and rear surface of the RSOA active device to reflect the light back across the amplification region as shown in figure 3.3. Thus, the device input and output are at a single facet and the device becomes more compact due do the backward and forward amplifications. This highly reflective structure in the active region conveniently admits an optical input signal travelling from forward propagating direction to increase its intensity and simultaneously its input optical signal is amplified also transformed as it exits out of the backward propagating direction without utilizing any added mechanism. The upstream modulation and downstream amplification of RSOAs dual processing together along with their capability of producing higher gain with lesser injection currents, increased modulation linearity, lower polarization dependency, less noise figure, and lower temperature sensitivity than conventional SOAs have rendered them indispensable for modern access realization for various applications that rely critically on the handling of bidirectionally propagating data and where colorless (i.e. wavelength-independent process) is highly desired.

3.4.2 Principal operation of SOA and design of XOR gate

The SOA-MZI is one of the most effective configurations for scheming nonlinear logic gates for signal retiming, reamplification, and regeneration applications. These regenerators consist of SOAs placed over one or both arms of the interferometer arrangement to permit high bit rate operations. This work considers the design of all-optical XOR and OR gates. Both all-optical gates are designed using SOA-assisted MZI configuration with the XPM nonlinearity. The phase change of the probe signal determines the output logic of the device. When both the signals propagate parallel into SOA, the refractive index inside the active region is altered and controls the phase change of the input probe signal. The SOA-MZI arrangement used for designing the intended XOR logic operation is given in figure 3.4.

SOAs are highly nonlinear, require low power, and are suitable for wavelength conversion and all-optical signal processing applications. They are simple to interface with other optical devices, which is essential for device construction when implementing all-optical logic gates. The SOA-MZI configuration is also called the symmetric MZI arrangement in the general SOA-MZI configuration; the optical pump signal is applied to the 3 dB coupler to divide the control signal equally between the upper and lower arms of the MZI structure. The probe signal is subjected to the effects of optical nonlinearity, while the optical control signal

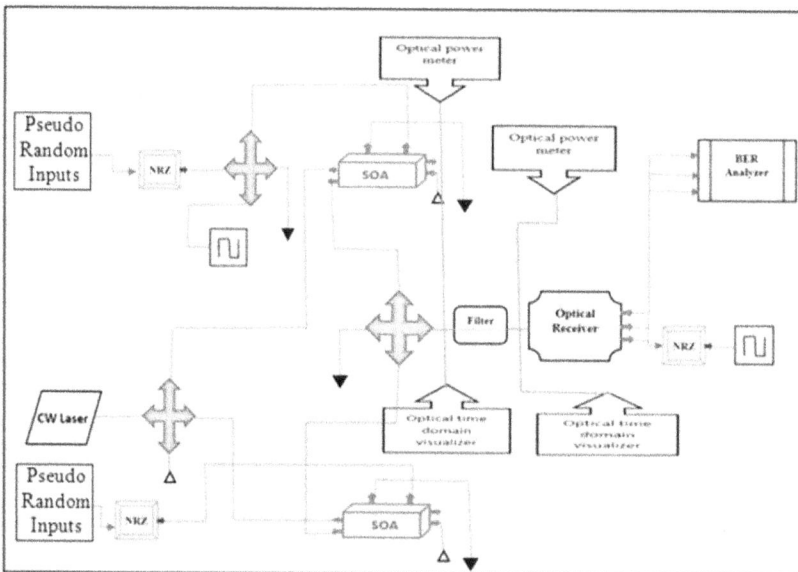

Figure 3.4. SOA-MZI schematic for XOR logic gate.

regulates the dynamics of the nonlinear optical effects. In both SOAs, the pulse of the control signal causes carrier depletion and controls the gain and phase of the input probe signal pulse, which are referred to as XGM and XPM, respectively. These nonlinear optical effects are incredibly efficient, allowing miniature devices to be controlled with a low-power signal. Furthermore, the SOA amplifies the control signal that depletes the carriers, lowering the needed power level of the input control signal.

If the MZI structure is well aligned, then the total power can be fed into one of the outputs by adjusting the path-length difference for a specific optical frequency, but fringe patterns are formed in both the outputs for the misaligned beams. The variations in path differences affect these interference pattern shapes. When the SOA gets input data signals, the refractive index of the medium varies and density of the carrier gets modulated. This leads to a phase shift over the counter propagating pump signal through the SOAs with respect to the intensity variations of an input signal. The amount of phase modulation accounted for by the probe signal during propagation inside the SOA is explained with the following equation (3.11):

$$\Delta\varphi = 2\pi n_0 \frac{L}{\lambda} + \alpha(\log G - \log G_0) \tag{3.11}$$

where λ is the data input signal wavelength propagating through the SOA, L is the length of the SOA active region, n_0 is medium refractive index, α is the SOA line-width enhancement factor, and G and G_0 are the linear and saturated device gain, respectively.

This equation is based on the analytical model of the wavelength converter under adiabatic approximation and the instantaneous response of the SOA. The hypothesis is that the XPM converter-based device immediately responds to the modulating input signal holds up to 10 Gbps [46].

By appropriately setting the bias currents, optical powers, and by SOA parameters design, the SOA control signals from the MZI arrangement interfere constructively or destructively at the interferometer and give the intended logic output based on the applied two input signals. In general, various logic outputs are achieved using XPM amid the SOA-MZI arms and are enhanced by optimizing SOA-MZI structure parameters. The amount of phase shift produced between the arms determines the output logic of the device and is defined by the following equation (3.12):

$$\Delta\varphi(t, z) = \sum_{i=1}^{k} \beta L \left[\frac{a\Gamma I}{\omega q t_a P_L} \exp\left(\frac{\Delta P_{1i}}{\tau}\right) + \frac{1}{\tau} \sum_{n=1}^{M} \frac{(-1)^n}{P_L^{\overline{n}}} \exp\left(\frac{\Delta P_{ni}}{\tau}\right) \left[a\Gamma L N_i^n(t_{ni}, Z_{ni},) - g_i P_i^{n-1} (t_{(n-1)i}, Z_{(n-1)i}) \right] \right] \tag{3.12}$$

where t is the time, τ is the carrier recombination time, z is the position towards the light propagation direction, t_a is the active region thickness, a is the differential gain, P_i (t,z) is the ith channel power, N_t is the transparency carrier density, ω is the waveguide width, $\Gamma = t_a/t_{eff}$ is confinement factor, and t_{eff} is the waveguide effective thickness. $N(t_1, z_1)$ and ΔP_1 are the time variation of carrier density and power. In order to reduce the system complexity, the derivative power coefficient P_L and the derivative carrier density N are considered as $\partial P_L (t_1, z_1)/\partial t_1$ and $\partial N (t_1, z_1)/\partial t_1$ respectively.

The refractive index in the active region of the SOA depends on the density of carriers N. This carrier density decreases with control signal propagation and modifies the refractive index, which sequentially increases the phase modulation of the data signal. The following first-order differential equations [47, 48] describe the nonlinear features of an ideal SOA that cause the phase change in the input probe signal:

$$\frac{dh_{CD}(t)}{dt} = \frac{(h_0 - h_{CD}(t))}{\tau_c} - (\exp[h_{CD}(t) - h_{CH}(t) - h_{SHB}(t)] - 1)\left(\frac{P_{in}(t)}{E_{sat}}\right) \quad (3.13)$$

$$\frac{dh_{CH}(t)}{dt} = -\frac{h_{CH}(t)}{\tau_{CH}} - \frac{\varepsilon_{CH}}{\tau_{CH}}(\exp[h_{CD}(t) + h_{CH}(t) + h_{SHB}(t)] - 1)P_{in}(t) \quad (3.14)$$

$$\frac{dh_{SHB}}{dt} = -\frac{h_{SHB}(t)}{\tau_{SHB}} - \frac{\varepsilon_{SHB}}{\tau_{SHB}}(\exp[h_{CD}(t) + h_{CH}(t) + h_{SHB}(t)] - 1)P_{in}(t) - \frac{dh_{CD}(t)}{dt} - \frac{dh_{CD}(t)}{dt} \quad (3.15)$$

The SOA output gain and corresponding phase change in the MZI structure are [49] calculated from:

$$G_{SOA}(t) = \exp(h_{CD}(t) + h_{CH}(t) + h_{SHB}(t) \quad (3.16)$$

$$\varphi_{SOA}(t) = -0.5(\alpha h_{CD}(t) + \alpha h_{CH}(t) + \alpha h_{SHB}(t) \quad (3.17)$$

where τ_c, τ_{CH}, and τ_{SHB} are the carrier lifetime, temperature relaxation rate, and carrier scattering rate; ε_{CH} and ε_{SHB} are the nonlinear gain suppression factors; α_{CH} and α_{SHB} are the LWF values; and $h_{CH}(t)$, $h_{CD}(t)$, $h_{SHB}(t)$ are the carrier heating, gain carrier depletion, and spectral hole burning of SOA, respectively.

The SOA electric field [50] with respect to phase and gain of the electric input filed is

$$E_{out}(t) = \sqrt{0.5P_{cw}} \times \exp(-0.5\log(G_{SOA}(t)) + j\varphi_{SOA}(t)) \quad (3.18)$$

where P_{cw} is the power of the CW input signal and G and φ are the gain and phase of the SOA.

Reproduced with permission from [52].

3.5 Design of OR gate using RSOA

The RSOA-MZI based OR gate is designed as shown in the schematic diagram in figure 3.5. A 3 dB cross coupler is used to deliver the pump signal, which has a frequency of 1558 nm, to the MZI structure. The upper arm of the structure contains the Pseduo input bits A and B. When the input signals are logic 0, the upper arm RSOA and lower arm RSOA of the OR gate remain the same, causing the probe signal to perceive the same gain. Both signals are cancelled by destructive interference when they are merged at the couplers, and the OR gate's output will be logic 0. The upper arm RSOA in the MZI experiences a drastic decrease in gain when compared to the lower arm RSOA when any one of the input signals is logic 1.

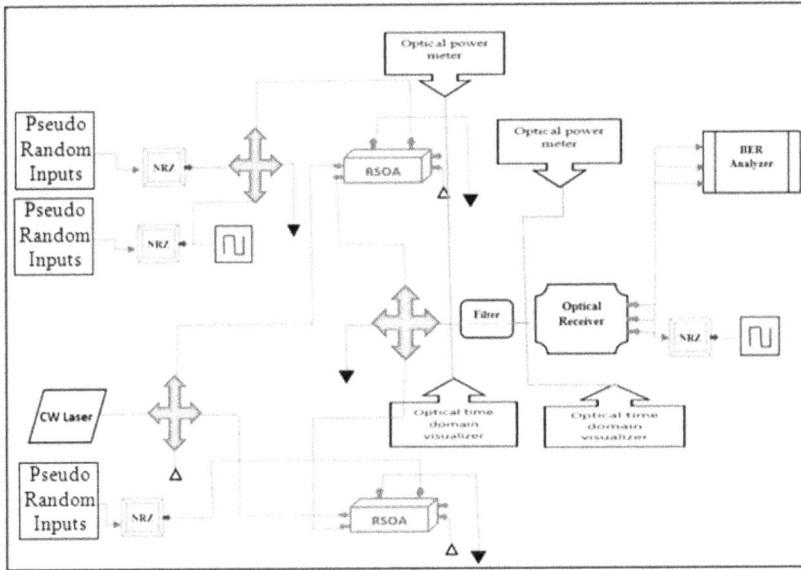

Figure 3.5. Schematic of all-optical OR gate using RSOA.

The XPM is then placed against the lower arm of the SOA, causing a differential phase difference, and the control then acquires a phase shift. The output power is increased and logic 1 is reached at the output with radian phase shift at the active region.

3.6 All-optical device design using SOA and RSOA

In order to verify the performance of all-optical gates, a type of all-optical device named octal to binary code converter was designed using RSOA and its performance compared with the SOA-based all-optical octal to binary code converter from our previous works [18, 51]. The XPM nonlinearity is used to construct the intended converter, which consists of the three OR gates as mentioned in figure 3.6. The inputs O_0, O_1, O_2, O_3, O_4, O_5, O_6, and O_7 are used to apply octal data at a frequency of 1552.52 nm. Three equally sized portions of bit O_7 power are divided and provided as input to the planned gates OR_1, OR_2, and OR_3. The gates OR_1 and OR_2 each receive an equal share of the power of bit O_6. The designed gates OR_1 and OR_3 are given the power of bit O_5 split into three equally sized pieces as input. The planned gates OR_2 and OR_3 are given the power of bit O_3 split into two equal portions. Gates OR_3, OR_2, and OR_1 are each given the power of bits O_4, O_2, and O_1, respectively. The binary code values shown as X, Y, and Z, respectively, illustrate the converted code produced from the octal bits O_0, O_1, O_2, O_3, O_4, O_5, O_6, and O_7. Table 3.2 lists the code converter's simulated findings, which are validated by the output powers of -100 dBm for logic 0 and 0.631 dBm for logic 1. In our previous work the SOA-based octal to binary code converter yielded a output power of

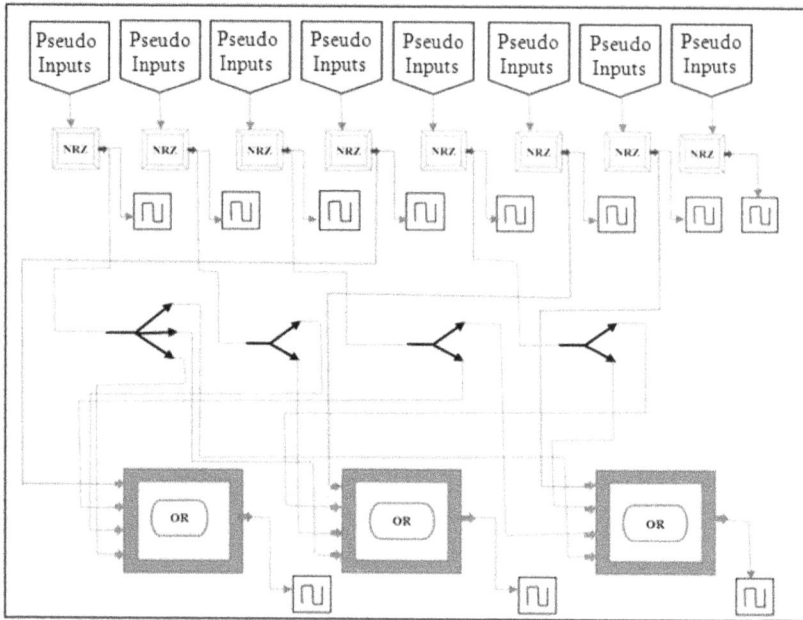

Figure 3.6. Schematic of all-optical octal to binary converter using RSOA-based OR gates.

Table 3.2. Truth table of octal to binary converter.

			Octal input					Output power in dBm		
O_0	O_1	O_2	O_3	O_4	O_5	O_6	O_7	A	B	C
1111	0000	0000	0000	0000	0000	0000	0000	−100	−100	−100
0000	1111	0000	0000	0000	0000	0000	0000	−100	−100	0.631
0000	0000	1111	0000	0000	0000	0000	0000	−100	0.631	−100
0000	0000	0000	1111	0000	0000	0000	0000	−100	0.631	0.631
0000	0000	0000	0000	1111	0000	0000	0000	0.631	−100	−100
0000	0000	0000	0000	0000	1111	0000	0000	0.631	−100	0.631
0000	0000	0000	0000	0000	0000	1111	0000	0.631	0.631	−100
0000	0000	0000	0000	0000	0000	0000	1111	0.631	0.631	0.631

1.92 dBm for logic 1 and −76.29 dBm for logic 0. The corresponding truth table for the octal to binary converter is given in table 3.2. The input–output timing diagram for the designed OR gate is given in figure 3.7.

Table 3.3 compares the obtained RSOA-based and SOA-based code converter and it is clear that the performance of the designed converter is better when implemented using RSOA due to its higher gain characteristics. The eye diagram in figure 3.8 calculated from the utilized OptiSystem software is beneficial to understand the performance of the system in terms of its performance metrics such

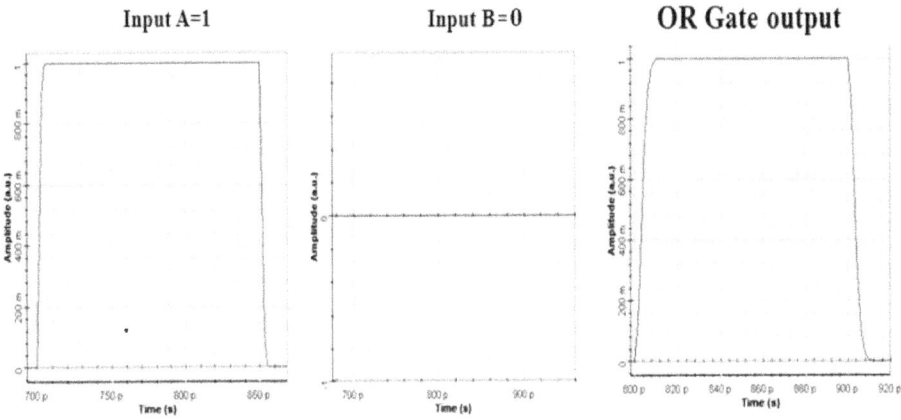

Figure 3.7. Input output timing diagram.

Table 3.3. Comparison of SOA-based and RSOA0based designs.

Output bits	SOA-based design			RSOA-based design		
	Q-factor	BER	ER (dB)	Q-factor	BER	ER (dB)
X	56.91	0	78.31	47.8941	0	100.64
Y	59.63	0	78.24	49.6634	0	100.64
Z	58.34	0	78.34	46.3539	0	100.64
QF		55			49	
ER (dB)		87			100.64	

as quality factor, extinction ratio, and bit error rate. The standard equation used to calculate QF, ER, and BER are given in equations (3.19)–(3.21):

$$ER = 10\text{Log}\frac{P_0}{P_1} \text{ (dB)};$$ (3.19)

$$BER = 0.5 \text{ erfc}\frac{QF}{\sqrt{2}};$$ (3.20)

$$QF = \omega_0\frac{\omega}{P_l};$$ (3.21)

where P_0 is logic '0' power and P_1 is logic '1' power and ω_0, ω, P_1 are angular frequency, stored energy, and loss of total power, respectively.

Table 3.4 gives a comparison of the proposed design with the existing work and shows that the designed logic gate and its corresponding optical device outperform the existing models.

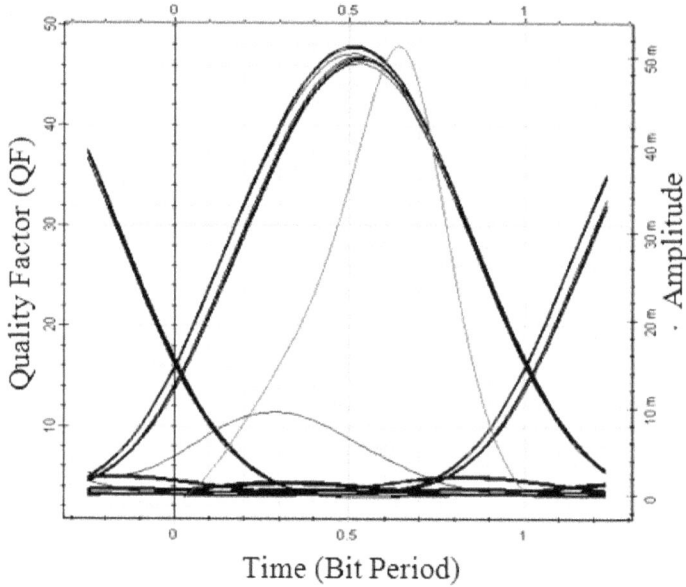

Figure 3.8. Eye diagram of designed circuit.

Table 3.4. Comparison with existing works.

Structure	Quality factor	Data rate (Gb s^{-1})	Extinction ratio (dB)
Proposed work	47.8941 (X)	1000	100.64
	49.6634 (Y)		100.64
	46.3539 (Z)		100.64
Soto *et al* 2002 [53]	—	80	15
Kim *et al* 2006 [54]	—	2.5 and 10	15
Zoe *et al* 2017 [33]	72	200	63
Nair *et al* 2021 [27]	23.87	100	23.65
Kotb *et al* 2021 [55]	18.5	100	—

3.7 Simulation setup of encoded inputs over wired and wireless optical channel

In this section, the performance of the designed gates are verified over wired and wireless optical channels. The simulation setups of encoded inputs with designed SOA-based XOR gate to transmit over wired and wireless optical channels are given respectively in figures 3.9 and 3.10.

The sample timing diagrams of the received binary bits over wired and wireless optical links are given in figure 3.11 to visualize the quality of the channel. Output power is 10.932 for both the channels at 10 Gbps and is given in table 3.5. The eye diagram illustration is useful in assessing the proposed design performance in terms

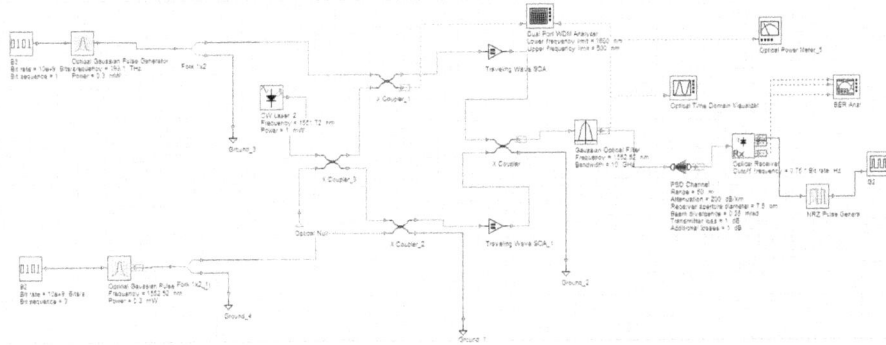

Figure 3.9. The encoded inputs with designed SOA-based XOR gate are transmitted over wireless optical channel.

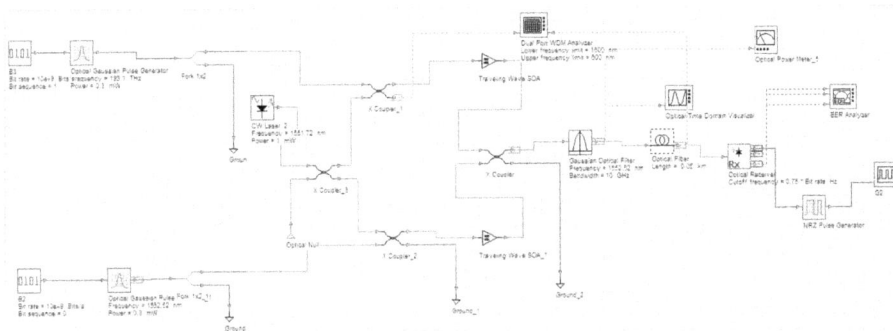

Figure 3.10. The encoded inputs with designed SOA-based XOR gate are transmitted over wired optical channel.

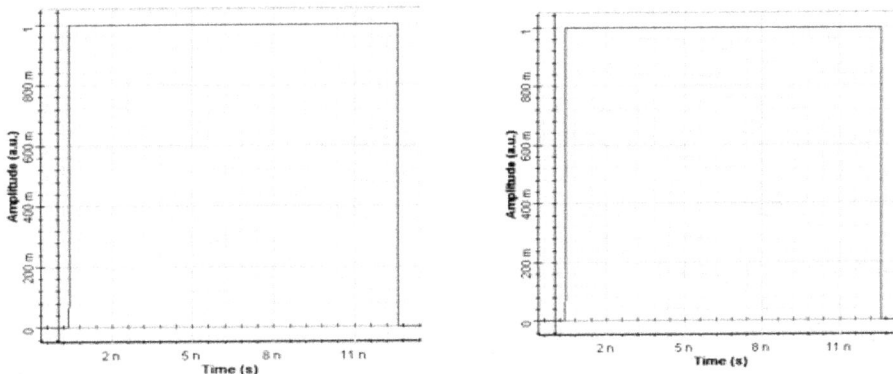

Figure 3.11. Timing diagram of observed output for (a) wired and (b) wireless systems.

of QF and BER. QF is calculated from the transmitted and received output power. BER value is calculated from the eye diagram using the formula. The standard BER formula gives the equation for the QF calculation, and the experiential formulas to calculate the ER, QF, and BER are given in equations (3.19)–(3.21). In these

Table 3.5. Performance metrics.

S.No.	Parameters	Wired optical system	Wireless optical system
1	Output power	10.932 dBm	10.932 dBm
2	Maximum QF	60.13	60.48
3	Minimum BER	0	0
4	Eye height	0.0133	0.001 21
5	Threshold	0.0054	0.0004
6	Decision Ins.	0.390	0.398

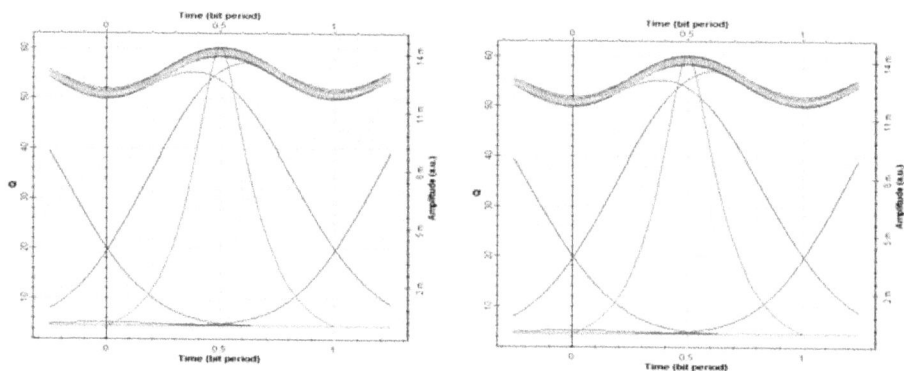

Figure 3.12. Eye diagram of received output bit after transmitting for (a) wired and (b) wireless systems.

equations, P_1 is logic '0' power and P_2 is logic '1' power. ω_0, ω, P_{loss} are angular frequency, energy stored, and total power loss, respectively. An eye diagram helps to validate the transmission efficiency of any communication channel, and the obtained graphs of the design are shown in figure 3.12. The obtained vertical and horizontal eye openings are adequate to ensure acceptable performance. The pulse sampling instant is indicated by vertical eye opening, and interference is nil. In the horizontal eye openings, time interval value determines the decision to be made. Eye diagrams generated using OptiSystem software are a visual representation of the quality and integrity of digital optical signals. The eye diagram is represented using different colors to make it easier to visualize and analyze the signal's quality and its various components. The main eye opening, typically shown in blue, represents the ideal signal or the desired signal level. This is the region where the receiver ideally samples the incoming data. The red region often represents the transition between signal levels. It indicates the points in time where the signal changes from high to low or vice versa. These transitions can be affected by signal distortion and jitter. The yellow regions in the eye diagram typically represent the noise and jitter in the signal. Jitter is the variation in the timing of signal transitions. The wider the yellow region, the more significant the jitter. The green areas usually indicate inter-symbol interference (ISI). ISI occurs when symbols from previous bits overlap with the current bit due to signal dispersion or distortion. Green areas show where the

Table 3.6. Performance metrics.

Parameter	Frequency (THz)	Gain (dB)	Noise figure (dB)	Input signal (dBm)	Input noise (dBm)	Input OSNR (dB)	Output signal (dBm)	Output noise (dBm)	Output OSNR (dB)
Fiber	193.1	21.90	−21.90	−10.98	−100	89.012	10.91	−100	110.914
system	193.2	−72.97	72.97	−6.02	−100	93.97	−79.001	−100	20.998
FSO	193.1	21.89	−21.89	−10.98	−100	89.0119	10.901	−100	110.904
system	193.2	−72.98	72.98	−6.02	−100	93.97	−79.001	−100	20.989

received signal may have multiple signal levels. Gray or black colors are sometimes used to represent additional information, such as annotations, labels, or other details related to the eye diagram. By observing the eye diagram, one can assess the extent of jitter, noise, and other impairments that can affect the ability to accurately receive and interpret the optical signal. This information is crucial for optimizing the design and performance of optical communication systems and ensuring reliable data transmission. Performance of the system is verified with various channel metrics such as gain in dB, noise figure, input signal input noise input OSNR, Output noise and output OSNR with 193.1 193.2 THz. Both designs exhibit good performance for the transmitted signal as can be seen from table 3.6. FSO has many benefits and good performance as seen in tables 3.4 and 3.5. The output SNR is lower for 193.2 THz when compared to 193.1 THz, as higher-frequency signals carry more information but are more susceptible to impairments, such as dispersion effects, optical losses, nonlinear effects, amplifier gain, filtering and crosstalk which can lead to lower SNR values.

3.8 Conclusion

The foremost objective in any transmission network is the efficient and secure transfer of data. FSO systems present a promising solution for several broadband networking applications. Still, there are few restrictions in the form of bad atmospheric conditions that may impact entire system performance. In this chapter, performance of a free space and fiber-optic communication system was analyzed with the help of secured encoded inputs with the designed all-optical XOR and OR gates based on SOA and RSOA type optical amplifiers, respectively. The all-optical gates were designed with SOA due its significant advantages such as good integration capability with other photonic integrated circuits devices. The perform-ance of the design was verified in both wired and wireless optical channels. The results were validated with the calculated performance metrics such as output power, BER, and QF. The outcomes obtained through simulations conducted using OptiSystem 7 software show that the SOA-based logic gate designs exhibit improved performance metrics within FSO system applications. These enhancements encom-pass a notable output power of 10.9 dBm, an almost imperceptible BER, and a robust QF measuring approximately 60. Since air serves as the FSO transmission medium and the light must travel through it, some environmental difficulties cannot be avoided as most atmospheric occurrences happened in the regions of the troposphere. Hence, careful selection of channel parameters is important for FSO systems. The sort of factors to take into account before setting up the system can be guided by extra caution and pre-study of the medium. Numerous studies are being conducted in this direction with the goal of bringing novel system designs, such as WDM-based FSO systems, to reduce the impact of attenuation. Researchers are actively exploring advanced multiplexing and modulation techniques to elevate the overall communication channel's quality. The era of all-optical communication, bolstered by the implementation of all-optical logic gates, serves as a significant enhancer of FSO system performance.

References

[1] Cotter D, Manning R J, Blow K J, Ellis A D, Kelly A E, Nesset D, Phillips I D, Poustie A J and Rogers D C 1999 Nonlinear optics for high-speed digital information processing *Science* **286** 1523–8

[2] Suzuki Y, Shimada J and Yamashita H 1985 High-speed optical–optical logic gate for optical computers *Electron. Lett.* **21** 161–2

[3] Cheng J and Zhou P 1992 Integrable surface-emitting laser-based optical switches and logic gates for parallel digital optical computing *Appl. Opt.* **31** 5592–603

[4] Sridarshini T, Geerthana S, Balaji V R, Thirumurugan A, Sitharthan R and Dhanabalan S S 2023 Ultra-compact all-optical logical circuits for photonic integrated circuits *Laser Phys.* **33** 076207

[5] Meindl J D 1995 Low power microelectronics: retrospect and prospect *Proc. IEEE* **83** 619–35

[6] Yavuz D D 2006 All-optical femtosecond switch using two-photon absorption *Phys. Rev. A* **74** 053804

[7] Caroline E, Michael M and Christina S 2021 Design and performance analysis of SOA-MZI-based Tbps all-optical gray converters with M-ary DPSK coded binary, gray, and octal inputs *Opt. Eng.* **60** 015103

[8] Michael M, Caroline B E and Xavier S C 2020 M-ary DPSK coded binary to gray, BCD to gray, and octal to binary all-optical code converters based on SOA-MZI configuration at 500 Gb/s *Appl. Opt.* **59** 8126–35

[9] Amirabadi M A and Vakili V T 2018 A new optimization problem in FSO communication system *IEEE Commun. Lett.* **22** 1442–5

[10] Padhy J B, Satarupa A and Patnaik B 2020 The effect of atmosphere on FSO communication at two optical windows under weather condition of Bhubaneswar city *Optical and Wireless Technologies* (Singapore: Springer) pp 417–25

[11] Kumar L B, Naik R P, Krishnan P, Raj A A B, Majumdar A K and Chung W Y 2022 RIS assisted triple-hop RF-FSO convergent with UWOC system *IEEE Access* **10** 66564–75

[12] Moghaddasi M, Mamdoohi G, Noor A S M, Mahdi M A and Anas S B A 2015 Development of SAC–OCDMA in FSO with multi-wavelength laser source *Opt. Commun.* **356** 282–9

[13] Lounis B and Moerner W E 2000 Single photons on demand from a single molecule at room temperature *Nature* **407** 491–3

[14] Nadeem F, Kvicera V, Awan M S, Leitgeb E, Muhammad S S and Kandus G 2009 Weather effects on hybrid FSO/RF communication link *IEEE J. Sel. Areas Commun.* **27** 1687–97

[15] Xu G and Zhang Q 2021 Mixed RF/FSO deep space communication system under solar scintillation effect *IEEE Trans. Aerosp. Electron. Syst.* **57** 3237–51

[16] Gebhart M, Leitgeb E, Muhammad S S, Flecker B, Chlestil C, Al Naboulsi M, de Fornel F and Sizun H 2005 Measurement of light attenuation in dense fog conditions for FSO applications *Atmospheric Optical Modeling, Measurement, and Simulation* **vol 5891** (Bellingham, WA: SPIE) pp 175–86

[17] Kaushal H and Kaddoum G 2016 Optical communication in space: challenges and mitigation techniques *IEEE Commun. Surv. Tutorials* **19** 57–96

[18] Agrawal G P and Olsson N A 1989 Self-phase modulation and spectral broadening of optical pulses in semiconductor laser amplifiers *IEEE J. Quantum Electron.* **25** 2297–306

[19] Jaz M 2013 *Experimental Characterisation and Modelling of Atmospheric Fog and Turbulence in FSO* (University of Northumbria at Newcastle (United Kingdom))

[20] Rashidi F, He J and Chen L 2018 Performance investigation of FSO–OFDM communication systems under the heavy rain weather *J. Opt. Commun.* **39** 37–42

[21] Rouissat M, Borsali A R and Chikh-Bled M E 2012 Free space optical channel characterization and modeling with focus on Algeria weather conditions *Int. J. Comput. Netw. Inf. Secur.* **4** 17

[22] Jeyarani J, Sriramkumar D and Caroline B E 2018 Performance analysis of free space optical communication system employing WDM-PolSK under turbulent weatherconditions *J. Optoelectron. Adv. Mater.* **20** 506–14

[23] Singh P, Singh A K, Arun V and Dixit H K 2016 Design and analysis of all-optical half-adder, half-subtracter and 4-bit decoder based on SOA-MZI configuration *Opt. Quantum Electron.* **48** 1–14

[24] Yang X, Weng Q and Hu W 2012 High speed all-optical long-term memory using SOA MZIs: simulation and experiment *Opt. Commun.* **285** 4043–7

[25] Data K, Chattapadhyay T and Sengupta I 2015 All-optical design of binary adders using semiconductor optical amplifier assisted Mach–Zehnder interferometer *Microelectron. J.* **46** 839–47

[26] Ramachandran M, Prince S and Maruthi N 2020 Design and simulation of all-optical shift registers using D flip flop *Microwave Opt. Technol.* **62** 2427–38

[27] Nair N, Kaur S and Singh H 2021 All-optical ripple carry adder based on SOA-MZI configuration at 100 Gb/s *Optik* **231** 166325

[28] Kaur S and Shukla M K 2017 All-optical parity generator and checker circuit employing semiconductor optical amplifier-based Mach–Zehnder interferometers *Opt. Appl.* **47** 263–71

[29] Jiao S M, Liu J W, Zhang L W, Yu F H and Zuo G M 2022 All-optical logic gate computing for high-speed parallel information processing *Opto-Electron. Sci.* **1** 220010

[30] Kanellos G T, Stampoulidis L, Pleros N, Tsiokos D, Kehayas E and Avramopoulos H 2003 Clock and data recovery circuit for 10 Gb/s asynchronous optical packets *IEEE Photonics Technol.* **15** 1666–8

[31] Kotb A, Zoiros K E and Guo C 2018 Performance investigation of 120 Gb/s all-optical logic XOR gate using dual-reflective semiconductor optical amplifier-based scheme *J. Comput. Electron.* **17** 1640–9

[32] Mandal G C, Mukherjee R, Das B and Patra A S 2018 Next-generation bidirectional triple-play services using RSOA based WDM radio on free space optics PON *Opt. Commun.* **411** 138–42

[33] Rizou Z V and Zoiros K E 2017 Performance analysis and improvement of semiconductor optical amplifier direct modulation with assistance of micro ring resonator notch filter *Opt. Quantum Electron* **49** 1–21

[34] Maji K, Mukherjee K, Raja A and Roy J N 2020 Numerical simulations of an all-Optical parity generator and checker utilizing a reflective semiconductor optical amplifier at 200 Gbps *J. Comput. Electron.* **19** 800–14

[35] Kotb A, Zoiros K E and Guo C L 2019 All-optical XOR gate using semiconductor optical amplifier Mach–Zehnder interferometer and delayed interferometer *Photonic Netw.*

[36] Nair N, Kaur S and Goyal R 2018 All-optical integrated parity generator and checker using SOA-based optical tree architecture *Curr. Opt. Photonics* **2** 400–6

[37] Michael M, Britto E C, Mohan K and Darshini S 2023 Design of RSOA-based all-optical parity generator and checker *J. Opt.* 1–14

[38] Leuthold J, Bonk R, Vallaitis T, Marculescu A, Freude W, Meuer C, Bimberg D, Brenot R, Lelarge F and Duan G H 2010 Linear and nonlinear semiconductor optical amplifiers *2010 Conf. on Optical Fiber Communication (OFC/NFOEC), collocated National Fiber Optic Engineers Conf.* pp 1–3

[39] Mokkapati S and Jagadish C 2009 III–V compound SC for optoelectronic devices *Mater. Today* **12** 22–32

[40] Rashed A N Z 2013 High reliability optical interconnections for short range applications in high performance optical communication systems *Opt. Laser Technol.* **48** 302–8

[41] Rashed A N Z, Kumar K V, Tabbour M S F and Sundararajan T V P 2019 Nonlinear characteristics of semiconductor optical amplifiers for optical switching control realization of logic gates *J. Opt. Commun.* **43**

[42] Said Y, Rezig H and Urquhart P 2011 Semiconductor optical amplifier nonlinearities and their applications for next generation of optical networks *Advances in Optical Amplifiers* (IntechOpen) pp 27–52

[43] Haridim M, Lembrikov B I and Ben-Ezra Y 2011 Semiconductor optical amplifiers *Advances in Optical Amplifiers* (IntechOpen) pp 3–26

[44] Rani A and Dewra M S 2013 Semiconductor optical amplifiers in optical communication system-review *Int. J. Eng. Res. Technol.* **2** 2710–9

[45] Minzioni P *et al* 2019 Roadmap on all-optical processing *J. Opt.* **21** 63001–55

[46] Durhuus T, Mikkelsen B, Joergensen C and Stubkjaer K E 1994 Semiconductor optical amplifiers for wavelength conversion *OSA Technical Digest Series* **vol 3**

[47] Qiang W and Dutta N K 2013 *Semiconductor Optical Amplifiers,* (Singapore: World Scientific)

[48] Kotb A 2016 Simulation of soliton all-optical logic XOR gate with semiconductor optical amplifier *Opt. Quantum Electron.* **48** 307

[49] Kotb A, Zoiros K E and Guo C 2018 160 Gbps photonic crystal semiconductor optical amplifier-based all-optical logic NAND gate *Photonic Netw. Commun.* **36** 246–55

[50] Cassioli D, Scotti S and Mecozzi A 2000 A time-domain computer simulator of the nonlinear response of semiconductor optical amplifiers *IEEE J. Quantum Electron.* **36** 1072–80

[51] Alzenad M, Shakir M Z, Yanikomeroglu H and Alouini M S 2018 FSO-based vertical backhaul/fronthaul framework for 5G+ wireless networks *IEEE Commun. Mag.* **56** 218–24

[52] Dhanabalan S S, Thirumurugan A, Raju R, Kamaraj S K and Thirumaran S (ed) 2023 *Photonic Crystal and Its Applications for Next Generation Systems.* (London: Springer Nature)

[53] Soto H, Erasme D and Guekos G 2001 5-Gb/s XOR optical gate based on cross-polarization modulation in semiconductor optical amplifiers *IEEE Photon. Technol. Lett.* **13** 335–7

[54] Kim J Y, Kang J M, Kim T Y and Han S K 2006 All-optical multiple logic gates with XOR, NOR, OR, and NAND functions using parallel SOA-MZI structures: theory and experiment *J. Lightw. Technol.* **24** 3392

[55] Kotb A and Zoiros W 2021 Numerical study of carrier reservoir semiconductor optical amplifier-based all-optical XOR logic gate *J. Mod. Optics* **68** 161–8

IOP Publishing

Advances in All-optical Communication

Shanmuga Sundar Dhanabalan, Arun Thirumurugan and Sridarshini Thirumaran

Chapter 4

Switching characteristics of optical solitons through inelastic interactions

M S Mani Rajan and P Nithya

Optical solitons are potential candidates in long-haul communication systems for ultrafast data transmission. In the context of soliton theory, soliton transmission in the nonlinear fiber medium is described by the well-known nonlinear Schrödinger equation. Especially for an inhomogeneous fiber medium, the nonlinear Schrödinger equation with variable coefficients can be employed to describe optical soliton transmission. Also, the optical soliton is a precursor to constructing ultrafast optical switches via inelastic interactions among solitons. Soliton interaction can be controlled by pulse parameters in a desirable manner for constructing various logic gate devices. This chapter also investigates the switching characteristics of optical solitons. The proposed technique may be a potential key component for the future of photonic devices and optical computing. In this work, obtained results are not only helpful for understanding the collisional properties of optical solitons but also for designing ultrafast optical switches.

4.1 Introduction

In the development of information technology, optical fibers play a vital role because of their large information-carrying capacity for long-haul data transmission [1]. The optical soliton is regarded as the natural data-bits and an important alternative for the next generation ultra-high speed optical fiber network [2]. The development of optical solitons is the subject of intense studies. A soliton transmission is a very efficient and attractive type of transmission in a fiber-optics telecommunication system as it does not change pulse shape during propagation over long distances and remains unaffected after collision with other solitons [3]. In advanced optical communication systems, propagation of short optical pulses in the order of a few picoseconds are signified by the nonlinear Schrödinger equation (NLSE) [4].

doi:10.1088/978-0-7503-5623-7ch4

Optical solitons have received immense research attention because of their robustness. In a nonlinear medium like optical fiber, optical solitons can be formed by the balancing of positive chirping produced by self-phase modulation and negative chirping generated by group velocity dispersion in the anomalous dispersion regime [5]. In nonlinear fiber optics, soliton switching is the subject of intense theoretical and experimental studies due to their potential applications in the construction of ultrafast switching devices [6]. The optical switches can be designed to achieve integrated photonic switches for ultrafast optical switches through the process of energy-exchange interactions. An optical soliton's particle- like characteristics enable the possibility of all-optical processing, which includes optical computing [7]. Various methods are employed to manipulate the soliton trajectories through the process of interaction by properly selecting input pulse parameters [8].

The interactions among nonlinear waves such as rogue waves [9], solitary waves [10], breather waves [11], cnoidal waves [12], etc., have been reported using the NLSE. In contrast to scalar NLS systems, the multicomponent NLSE contains additional coupling terms and interactions among them yield some significant interaction scenarios [13].

The NLSE is adequate to describe the picosecond optical pulse propagation in an inhomogeneous nonlinear fiber medium [14]. However, such models will not provide a realistic description if the medium is inhomogeneous. On the other hand, optical fiber medium is practically inhomogeneous, which can be theoretically signified by including variable coefficients in the standard NLSE [15]. Thus, variable coefficients are directly incorporate the inhomogeneous nature of optical fibers which arise due to the imperfection during the fabrication, variation in the fiber diameters, changes in the lattice parameters [16].

When a femtosecond optical soliton propagates through nonlinear optical fiber, the standard NLSE is inadequate to describe the optical pulse transmission [17]. In order to study the propagation of femtosecond soliton propagation, third-order dispersion (TOD), self-steepening (SS), and stimulated Raman scattering (SRS) are used with the standard NLSE [18]. Moreover, recent publications have confirmed that higher-order NLSEs with variable coefficients provide some novel characteristics such as dispersion management, soliton shaping, and soliton control technology through adopting suitable inhomogeneous profiles [19].

4.1.1 Telecommunication windows

Fiber-optic communications typically operate in any one of the following 'telecom windows':

1. The first window utilizes wavelengths from 800 to 900 nm. In this wavelength span, absorption loss is comparatively very high and efficient optical amplifiers are not available in this spectral range. Hence, this wavelength range is not suitable for long- distance optical communication.

2. The wavelength around 1300 nm occupies the second telecom window and attenuation is relatively less and chromatic dispersion of silica glass is also

low. This window is used for long-haul optical communication systems. However, the effect of fiber nonlinearity is significantly large, which is undesirable for long-haul communication links. Furthermore, fiber amplifiers not available for 1300 nm while erbium fiber amplifiers available for 1500 nm.

3. Currently, the third telecom window is extensively used with a wavelength around 1500 nm. In this window, absorption loss of optical fiber is significantly low and erbium-doped fiber amplifiers are employed for the amplification process. The 1500-ms wavelength directly implies the dispersion regime is anomalous where group velocity dispersion and self-phase modulation are opposite in nature.

The second telecom window and third telecom window are divided into six wavelength bands:

Band	Wavelength range (nm)
O band (O—Original)	1260–1360
E band (E—Extended)	1360–1460
S band (S—Short wavelengths)	1460–1530
C band (C—Conventional)	1530–1565
L band (L—Long wavelengths)	1565–1625
U band (U—Ultra long wavelengths)	1625–1675

4.1.2 Optical soliton formation in an optical fiber

In 1960, the concept of 'soliton' was theoretically proposed for the first time and solitary wave was observed by John Scott Russell in a canal near Edinburgh. During that century, there was vast discussion regarding the existence of a special kind of solitary wave. In the context of nonlinear fiber optics, a complex envelope of optical field does not suffer along the propagation due to the exact balance of nonlinear and linear effects, referred to as optical soliton. In physics, a solution is a self-reinforcing solitary wave that maintains its shape while it travels at a constant speed. Theoretical calculations and numerical solutions show that the nonlinear dependence of refractive index on intensity makes it possible to transmit ultra-short optical pulses without distortion [20]. The NLSE describes soliton propagation in nonlinear dispersive optical fibers. Applications of optical solitons in fiber- optics communication systems is part of a new technology called photonic networks for high-speed logic gate devices. Solitons are very stable solitary waves and they are behaving like 'particles'. Hence, they can travel after the collision without change in the pulse parameters like amplitude, velocity and phase.

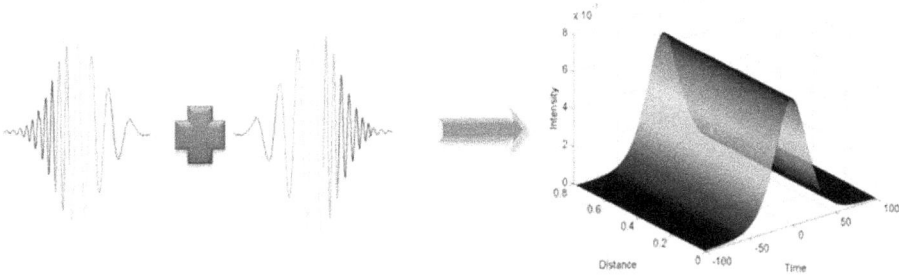

In the scenario of nonlinear science, especially in fiber optics, soliton pulse formed through the combined effects of Group Velocity Dispersion (GVD) and Self-Phase Modulation (SPM) which are arise due to second-order dispersive effect and non-linear Kerr effect. The combined properties of linear effect of anomalous GVD and nonlinear SPM effect are causing the formation of optical solitons in a dispersive nonlinear optical fiber. In practical, it is very difficult to generate fundamental optical soliton due to fiber absorption. Because of absorption in the fiber medium, nonlinearity falls to weaken effect which leads to greater dispersion effect and overlap will occur with the adjacent pulses. In order to overcome these issues, there are two solutions are possible. Firstly, the attenuation loss can be compensated by means of optical amplification process through optical/electrical amplifiers. The second solution is a dispersion management scheme to manage the dispersion.

4.1.3 Nonlinearity

Mathematically, nonlinear is a term that indicates the variation where two variables are varied with a nonlinear relationship. It means the graphical representation of two variables is not a straight- ine. In a nonlinear relationship, there is a change in the output do not change in direct proportion to changes in any of the inputs [21].

The carrier frequency of the optical is mainly affected by the time factor, which leads to pulse shaping by means of GVD effect. Since the power dependence of the refractive index of the medium, it is a limiting factor in the fiber optic communication systems [22]. On the other hand, the SPM effect is mainly caused by the Kerr effect, which is responsible for the nonlinear phase shift of the optical field by its own intensity. SPM can lead to considerable spectral broadening of pulses propagating inside an optical fiber [23]. Nonlinear science can be impossibly broad, too interdisciplinary, or 'the study of everything'. It can also be said as a system that lacks a systematic mathematical framework, and the complexity of natural non-linear phenomena suggest that nonlinearity can be best understood by classifying various manifestations in various different systems and to observe and study the common features. The concept of nonlinearity has gained interest due to their similarities of their nature in numerous fields such as nonlinear science, mathematical physics, plasma physics, biophysics, nonlinear fiber optics, Bose–Einstein condensation etc. These common concepts give insight into nonlinear problems. In various disciplines and by understanding these paradigms one can understand the essence of nonlinearity and its consequences in all the fields. The first and most crucial development is the high-speed electronic computers that allow us quantitative numerical simulations of nonlinear systems. In some cases, nonlinear behavior of physical problem may require computational investigations where analytic methods are inactive. The origin of nonlinear response is related to the non-harmonic motion of bound electrons under the influence of an applied field. Hence, the induced polarization P resulting from the electric dipoles [24].

$$P = \varepsilon_0(\chi^{(1)}E + \chi^{(2)}E^2 + \chi^{(3)}E^{(3)} + \cdots) \tag{4.1}$$

In the above expression (4.1), ε_0 represents the permittivity of vacuum and $\chi^{(j)}$ is order of susceptibility tensor with rank $j + 1$. For example, $\chi^{(1)}$ determines the linear susceptibility tensor of nonlinear dispersive medium. In the case of silica glass, even power terms of E will not exist in the expression for P due to the centrosymmetric nature of SiO_2 molecules. Therefore, the lowest order of nonlinearity mainly occurs due to the third-order susceptibility tensor $\chi^{(3)}$, which is known as Kerr nonlinearity or cubic nonlinearity. Since SiO_2 exhibits inversion symmetry nature, second-order susceptibility $\chi^{(2)}$ vanishes for silica glass. For silica, the third-order susceptibility value is smaller by a factor of 100 or more when compared with other kind of nonlinear materials. In an optical fiber, when an optical pulse is propagated through the nonlinear dispersive medium, the real part of $\chi^{(3)}$ is responsible for the Kerr effect while the imaginary part causes the Raman effect.

In a nonlinear optical fiber medium, fiber nonlinearities limit the amount of data transfer when optical signal is transmitted through an optical fiber. Photonic network developers always consider these limitations to reduce the harmful effects due to fiber nonlinearities. There are two kinds of mechanisms that cause nonlinearity: optical power-dependent nonlinearity and scattering nonlinearity. In the optical fiber, nonlinearity of silica glass significantly varies due to the intensity-dependent refractive index. This response contributes to the nonlinear modification of refraction $n(\omega, |E|^2)$ given as [25]

$$n(\omega, |E|^2) = n_0(\omega) + n_2(\omega)|E|^2 \tag{4.2}$$

$n_2(\omega)$ is related to $\chi^{(3)}$ through the relation [26]

$$n_2(\omega) = \frac{3}{4n_0}\chi^{(3)} \tag{4.3}$$

In a single-mode fiber, SPM occurs due to the changing of refractive index with intensity of the incident laser pulse. As a result of optical nonlinearity, intensity of input optical signal is modulated and nonlinear phase shift occurs, which causes the phase modulation. The positive chirping exists in the input optical pulse, which leads to spectral broadening in the anomalous dispersion region of the optical fiber. When a complex envelope of optical signal is time-dependent, nonlinear phase shift occurs in the transmitted optical pulse. The induced positive chirp due to SPM is exactly balanced with the chirp arises due to GVD where chirping is negative. Thus, the type of chirping is depending on the sign of the GVD coefficient i.e., dispersion coefficient is negative for anomalous dispersion regime, and the chirping caused by SPM is positive. Hence, pulse broadening due to GVD is greatly reduced by SPM effect. When these two effects are balanced exactly, soliton pulses propagate with more stability.

With cubic nonlinear effect, phase shift is induced by the optical pulse itself and is referred to as SPM. As a result of Kerr effects [27], the refractive index of the fiber depends upon the intensity component. The leading edge of the pulse attains a positive refractive index gradient due to the higher intensity portion of the experiences high refractive index and the trailing edge of the pulse gets negative refractive index gradient. Due to this change in refractive index, the pulse experiences a phase change. As the pulse modulates its own phase according to its intensity profile, it is called SPM [28]. SPM induces spectral compression, which is an effect opposite to pulse broadening due to GVD. SPM can be used for the fast optical switching that is used for passive mode locking.

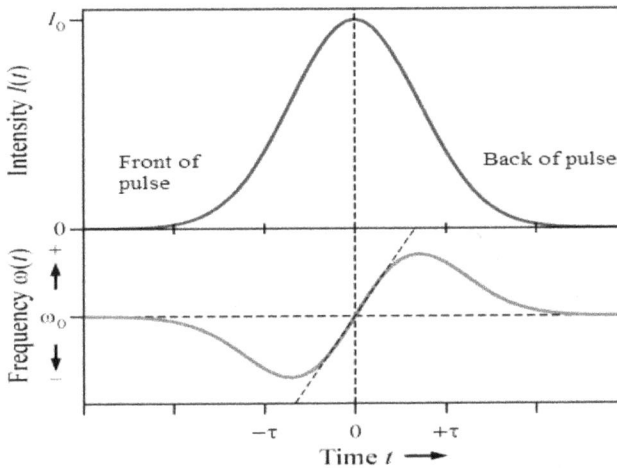

If phase modulation of an input signal is caused by a copropagated signal within the same fiber, then the effect is known as cross-phase modulation (XPM). For example, when multiple signals with different wavelengths are transmitted in the same fiber, Kerr effect is induced by one signal as a result of phase modulation of the other signals.

The cross-phase modulation can be significantly reduced in the wavelength division multiplexing (WDM) channels by carefully selecting unequal bit rates for copropagating signals. In an ideal cylindrical fiber core, there would be no coupling between two polarization modes of single-mode fiber. But in real fiber, manufacturing defects lead to deviation from cylindrical symmetry and result in birefringence. The birefringence is mainly responsible for polarization mode dispersion.

4.1.4 Group velocity dispersion

Group velocity dispersion (GVD) is the phenomenon of dependence of the group velocity of light in a transparent medium on the optical frequency or wavelength.

The term can also be said as a precisely defined quantity, namely the derivative of the inverse group velocity with respect to angular frequency [29].

$$\text{GVD} = \frac{\partial}{\partial \omega} \frac{1}{v_g} = \frac{\partial}{\partial \omega}\left(\frac{\partial k}{\partial \omega}\right) = \frac{\partial^2 k}{\partial \omega^2} \tag{4.4}$$

The GVD can be defined as group delay dispersion per unit length. The basic units are $s^2\, m^{-1}$. The GVD of fused silica is 35 $fs^2\, mm^{-1}$ at 800 nm when 26 $fs^2\, mm^{-1}$ at 1500 nm. In particular, zero-dispersion wavelength (ZDW) for an optical fiber is 1.3 μm where GVD does not occur.

In the context of a fiber-optic communication system, mathematically the GVD is defined as a derivative with respect to wavelength. GVD can be expressed as

$$D_\lambda = \frac{\partial}{\partial \lambda} \frac{1}{v_g} = \frac{-2\pi C}{\lambda^2} \tag{4.5}$$

$$\text{GVD} = \frac{-2\pi C}{\lambda^2} \cdot \frac{\partial^2 k}{\partial \omega^2} \tag{4.6}$$

where c is the vacuum velocity of light, and its quantity is usually specified with units of ps $(nm\ km)^{-1}$ (picoseconds per nanometer wavelength change and kilometer propagation distance). For instance, 20 ps $(nm\ km)^{-1}$ at 1550 nm (a typical value for telecom fibers) corresponds to $-25\ 509\ fs^2\ m^{-1}$. It is very critical to understand the different signs of GVD and D_λ which are related through shorter wavelengths and higher frequencies. In order to avoid this controversy, the terms normal and anomalous dispersion regions are used instead of positive and negative dispersion. In normal dispersion regime, the group velocity is decreasing with increasing optical frequency.

A soliton or solitary wave is otherwise called a self-reinforcing wave packet that maintains the shape while propagating at constant velocity. The term soliton is used in mathematics or physics—any cancellations of nonlinear and dispersive effects in the medium cause solitons. Solitons in optical fibers are formed because of a balance between the chirps induced by GVD and SPM [30]. When GVD and SPM act independently, they will limit the system performance. In this chapter, analysis of dispersion-induced pulse broadening and nonlinear optical effects is used. Inside an optical fiber, GVD broadens the optical pulse at the time of propagation. If the pulse is initially chirped in the right way, then broadening will not occur. A compressed chirped pulse is used in initial propagation. If β_2 and the chirp parameter C is changed to have the opposite sign, the term $\beta_2 C$ will be negative. In nonlinear effects, SPM induced a chirp on the optical pulse such that $C > 0$. The condition $\beta_2 C < 0$ is satisfied if β_2 is less than 0. Since the chirp induced by SPM is dependent on power, the effect of SPM is varying with respect to propagation distance. The exact balance between the SPM and GVD means that SPM-induced positive chirping cancels the GVD induced negative chirping. In such a situation, the propagation of optical pulse is stable and without distortion.

4.1.5 Attenuation

The typical value of attenuation loss of an optical fiber is 0.2 dB km^{-1} while the loss for copper cable is 5 dB km^{-1}. In copper cables, signal loss varies nonlinearly with modulation frequency. Due to attenuation effect, signal strength is significantly reduced and can be determined by comparing output power with input power as

$$\text{Loss in dB} = -10 \log_{10}(P_{\text{out}}/P_{\text{in}}) \qquad (4.7)$$

According to expression (4.7), when optical signal travels for a long distance, input optical power decreases significantly because of attenuation. In real optical fiber, the attenuation process can be understood clearly through the effective length (L_{eff}) of an optical fiber. In practice, power is constant over the specific distance or length of optical fiber, which may be called the effective length. Therefore, effective length is a powerful parameter to estimate the effect of nonlinearities in an optical fiber medium. In an optical fiber, attenuation parameter α is the most important one since input optical power is attenuated during the signal propagation. It is related to effective length and expressed as

$$L_{\text{eff}} = \frac{1 - e^{-\alpha L}}{\alpha} \qquad (4.8)$$

where α is the attenuation constant. The effective length L_{eff} means that where the nonlinear shift becomes significant.

Physically, the nonlinear length (L_{eff}) indicates the distance at which the nonlinear phase shift reaches 1 radian, and it provides a length scale over which the nonlinear effects become relevant for optical fibers. After travel through an optical fiber, the output power of transmitted signal is P_{out} and can be measured by initial power (P_{in}) through

$$P_{\text{out}} = P_{\text{in}} e^{-\alpha L} \qquad (4.9)$$

In the measurement of attenuation constant α, attenuation due to all sources must be taken into account. Like GVD, attenuation losses of the optical fiber also vary with wavelength. Hence, different frequency components of incident optical pulse are differently attenuated in various magnitudes. In a silica optical fiber, Rayleigh scattering, water (OH−) absorption, and metal-oxide absorption peaks have a significant contribution to loss spectrum. As an example, electronic resonances occur in the UV region when vibrational resonances lie in the FIR (far infrared) for a silica glass fiber. Hence, silica glasses can transmit optical signal in the wavelength span of 0.5–2.2 μm. During the manufacturing process of optical fibers there are always density fluctuations that cause a Rayleigh type of scattering, which is scattering in all directions. According to Rayleigh law, loss varies with $1/\lambda^{4}$, which is more dominant at shorter wavelength range. On the other hand, bending loss is another scattering loss for signal degradation that occurs at the interface of core and cladding. Furthermore, splicing loss and connector losses also contribute to the attenuation of input optical signal, but with significantly less magnitude.

4.2 Governing theoretical model

In this chapter, we investigate the femtosecond soliton switching characteristics via optical soliton interactions by considering the coupled higher-order nonlinear Schrödinger equation with inhomogeneous coefficients. Also, the influence of GVD, third-order dispersion, and nonlinear coefficients with their control parameters on the characteristics of femtosecond soliton switching are discussed in detail. Thus, the coupled theoretical model is considered as follows [31]:

$$i\, q_{jz} - \frac{d_2(z)}{2} q_{jtt} - \chi(z)\left(\sum_{n=1}^{2}|q_n|^2\right)q_j + i\, d_3(z)q_{jttt} + i\, R(z)\left(\sum_{n=1}^{2}|q_n|^2\right)q_{jt} + i\, \delta(z)\left(\sum_{n=1}^{2} q_{nt}\, q_j{}^*\right)q_j + i\, G(z)q_j = 0 \quad (4.10)$$

In equation (4.10), variable coefficients $d_2(z)$, $\chi(z)$, $d_3(z)$, $R(z)$, $\delta(z)$, and $G(z)$ are denoting the second-order dispersion, cross-phase modulation, third-order dispersion, cubic nonlinearity, stimulated Raman scattering, and amplification or attenuation, respectively, where q_j ($j = 1,2$) is the two bimodal components of propagating fields. In the complex envelope, subscripts z and t imply the derivative of $q(z,t)$ with respect to z and t. The symbol * indicates the complex conjugate. By incorporating higher-order effects in the standard NLSE, resultant equation (4.10) can be used to describe the femtosecond soliton propagation in a birefringent fiber. By properly selecting the suitable parameters for control functions, one can achieve the soliton switching characteristics.

The coupled nonlinear Schrödinger (CNLS) type equations are used to describe the optical soliton propagation in birefringent fibers, multi-mode fibers [32]. In a single-mode fiber, birefringence is very weak whereas very strong birefringence occurs in the multi-mode fibers. In the case of birefringence, there are two different transmission principal axes: fast and slow axes. In a birefringent fiber, different frequencies copropagate along the fiber when two or more optical fields exist that can interplay through the process of XPM [33]. The expression (4.10) can be applicable to WDM systems where multi-wavelength components are propagated simultaneously that enhance the transmission capacity of a fiber-optic communication system. Theoretically and experimentally, it has been confirmed that the optical soliton-based transmission system with WDM is a potential candidate to enhance the information-carrying capacity of a fiber-optic communication network.

Soliton control technology has received huge attention due to its potential applications in the field of construction of optical switches based on an optical fiber system [34]. By using the control parameters with the freedom of choosing them in the obtained solutions, soliton control technology is practically feasible [35]. The interactional behavior of optical solitons in an inhomogeneous optical fiber have been reported [36]. In the solitonic communication systems, optical solitons have been managed through the process of nonlinear tunneling [37]. Collisional dynamics of optical solitons for the modified Hirota equation have been reported for constructing all-optical devices [38]. By means of symbolic computation, NLSEs with variable coefficient have been investigated with some novel soliton solutions [39]. With the nature of birefringence, single-mode fiber is not only supported for single-mode operation but will also weakly support bimodal propagation [40].

In practical terms, orthogonal polarization modes supported by a real optical fiber system propagate in the orthogonally perpendicular direction. Hence, while we consider the muti-mode fibers, XPM must be taken into account with self-phase modulation where the phase is changed in each mode of the copropagating mode. On the other hand, standard NLSEs are insufficient to describe the optical soliton transmission in multi-mode optical fiber or multi-core optical fiber. The combined effects of variable coefficients with coupled NLSEs are more adequate to describe the physical significance of inhomogeneous fiber when compared with constant coefficients in practical applications. For an inhomogeneous multi-mode optical fiber, the coupled nonlinear Schrödinger equations (CNLSEs) with distributed coefficients have been discussed and reported in [41]. Exclusively, rogue wave soliton solutions obtained for three coupled NLSEs [42].

The CNLSEs have many advantages including shape-changing collisions, energy-exchange interaction among intermodals, etc. [43]. The propagated optical solitons in the coupled NLS systems are called vector solitons, which have potential applications in the construction of ultrafast optical switching devices, in the design of soliton-controlled logic gates [44], and in optical pulse amplification through tapered erbium-doped fiber [45].

In the design of optical devices, to ensure stability of solitons, optical solitons are guided under interaction between one or more solitons with signal solitons [46]. In order to improve performance of soliton-based photonic integration devices, soliton interaction can be controlled with minimum spacing between the solitons for designing optical devices that can be used to construct the photonic switching circuits. Interactions of solitons have provided a relevant solution to use one pump light to manipulate the direction, intensity, phase, polarization, and other parameters of another signal light without electrical signal [47]. On the other hand, photonic crystal (Ph C)-based optical gates in the development of all-optical half adders have been reported [48]. In the field of photonics, optics have a significant impact in environmental applications and supports for the development of a green future [49]. Using 2D photonic crystal, various logic gates such as AND, OR, and EX-OR can be constructed [50, 52].

4.3 Lax pair for the system (4.10)

In this section, using symbolic computation, we will construct a new Lax pair of equation (4.10) via the Ablowitz–Kaup–Newell–Segur scheme [53]. For system (4.10), we present the linear eigenvalue problem as follows:

$$\psi_t = U\psi; \quad \psi_z = V\psi;$$
$$\psi = (\varphi_1, \varphi_2, \varphi_3)^T \tag{4.11}$$

$$U = i\,\lambda \begin{pmatrix} -1 & 0 & 0 \\ 0 & 1 & 0 \\ 0 & 0 & 1 \end{pmatrix} + \begin{pmatrix} 0 & T\,q_1(z,\,t) & T\,q_2(z,\,t) \\ -T\,\bar{q}_1(z,\,t) & 0 & 0 \\ -T\,\bar{q}_2(z,\,t) & 0 & 0 \end{pmatrix}$$

$$V = i\beta_2(z)(4\lambda^3 + \lambda^2)\begin{pmatrix} -1 & 0 & 0 \\ 0 & 1 & 0 \\ 0 & 0 & 1 \end{pmatrix} - i\beta_2(z)(4\lambda^3 + \lambda)\begin{pmatrix} 0 & -iT\,q_1(z,\,t) & -iT\,q_2(z,\,t) \\ iT\,\bar{q}_1(z,\,t) & 0 & 0 \\ iT\,\bar{q}_2(z,\,t) & 0 & 0 \end{pmatrix}$$

$$+ \frac{1}{2}i\beta_2(z)(4\lambda + 1)\begin{pmatrix} -T^2\,|q_1|^2 - T^2\,|q_2|^2 & -T\,q_{1t} & -T\,q_{2t} \\ -T\,\bar{q}_{1t} & T^2\,|q_1|^2 & T^2\bar{q}_1 q_2(z,\,t) \\ -T\,\bar{q}_{2t} & T^2 q_1(z,\,t)\bar{q}_2 & T^2\,|q_2|^2 \end{pmatrix}$$

$$- i\beta_2(z)\begin{pmatrix} M_{11} & M_{12} & M_{13} \\ M_{21} & M_{22} & M_{23} \\ M_{31} & M_{32} & M_{33} \end{pmatrix}$$

$$M_{11} = -i\,T^2\big(q_1(z,\,t)\bar{q}_{1t} - q_{1t}\,\bar{q}_1(z,\,t) + q_2(z,\,t)\bar{q}_{2t} - q_{2t}\,\bar{q}_2(z,\,t)\big)$$

$$M_{12} = i\,T\,q_{1tt} + 2i\,T^3\big(|q_1|^2 + |q_2|^2\big)q_1(z,\,t)$$

$$M_{13} = i\,T\,q_{2tt} + 2i\,T^3\big(|q_1|^2 + |q_2|^2\big)q_2(z,\,t)$$

$$M_{21} = -i\,T\,\bar{q}_{1tt} - 2i\,T^3\big(|q_1|^2 + |q_2|^2\big)\bar{q}_1(z,\,t)$$

$$M_{22} = i\,T^2\big(q_1(z,\,t)\bar{q}_{1t} - q_{1t}\,\bar{q}_1(z,\,t)\big)$$

$$M_{23} = i\,T^2\big(q_2(z,\,t)\bar{q}_{1t} - q_{2t}\,\bar{q}_1(z,\,t)\big)$$

$$M_{31} = -iT\,\bar{q}_{2tt} - 2i\,T^3\big(|q_1|^2 + |q_2|^2\big)\bar{q}_2(z,\,t)$$

$$M_{32} = i\,T^2\big(q_1(z,\,t)\bar{q}_{2t} - q_{1t}\,\bar{q}_2(z,\,t)\big)$$

$$M_{33} = i\,T^2\big(q_2(z,\,t)\bar{q}_{2t} - q_{2t}\,\bar{q}_2(z,\,t)\big)$$

where $M = \sqrt{\frac{\gamma(z)}{\beta_2(z)}}$. One can attain the system (4.10) through the compatibility condition $\frac{\partial U}{\partial z} - \frac{\partial V}{\partial t} + UV - VU = 0$ with the following relations between the inhomogeneous coefficients:

$$\beta_3(z) = -\beta_2(z),$$

$$\delta(z) = \chi(z) = -3\gamma(z)$$

and

$$\Gamma(z) = \frac{1}{2}\frac{W[\gamma(z),\,\beta_2(z)]}{\beta_2(z)\gamma(z)} \quad \text{with} \quad W[\gamma(z),\,\beta_2(z)] = \gamma(z)\beta_{2z} - \beta_2(z)\gamma_z.$$

In equation (4.11), it is noted that the parameter λ is called the iso-spectral parameter, which means it is considered as a constant.

4.4 Two soliton solutions through Darboux method

According to soliton dynamics, the Darboux transformation [54] is a very efficient tool for constructing the analytical solutions of integrable systems. Generally, Darboux transformation can be applied to keep the linear eigenvalue problem corresponds to a nonlinear Schrödinger system and to obtain the new eigenfunction which relates the original eigenfunction.

$$\psi = D\psi = (\lambda I_{3X3} - S)\psi \qquad (4.12)$$

where D is the Darboux matrix.

$$S = -H \Lambda H^{-1} \qquad (4.13)$$

$$H = \begin{pmatrix} \phi_{11} & -\phi_{21} & \phi_{31} \\ \phi_{21} & -\phi_{11} & 0 \\ \phi_{31} & 0 & -\phi_{11} \end{pmatrix}, \Lambda = \begin{pmatrix} \lambda_1 & 0 & 0 \\ 0 & \lambda_1^* & 0 \\ 0 & 0 & \lambda_1^* \end{pmatrix} \qquad (4.14)$$

By substituting equation (4.14) in equation (4.13), we estimate the value of matrix 'S' and get the fundamental Darboux transformation to arrive at the following generalized soliton solution:

$$q'_1 = q_1 - 4\sqrt{\frac{\beta_2}{\gamma}} S_{12}$$

$$q'_2 = q_2 - 4\sqrt{\frac{\beta_2}{\gamma}} S_{13} \qquad (4.15)$$

With $S_{12} = -S^*{}_{12}$ and $S_{13} = -S^*{}_{13}$. Corresponding to this procedure and taking the Darboux transformation n times, we find the following formula:

$$q'_1 = q_1 - 4\sqrt{\frac{\beta_2}{\gamma}} \sum_{m=1}^{n} \frac{\text{Im}(\lambda_m)\varphi_{1,m}(\lambda_m)\varphi_{2,m}^*(\lambda_m)}{A_m}$$

$$q'_2 = q_2 - 4\sqrt{\frac{\beta_2}{\gamma}} \sum_{m=1}^{n} \frac{\text{Im}(\lambda_m)\varphi_{1,m}(\lambda_m)\varphi_{3,m}^*(\lambda_m)}{A_m}$$

$$\varphi_{k,m+1}(\lambda_{m+1}) = (\lambda_{m+1} + \lambda_m^*)\varphi_{k,m}(\lambda_{m+1}) - \frac{B_m}{A_m}(\lambda_m + \lambda_m^*)\varphi_{k,m}(\lambda_m)$$

$$A_m = |\varphi_{1,m}(\lambda_m)|^2 + |\varphi_{2,m}(\lambda_m)|^2 + |\varphi_{3,m}(\lambda_m)|^2$$

$$B_m = \varphi_{1,m}(\lambda_{m+1})\varphi_{1,m}^*(\lambda_m) + \varphi_{2,m}(\lambda_{m+1})\varphi_{2,m}^*(\lambda_m)$$

$$(4.16)$$

Substituting the zero solution of equation (4.10) as $q_1 = 0$ and $q_2 = 0$ with $\lambda_1 = \alpha_1 + i\rho_1$ into equation (4.16), one can derive the one soliton solution for equation (4.10). Using that one soliton solution as the seed solution with $\lambda_2 = \alpha_2 + i\rho_2$ in equation (4.16), we can derive the two-soliton solution. Thus, in recursion, one can generate up to n-soliton solution. Here we present only the two-soliton solutions in explicit forms. By putting $n = 2$ in equation (4.16), we get two-soliton solution as follows:

$$q'_1(2) = -4\sqrt{\frac{\beta_2(z)}{\gamma(z)}}\frac{X}{Y}$$

$$q'_2(2) = -4\sqrt{\frac{\beta_2(z)}{\gamma(z)}}\frac{X}{Y}$$

(4.17)

The values of X and Y are

$$X = a_1 \mathrm{Cosh}(2\theta_2)\exp(2i\varphi_2) + a_2 \mathrm{Cosh}(2\theta_1)\exp(2i\varphi_2) + i\, a_3(\mathrm{Sinh}(2\theta_2)\exp(2i\varphi_1) - \mathrm{Sinh}(2\theta_1)\exp(2i\varphi_2))$$

$$Y = b_1 \mathrm{Cosh}(2\theta_1 + 2\theta_2)\exp(2i\varphi_2) + a_2 \mathrm{Cosh}(2\theta_1 - 2\theta_2)\exp(2i\varphi_2) + b_3 \mathrm{Cos}(2\varphi_2 - 2\varphi_1)$$

$$a_1 = \frac{\alpha_1}{2}\left(\alpha_1^2 - \alpha_2^2 + (\rho_1 - \rho_2)^2\right)$$

$$a_2 = \frac{\alpha_1}{2}\left(\alpha_2^2 - \alpha_1^2 + (\rho_1 - \rho_2)^2\right)$$

$$a_3 = \alpha_1 - \alpha_2(\rho_1 - \rho_2)$$

$$b_1 = \frac{1}{4}(\alpha_1 - \alpha_2)^2 + (\rho_1 - \rho_2)^2$$
$$b_2 = \frac{1}{4}(\alpha_1 - \alpha_2)^2 + (\rho_1 - \rho_2)^2$$
$$b_3 = -\alpha_1\alpha_2$$

$$\theta_j = \rho_j t - \frac{1}{2}\rho_j\left(3\alpha_j^2 - \rho_j^2\right)\int \beta_2(z)dz + \frac{1}{2}\alpha_j\rho_j\int \beta_2(z) + \theta_{j0}$$

$$\phi_j = -\frac{1}{2}\alpha_j t - \frac{1}{2}\alpha_j\left(3\rho_j^2 - \alpha_j^2\right)\int \beta_2(z)dz + \frac{1}{4}\left(\alpha_j^2 - \rho_j^2\right)\int \beta_2(z) + \phi_{j0}$$

Where θ_{j0} and φ_{j0} are called integration constant.

4.5 Discussion on switching characteristics of femtosecond solitons

In [55], the authors explicitly attained bright two-soliton solutions for the saturable Schrödinger equation more general than previously, and derived explicit asymptotic results for collisions in the anomalous dispersion region. In a multi-component system, the remarkable interaction properties of vector optical solitons are used to construct logic gate switching devices where energy is exchanged between solitons for switching characteristics. In this work, we demonstrate that switching characteristics of vector solitons by properly controlling the soliton parameters to display the switching possibilities of optical solitons which are opposing the earlier estimation that this was not possible in integrable systems. Moreover, these investigations are used in the implementation of optical soliton interaction-based logic operations for optical soliton-assisted ultrafast switching devices in nonlinear photonic circuits where radiative losses are minimal.

The peculiar properties of vector solitons open a new window for physical applications that are not possible in the case of scalar solitons. In particular, energy-exchange interactions among bright vector solitons offer the technology to develop photonic circuits in an optical fiber communication system. References [56, 57] have pointed out the concept of simulating the quantum logic via collisions of vector solitons in the direction of designing a true quantum information processor. Thus, these results have a good impact on the investigation of optical soliton computing via energy-exchange interactions of vector solitons in some potential research areas like optical computing, WDM-employed soliton transmission system, all-optical switching devices, etc.

In an anomalous GVD regime, two-mode or two-core optical fiber is efficiently utilized to achieve the soliton inelastic collisions where the entire soliton pulse can be switched from one mode to another mode. These switching characteristics can be applied in the design of ultrafast logic gates devices. Current literature shows that asymmetric fiber couplers can also be used to attain the switching process in optical networks for ultrafast logic gates. In this chapter, the interactions between two solitons in birefringent fibers are studied.

For the obtained solution (4.17), we attain the inelastic soliton interactions as displayed in figure 4.1(a). Also, from figure 4.1(b), we infer that there is an energy exchange between optical solitons during the course of interactions. Furthermore, energy can be lost or transferred from one soliton to another through the process of energy exchange caused by inelastic interactions.

Due to the influence of inhomogeneous effects on the two-soliton interaction, an energy-exchange interaction is observed as shown in figure 4.2(a). In particular, in this interaction, two solitons with equal amplitudes are inelastically interacted at $z = 0$. After the interaction, one soliton gets well amplificated while another one is highly attenuated. This amplification process gives the amplified optical signal through the process of inelastic interaction by properly adopting control parameters.

As depicted in figure 4.3(a), soliton interactions are purely elastic where energy transfer does not occur and optical solitons simply pass through each other without

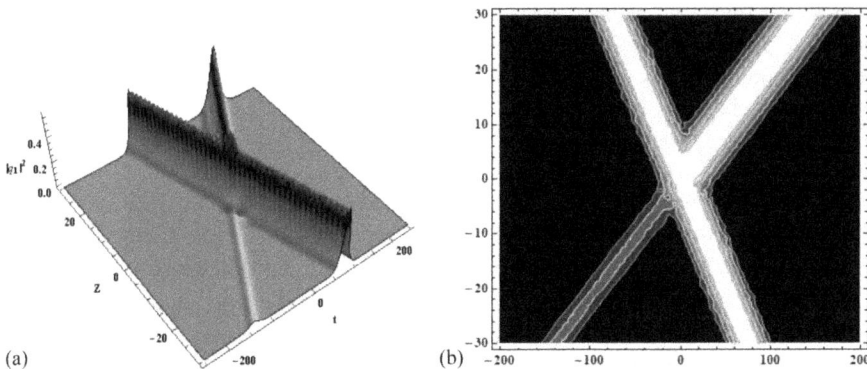

Figure 4.1. (a) 3D view of inelastic interactions between two bright solitons with unequal amplitudes. (b) Contour plot.

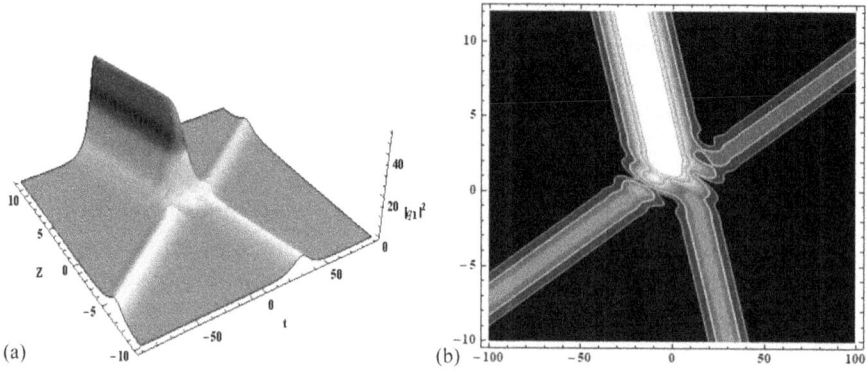

Figure 4.2. (a) 3D view of inelastic interactions between two bright solitons with equal amplitudes. (b) Contour plot.

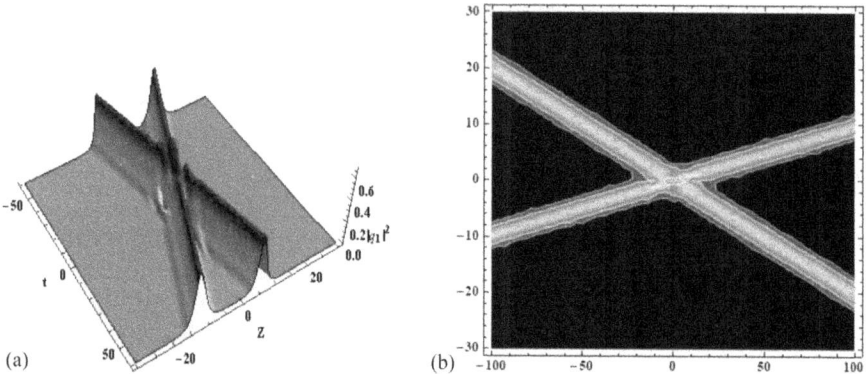

Figure 4.3. (a) 3D view of elastic interactions among two bright solitons with equal amplitudes. (b) Contour plot.

affecting each other except for phase shift. They remain unaffected by the collisional process when they undergo elastic interactions with each other. We infer that after the interactions among solitons, the intensity, velocity, and shape of optical spatial solitons are invariant as clearly shown in figure 4.3(b).

4.6 Conclusions

Two-component coupled NLSE was considered in this chapter and solved by the Darboux transformation method via constructed Lax pair. This theoretical model is adequate to describe the two-mode propagation in an inhomogeneous optical fiber. Using obtained solutions via Darboux procedure, elastic and inelastic interactions were investigated. In particular, the switching dynamics of two optical solitons were studied for inelastic type collision. We suggest the obtained results be used for the construction of all-optical switching devices in optical network systems via soliton interactions.

References

[1] Hasegawa A and Tappert F 1973 Transmission of stationary nonlinear optical physics in dispersive dielectric fibers *Appl. Phys. Lett.* **23** 142–4

[2] Mollenauer L F, Stolen R H and Gordon J P 1980 Experimental observation of picosecond pulse narrowing and solitons in Optical fibers *Phys. Rev. Lett.* **45** 1095–8

[3] Ablowitz M J and Clarkson P A 1991 *Soliton Nonlinear Evolution Equations and Inverse Scattering* (New York: Cambridge University Press) pp 313–44

[4] Radhakrishnan R, Lakshmanan M and Hietarinta J 1997 Inelastic collision and switching of coupled bright solitons in optical fibers *Phys. Rev. E* **56** 2213–6

[5] Serkin V N and Hasegawa A 2000 Novel soliton solutions of the nonlinear Schrödinger equation model *Phys. Rev. Lett.* **85** 4502–5

[6] Wanga S, Zhang X, Ma G and Zhu D 2023 All-optical switches based on three soliton inelastic interaction and its application in optical communication systems *Chin. Phys.* B **32** 30506–11

[7] Hasegawa A and Kodama Y 1995 *Soliton in Optical Communications* (London: Oxford University Press) pp 67–88

[8] Xie X Y and Liu X B 2020 Elastic and inelastic collisions of the semirational solutions for the coupled Hirota equations in a birefringent fiber *Appl. Math. Lett.* **105** 106291–7

[9] Wang L and Yan Z 2021 Rogue wave formation and interactions in the defocusing nonlinear Schrödinger equation with external potentials *Appl. Math. Lett.* **111** 106670–8

[10] Li Y and Liu J 2018 New periodic solitary wave solutions for the new (2+1)-dimensional Korteweg–de Vries equation *Nonlinear Dyn.* **91** 497–504

[11] Gelash A, Xu G and Kibler B 2022 Management of breather interactions *Phys. Rev. Res.* **4** 33197–221

[12] Fei J and Cao W 2020 Explicit soliton–cnoidal wave interaction solutions for the (2 + 1)-dimensional negative-order breaking soliton equation *Waves Random Complex Medium* **30** 54–64

[13] Xu Z W, Yu G F and Zhu Z N 2016 Soliton dynamics to the multi-component complex coupled integrable dispersionless equation *Commun. Nonlinear Sci. Numer. Simul.* **40** 28–43

[14] Kengne E 2014 Analytical solutions of nonlinear Schrödinger equation with distributed coefficients *Chaos, Solitons Fractals* **61** 56–68

[15] Angelin V and Mani Rajan M S 2018 Attosecond soliton shaping through dispersion tailoring technique in a monomode optical fiber *Optik* **167** 196–203

[16] Mani Rajan M S 2020 Transition from bird to butterfly shaped nonautonomous soliton and soliton switching in erbium doped resonant fiber *Phys. Scr.* **95** 105203

[17] Hosseini K, Mirzazadeh M, Baleanu D, Salahshour S and Akinyemi L 2022 Optical solitons of a high-order nonlinear Schrödinger equation involving nonlinear dispersions and Kerr effect *Opt. Quantum Electron.* **54** 177

[18] Tariq K U, Younis N and Rizvi S T R 2018 Optical solitons in monomode fibers with higher order nonlinear Schrödinger equation *Optik* **154** 360–71

[19] Xie X Y, Tian B, Liu L, Guan Y Y and Jiang Y 2017 Bright solitons for a generalized nonautonomous nonlinear equation in a nonlinear inhomogeneous fiber *Commun. Nonlinear Sci. Numer. Simul.* **47** 16–22

[20] Osman M S 2019 One-soliton shaping and inelastic collision between double solitons in the fifth-order variable-coefficient Sawada–Kotera equation *Nonlinear Dyn.* **96** 1491–6

[21] Li J, Xu T, Meng X H, Zhang Y X, Zhang H Q and Tian B 2007 Lax pair, Bäcklund transformation and N-soliton-like solution for a variable-coefficient Gardner equation from nonlinear lattice, plasma physics and ocean dynamics with symbolic computation *J. Math. Anal. Appl.* **336** 1443–55

[22] Lu X, Zhu H W, Yao Z, Meng X H, Zhang C, Zhang C Y and Tian B 2008 Multi-soliton solutions in terms of double Wronskian determinant for a generalized variable-coefficient nonlinear Schrödinger equation from plasma physics, arterial mechanics, fluid dynamics and optical communications *Ann. Phys.* **323** 1947–55

[23] Remoissenet M 1993 *Waves Called Solitons* (Heidelberg: Springer) pp 204–25

[24] Hasegawa A and Kodama Y 1987 Nonlinear pulse propagation in a monomode dielectric guide *IEEE J. Quantum Electron.* **23** 510–24

[25] Agrawal G P 2011 Nonlinear fiber optics: its history and recent progress *J. Opt. Soc. Am.* B **28** A1–A10

[26] Stolen R H and Ashkin A 1973 Optical Kerr effect in glass waveguide *Appl. Phys. Lett.* **22** 294–6

[27] Maker P D, Terhune R W and Savage C M 1964 Intensity dependent changes in the refractive index of liquids *Phys. Rev. Lett.* **12** 507–9

[28] Stolen R H and Lin C 1978 Self-phase-modulation in silica optical fibers *Phys. Rev.* A **17** 1448–53

[29] Weber H P and Hodel W 1988 Propagation of sub picosecond pulses and soliton formation in an optical fiber *Phys. Scr.* **23** 200–5

[30] Ohkuma K, Ichikawa Y H and Abe Y 1987 Soliton propagation along optical fibers *Opt. Lett.* **12** 516–8

[31] Mani Rajan M S, Hakkim J and Mahalingam A 2013 Dispersion management and cascade compression of femtosecond nonautonomous soliton in birefringent fiber *Eur. Phys. J.* D **67** 6150–7

[32] Han P Q and Shin H J 1999 Systematic construction of multicomponent optical solitons *Phys. Rev.* E **61** 3093–106

[33] Islam M N, Mollenauer L F, Stolen R H, Simpson J R and Shang H T 1987 Cross-phase modulation in optical fibers *Opt. Lett.* **12** 625–7

[34] Serkin V N and Hasegawa A 2000 Soliton management in the nonlinear Schrödinger equation model with varying dispersion, nonlinearity, and gain *JETP Lett.* **72** 89–92

[35] Yu W T, Zhou Q, Mirzazadeh M and Liu W J 2019 Phase shift, amplification, oscillation and attenuation of solitons in nonlinear optics *J. Adv. Res.* **15** 69–76

[36] Liu W J, Zhang Y J and Triki H 2019 Interaction properties of solitonics in inhomogeneous optical fibers *Nonlinear Dyn.* **95** 557–63

[37] Mahalingam A and Mani Rajan M S 2015 Influence of generalized external potentials on nonlinear tunneling of nonautonomous solitons: Soliton management *Opt. Fiber Technol.* **25** 44–50

[38] Mani Rajan M S and Mahalingam A 2015 Nonautonomous solitons in modified inhomogeneous Hirota equation: soliton control and soliton interaction *Nonlinear Dyn.* **79** 2469–84

[39] Liu W J, Tian B and Zhang H Q 2008 Types of solutions of the variable-coefficient nonlinear Schrödinger equation with symbolic computation *Phys. Rev.* E **78** 66613–8

[40] Mani Rajan M S, Mahalingam A and Uthayakumar A 2014 Nonlinear tunneling of optical soliton in 3 coupled NLS equation with symbolic computation *Ann. Phys.* **346** 1–13

[41] Mani Rajan M S and Bhuvaneshwari B V 2018 Controllable soliton interaction in three mode nonlinear optical fiber *Optik* **175** 39–48

[42] Chen Z L and Jie L 2013 Rogue-wave solutions of a three-component coupled nonlinear Schrödinger equation *Phys. Rev.* E **87** 13201–8

[43] Soljacic M, Steiglitz K, Sears S M, Segev M, Jakubowski M H and Squier R 2003 Collisions of two solitons in an arbitrary number of coupled nonlinear Schrodinger equations *Phys. Rev. Lett.* **90** 254102

[44] Jiang Y, Tian B, Liu W J, Sun K, Li M and Wang P 2012 Solitons interactions and complexes for coupled nonlinear Schrodinger equations *Phys. Rev.* E **85** 36605–22

[45] Mani Rajan M S 2016 Dynamics of optical soliton in a tapered erbium-doped fiber under periodic distributed amplification system *Nonlinear Dyn.* **85** 599

[46] Liu W, Yang C, Liu M, Yu W, Zhang Y and Lei M Effect of high-order dispersion on three-soliton interactions for the variable-coefficients Hirota equation *Phys. Rev.* E **96** 042201

[47] Wang Q, Yang J R and Liang G 2020 Controllable soliton transition and interaction in nonlocal nonlinear media *Nonlinear Dyn.* **101** 1169–79

[48] Sivaranjani R, Shanmuga Sundar D, Sridarshini T, Sitharthan R, Karthikeyan M, Sivanantha Raja A and Marcos Flores C 2020 Photonic crystal based all-optical half adder: a brief analysis *Laser Phys.* **30** 116205

[49] Geerthana S, Syedakbar S, Sridarshini T, Balaji V R, Sitharthan R and Shanmuga Sundar D 2022 Current and future horizon of optics and photonics in environmental sustainability *Sustain. Comput. Inform. Syst.* **36** 100815

[50] Geerthana S, Syedakbar S, Sridarshini T, Balaji V R, Sitharthan R and Shanmuga Sundar D 2022 2D-Ph C based all optical AND, OR and EX-OR logic gates with high contrast ratio operating at C band *Laser Phys.* **32** 106201

[51] Shanmuga Sundar D, Arun T, Ramesh R, Sathish kumar K and Sridarshini T 2023 *Photonic Crystal and Its Applications for Next Generation Systems* (Singapore: Springer)

[52] Sridarshini T, Geerthana S, Balaji V R, Arun T, Sitharthan R, Sivanantha Raja A and Shanmuga Sundar D 2023 Ultra-compact all-optical logical circuits for photonic integrated circuits *Laser Phys.* **33** 076207

[53] Ablowitz M J, Kaup D J and Newel A C Nonlinear-evolution equations of physical significance *Phys. Rev. Lett.* **31** 125–7

[54] Gu H, Hu H S and Zhou Z X 2005 *Darboux Transformation in Soliton Theory and Its Geometric Applications* (Shanghai: Shanghai Science and Technology)

[55] Jakubowski M H, Steiglitz K and Squier R K 1997 Information transfer between solitary waves in the saturable Schrödinger equation *Phys. Rev.* E **56** 7267–73

[56] Janutka A 2007 Error of quantum-logic simulation via vector-soliton collisions *J. Phys.* A **40** 10813–27

[57] Janutka A 2008 Quantum-like information processing using vector solitons *J. Phys.* A **41** 375202–19

IOP Publishing

Advances in All-optical Communication

Shanmuga Sundar Dhanabalan, Arun Thirumurugan and Sridarshini Thirumaran

Chapter 5

Silicon photonic modulators for high-speed applications—a review

R G Jesuwanth Sugesh, V R Balaji, A Sivasubramanian, Gopalkrishna Hedge, M A Ibrar Jahan and Richards Joe Stanislaus

Internet traffic is rising at an alarming rate, taking a toll on data centres. The recent pandemic has stressed the importance of advancement in networks. The advantage of optical communication is brought down to the chip level to meet futuristic network demands with the help of Silicon Photonics technology. The research aims toward greater transmission rates with compact footprint devices on the chip level. The effective modulation of an optical modulator is the prerequisite for high-speed transmission. A paradigm shift in the selection of new materials to increase the modulator's effectiveness has been witnessed in recent years. The chapter aims to provide a review of the Electro-optic Silicon Photonic modulators, novel materials that are set to revolutionize the modulator and the performance metrics to be met for inter and intra-high speed data centre applications.

5.1 Introduction

Affordable internet and technological advancements such as artificial intelligence, cloud computing, and IoT have propelled the use of the internet. Internet traffic is increasing at an alarming rate due to the increase in business transactions, video streaming, web conference, file sharing, etc. With the introduction of 5G, the transmission speed has increased 100 times faster than the present networks. It is predicted that by 2025 there will be 75 billion IoT endpoints [1]. The biggest lesson that COVID-19 has taught economies is that digitisation is paramount. Network traffic within a data centre is higher than the number of bits transmitted. The never-ending increase in network traffic has led to the development of warehouse-scale data centres, which comprise many servers interconnected by optical links.

For long-haul communications, optical communications are through optical fibres. Single-mode fibre (SMF) has been preferred due to its minimal loss

doi:10.1088/978-0-7503-5623-7ch5

(0.2 dB km^{-1}), providing data transmission for long distances. The cost of optical fibre installation and maintenance is expensive. This creates the requirement for efficient bandwidth management. The simple way of increasing spectrum efficiency is by transmitting at multiplexed wavelengths. Short-haul communications between and within data centres have improved with the help of optical communication. Research is being undertaken to produce energy-efficient optical interconnects in the order of a few pico-Joules per bit transmitted for fast and reliable data communications. Multi-mode fibres (MMF) provide short distance transmission of multiple modes through a single fibre, improving bandwidth. Research in optical interconnects has increased due to their efficiency in communications such as rack-to-rack communications, between chip to chip, and even intra-chip communications. The core of any optical interconnect system is the optical transceiver.

Silicon photonics (SiPh) takes advantage of the benefits of sophisticated CMOS manufacturing technology employed in the microelectronics sector. The most significant advantage of SiPh is the ability to bring electronic and photonic components on a single platform, leading to monolithically integrated electronic-photonic integrated circuits (EPIC). There are other PIC platforms produced at specialised fabrication facilities (fabs), such as indium phosphide (InP) and lithium niobate (LiNbO$_3$) [2, 3], which employ 100 and 150 mm wafers, respectively. SiPh-integrated circuits are usually produced on wafers that have a 200 or 300 mm diameter. In modern R&D pilot lines and CMOS commercial fabs, the large wafer size enables a high density of compact dies per wafer at low-cost scaling to large commercial quantities. The commercial availability of SiPh-enabled transceivers demonstrates SiPh's potential as a photonic integration technique for high-speed modulators.

In Silicon on Insulator (SOI) wafer, there is a silicon dioxide (SiO$_2$) layer called the buried oxide (BOX) layer, which provides an excellent platform for developing photonic devices. The BOX layer prevents the light from getting coupled to the silicon substrate below. This allows sub-micrometre waveguides, bending the waveguide with a small radius and high-density integration. Due to the low refractive index of silicon, it is transparent at telecommunication wavelengths, and thus low-loss waveguides can be designed.

The modulator plays a vital role in any photonic communication link. In silicon, optical modulation is obtained with the help of the plasma dispersion effect (electro-optic modulators) [4] or thermal effect (thermo-optic modulators). It has been observed that the plasma dispersion effect provides high-speed data transmission operation. To compensate for the low modulation efficiency in silicon, either long devices or high doping concentrations should be used. By doing so, the capacitance increases, limiting the modulator's bandwidth. To reduce the capacitance, a travelling wave driving scheme is utilised in the Mach–Zehnder modulator (MZM). Bandwidth and modulation depth improves with the inclusion of a travelling wave electrode. This makes the travelling wave electrode Mach–Zehnder modulator (TW-MZM) among the most preferred modulator [5]. Ring modulator (RM) is preferred for its compact footprint and low power consumption and optical loss. The challenge in an RM is its low fabrication and temperature tolerance.

5.2 Phase shifters

Silicon optical modulators are the workhorse of silicon photonics-based optical interconnects. In recent years, modulator performance has increased considerably regarding footprint, modulation efficiency, and energy usage. However, there are still numerous challenges to overcome. The co-design of photonics and electronics is required for an ideal design, which implies a trade-off between multiple performance attributes. In this chapter, the two most preferred modulators for data communication (i.e., MZM and RM) are discussed, the trade-off conditions in designing the two modulators are briefed, and the performance metrics of the designed modulators and the data centre requirements are presented.

At telecommunication wavelengths, the principal EO effects used in other semiconductors, including the Kerr effect, Pockels effect, and Franz–Keldysh effect, are absent in bulk unstrained silicon. Soref and Bennett have determined the mathematical expression for the plasma dispersion effect using experiments. That relates to the relationship of free carrier density with refractive index (Δn) (5.1) and absorption coefficient ($\Delta \alpha$) is provided in (5.2) [4].

$$\Delta n = -8.8 \times 10^{-22} N_n - 8.5 \times 10^{-18} N_p^{0.8} \qquad (5.1)$$

$$\Delta \alpha = 8.5 \times 10^{-18} N_n + 6 \times 10^{-18} N_p \qquad (5.2)$$

The variations in the concentration of free electrons and holes are represented by N_n and N_p, respectively. These equations describe how the refractive index of silicon can be manipulated with free carrier concentrations. They also imply one essential trade-off in phase modulator design: optical loss and modulation efficiency. Increased free carriers cause a significant shift in refractive index, but they also cause higher optical loss.

5.2.1 Silicon-based phase shifter

The modulation efficiency (V cm) of a modulator utilising the Electro-optic (EO) effect is given by the voltage (V) that creates a phase shift in the phase shifter (PS) of unit length (cm). The voltage consumption and the phase shifter length of the PS have to be minimised to obtain high modulation efficiency. The performance of the PS to get high modulation efficiency is dependent on the waveguide dimensions, carrier manipulation technique, doping carrier concentration, and doping pattern. The PS is embedded in MZIs [6–8], RRs [9–11], Michelson interferometers [12], Bragg reflectors [13], Fabry–Perot cavities [14], and photonic crystal cavities [15] to convert phase modulation to intensity modulation. The waveguide dimensions play a vital role in confining the optical beam to reduce the propagation loss, bent loss, and the coupler's working.

Electrically altering the free carrier concentration can be accomplished in a variety of methods, including carrier accumulation [16], carrier injection [6], and carrier depletion [7] (figure 5.1). The various doping pattern configurations are found in the literature (figure 5.2).

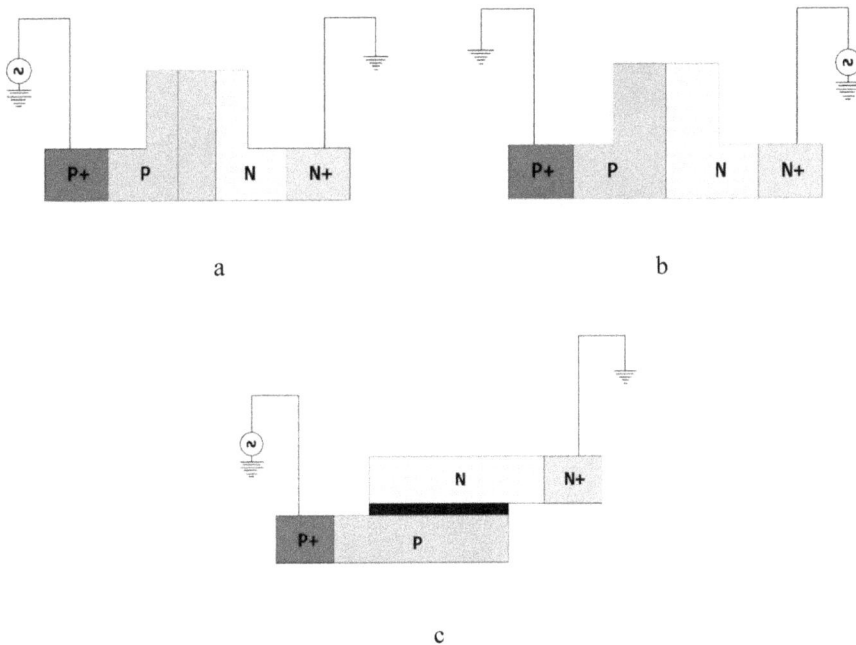

Figure 5.1. (a) Baseline architecture of carrier injection, (b) carrier depletion, and (c) carrier accumulation plasma dispersion phase shifters.

A forward-biased PN junction or PIN junction is commonly used for carrier injection. Because of the forward biassing, an excessive minority carrier accumulates in the junction, changing the refractive index of silicon [6]. Because of the large minority carrier in the PN junction in forward bias, the carrier injection approach has a high modulation efficiency in a small cross-section [17]. It has been reported that carrier injection-based modulators (CIM) use vertical PIN [18] and lateral PIN configurations [17]. CIMs commonly employ lumped drive schemes due to the practicality of a short PS (in μm) [17]. CIM's limited speeds are improved by using pre-emphasised electrical signalling techniques where a component of the electrical driving signal has a modulator voltage greater than V, with the predicted result being increased power consumption [19]. A silicon micro-ring modulator dependent on a forward-biased PIN junction was developed in 2005 [20]. To decrease doping-induced optical loss, a lateral PIN junction was inserted in the rib waveguide with intrinsic rib area and p+/n+ doping on the slab layer, as illustrated in figure 5.2(b). A carrier injection-based MZM was also demonstrated in [6]. This gadget was able to attain a $V\pi L$ of 0.29 V cm. In recent years, passive equalisation technologies have been developed to improve the high-speed operation of CIM. The bandwidth of the CIM can be increased by including an RC passive equaliser [21]. Passive RC equalisation recently enabled 70 Gbaud functioning of a carrier injection MZM [22]. The modulator had a 3 dB bandwidth of 37 GHz and used a 0.25 mm long PS with a loss of 28 dB cm^{-1} and a 2 V cm modulation efficiency. In [19], a side-wall corrugated MZM operating at 25 Gbps achieves extraordinary 0.274 V cm

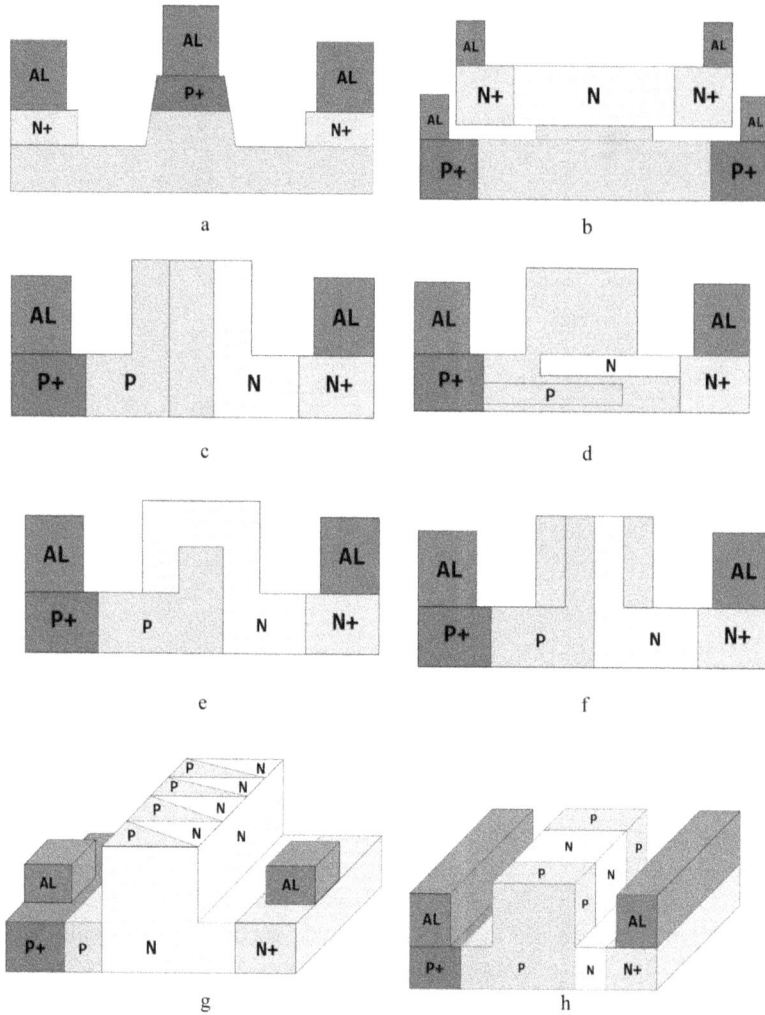

Figure 5.2. Various configurations of plasma dispersion phase shifters. (a) Three terminal junction, (b) vertical junction, (c) PIN junction, (d) epitaxial vertical junction, (e) wrapped junction, (f) PN junction with contour doping at corners, (g) zig-zag junction, and (h) interleaved junction.

modulation efficiency with a FIR filter employed for frequency compensation, which is required for broadband operation. The displayed modulator contained a phase shifter that was 250 μm long and had 5.3 dB mm^{-1} optical loss. The carrier injection devices have a slower operating speed than other devices that rely on other methods due to the lengthy life duration of the extra minority carriers in the silicon.

A reverse-biased PN or PIN junction-based rib waveguide is used in silicon optical modulators employing carrier depletion (figure 5.2(c)). A reverse bias was used to remove the free charge carrier from the waveguide's core. In carrier depletion modulators, the current greatest recorded data rates of 100 Gbps were also demonstrated utilising OOK modulation [23]. Carrier depletion PS have a low capacitance (0.2–0.8 pF mm^{-1}), which results in low modulation efficiency [5, 21]. The capacitance can be raised to enhance modulation efficiency by either shortening the mode size or reducing the breadth of the depletion area [5]. Higher doping doses are required for the latter, which results in an increase in free carrier absorption loss. Over a period of time, a number of PN junction designs have been described to handle the problem of simultaneously maximising the modulator's efficiency, speed, and loss. Below are the typical ways PN junctions are implemented.

(i) Vertical PN junction pattern

A vertical P(I)N junction is created by placing the vertical doped regions with respect to the substrate. Greater overlap between the doped region and the optical propagation path is known to lower power consumption and enhance modulation efficiency (higher capacitance) at these junctions [8].

(ii) Horizontal P(I)N or lateral pattern

The PN junction is arranged side-by-side in the middle of the waveguide in a horizontal or lateral PN junction (laterally symmetric design) [24]. Usually, the idea is to take advantage of the dominantly *p*-doped active region's larger index change and lesser loss.

(iii) Interleaved or interdigitated P(I)N pattern

The junction is formed with P- and N-doped regions arranged alternatingly along the waveguide's length [11]. By balancing the power efficiency and speed, it is known to give high modulation efficiency at the expense of fabrication steps.

An Athermal micro-disk modulator with a 250 pm V^{-1} modulation efficiency of, operating at 25 Gbps while requiring 0.9 fJ/bit utilising OOK signalling is described using a vertical PIN junction architecture [9]. The modulator had ~1 dB IL and 6 dB ER at 25 Gbps. A symmetric PN junction was used to achieve operating speeds of 60 Gbps with a 3.8 dB ER [24]. The PN junction phase shifter implanted on both arms of the MZI functioned in a succession of push–pull modes. The net capacitance is cut in half by connecting the two-phase shifters in series, and the modulator's chirp is limited by the push–pull technique. Symmetric carrier depletion MZM with a sub-1 V drive signal performed at 20 Gbps had a power consumption of 200 fJ/bit [25]. An 0.5 mm^2 footprint RM with a symmetric PS demonstrated a modulation efficiency of 25 pm V^{-1} and consumed energy of 36 fJ/bit for 40 Gbps operation [10]. The carrier depletion PS is achieved using a photonic crystal structure with a length of 200 μm in a recent development of a 64 Gbps MZM [14]. The modulator needed a 5.5 V voltage swing to generate a 21 pJ/bit energy usage and 4.8 dB ER. A demonstration achieved a 1 dB mm^{-1} loss without reducing modulator speed but by improving the use of self-aligned PN junctions [26]. The self-aligning method is used to create a PIPIN PS. It operated at 40 Gbps, with a 3.5 V cm modulation efficiency for a PS of 0.95 mm incurring an optical loss of 4.5 dB [27]. A PIPN PS structure of 2.5 mm in an MZM achieved a modulation efficiency of 1 V cm supporting a high-speed operation of 70 Gbps with an extinction ratio of 6.5 dB and BER of 3.7 × 10^{-5} [28]. Apart from standalone high-speed PS implementations based on these three fundamental junction designs and their modifications, there have been reports of PN junction patterns that combine all three techniques in a single PN junction design [11, 27, 29]. The 'zig-zag' PS for an MZM, for example, delivers modulation efficiency of 1.6 V cm for a PS of 2-mm with loss of 4.4 dB [29]. The modulator had a bandwidth of 55 GHz and operated at 90 Gbps. A corrugated PN structure was obtained with the combination of vertical and horizontal PN structure. This introduced multiple PN junctions along the light propagation path thus improving the modulation efficiency of 0.7 V cm for a 1.5 mm long PS with an energy per bit transmission of 3.3 pJ/bit. The increased interaction of the carriers with the mode propagation path improved the modulation efficiency but at the cost of free carrier absorption loss. This increased the optical loss to 7.4 dB [30]. Another example is a phase shifter that combines horizontal, vertical, and interleaved techniques to provide 40 Gbps of operation for a racetrack RM. At 1 V reverse bias, this implementation had 0.76 V cm and loss 3.5 dB mm^{-1} [27].

Slotted rib structures were engineered to guide the light mode into the slot and the PN junctions were formed on the slides of the slot. The modulator supported a 70 Gbps modulation speed with 4.9 dB modulation depth for a 4 mm long PS [31]. A core-based slot PS was designed to achieve high modulation efficiency and strong coupling between RF and optical modes [32, 33]. The improved interaction of the propagating optical mode with the PN junctions formed on the top and bottom of the core influenced better modulation efficiency of 0.75 V cm for a PS length of 1.5 mm [33]. The device supported a modulation speed of 90 Gbps with modulation depth of 7.37 dB and BER of 1.6×10^{-12}. Better results were contributed by the symmetrical carrier depletion from the junctions. The modulation is improved by the light-matter interaction by the vertical PN junctions formed along the sides of the slot [34]. This design offered a high modulation efficiency of 0.74 V cm and supported an operating speed of 100 Gbps in NRZ OOK modulation. A zig-zag PN junction PS in RM of 9 μm radius offered a modulation efficiency of 38 pm v^{-1} operating at a modulation speed of 100 Gbps in the PAM4 modulation technique [11].

In order to meet the ever-increasing demands of the internet, research on Silicon modulators requires complex modulation formats. Plasma dispersion modulators exhibit chirp because the relationship between voltage and refractive index change is not linear. The plasma dispersion effect based MZM are larger in size (mm) and Ring modulators suffer from high temperature and fabrication sensitivity. Delivering future expectations for single-lane 100 Gbps modulation speeds and beyond in an energy-efficient way is a difficult task. These challenges have propelled researchers to look for alternative materials.

5.2.2 Hybrid phase shifters

There is a shift in the research, to incorporate materials with potent electro-optic effects into the SiPh modulators. The integration of novel materials opens up pathways for energy-efficient high-speed modulation [35]. While maintaining the advantages offered by the established SiPh technology, one can achieve outstanding modulator performance through the wafer-scale integration of these novel material systems with SiPh platforms [35]. These new materials can be roughly divided into organic (electro-optic) materials, ferroelectrics, (silicon-)germanium (GeSi), III–V semiconductors, and 2D materials.

Alternative methods in SiPh to create compact, high-speed, energy-efficient electro-absorption modulators (EAMs) are provided by the Franz–Keldysh (FK) effect and the quantum-confined Stark (QCS) effect with the inclusion of GeSi. Band engineering is necessary for both effects. As a result, their optical bandwidth is constrained (30 and 20 nm for the FK effect and QCS effect, respectively). FK-based modulators favour high-speed operation and have reached 100 Gbps data rate with OOK modulation [36]. Ferroelectrics are substances with spontaneously occurring polarisation that can be changed by means of an electric field that is sufficiently strong (Pockels effect). The combination of ferroelectric substances like lead zirconate titanate (PZT), barium titanate ($BaTiO_3$), and $LiNbO_3$ with SiPh offers a viable path for the construction of effective, high-speed efficient modulators due to their potent Pockels effect. $LiNbO_3$-based modulators deliver a high operating speed of 100 Gbps with low-loss whereas $BaTiO_3$-based modulators are selected for high modulation efficiency (<0.5 V cm) [2, 3, 37, 38]. Since the Pockels effect in organic electro-optic materials is electrical in nature as opposed to ferroelectric, it can result in an ultra-fast response. The organic electro-optic modulators have reported 100 Gbps data rate operation, high modulation efficiency, and fj/bit energy per bet transmission when compared with other Pockels effect modulators [39, 40]. III–V on Si modulators has demonstrated outstanding modulation efficiency and low-loss performance [41, 42]. New research involves atomically thick 2D materials, such as graphene-based modulators that transmit a graphene layer onto a SiN or Si waveguide. Other 2D materials, like molybdenum disulfide (MoS_2) and tungsten disulfide (WS_2), have also been merged with SiPh in response to the encouraging results seen with graphene. When these materials are combined with SiN, the results reveal phase change in relation to a change in absorption for telecom wavelengths. To achieve the highest performance, graphene needs more research and development. High carrier mobility has to be maintained by graphene to maximise the potential of graphene modulators in SiPh [43, 44]. Hybrid modulators are a way to overcome the limits of plasma dispersion modulators. Their success depends on the optimization of the complex fabrication process.

5.3 Mach–Zehnder modulator (MZM)

The Mach–Zehnder modulator (MZM) is a Mach–Zehnder interferometer (MZI)-configured optical intensity modulator. Two Y-junctions at both input and output, as well as two balanced or unbalanced arms, make up an MZM. The input power is shared evenly between two arms by the Y-junction at the input, and the phase of light moving in one or both arms is modified by the plasma dispersion effect. The Y branch combiner plays a major role in the constructive and destructive interference of the MZM. This forms the fundamental working of an MZM. When the phase difference is zero or in phase, then the power supply adds up and this denotes the constructive interference of MZM. As the phase difference increases the mode gets excited and the power leaks out through the walls. When the phase difference is π, it is observed that all the output power is dispersed into the cladding and this denotes the destructive interference of MZM.

Due to the sheer EO effect in silicon, the MZM tends to be rather large [4]. Thus, optimizing the doping pattern plays a vital role in reducing the footprint of MZM. Although device size may be lowered (0.5 mm) by doping optimization [7], this does not always equate to low load capacitance. Due to their independence from the RC time constant, travelling wave (TW) electrode designs are favoured over lumped designs in MZM. This improves the modulator's bandwidth. The figure represents a TWE on a PS with a PN junction-equivalent circuit. The following three aspects must be considered while designing a high-speed TWE [30, 31, 45]. To begin, the electrical and optical signals' velocities should be matched because a velocity mismatch would reduce modulation efficiency. Second, the TWE's radio frequency (RF) loss should be kept to a minimum. Finally, the phase shifter's characteristic impedance should be 50 or 25 Ω [25].

The TWE is used in the majority of the high-speed silicon MZMs that have been shown thus far. Silicon MZM is commonly utilised in silicon photonic transceiver circuits due to its benefits of enormous optical bandwidth, high temperature and process variation tolerance, and low cost. The footprint, modulation efficiency, optical loss, parasitic, power consumption, and bandwidth trade-offs are all part of the MZM design process. Figure 5.3 shows the PN junction-equivalent circuit with the device model. One of the most basic design factors, for example, is the doping concentration. More free charge carriers result from greater doping concentrations, which leads to increased modulation efficiency.

Higher doping concentrations, on the other hand, will result in increased optical loss owing to absorption. As a result, the doping profile in the active area may be

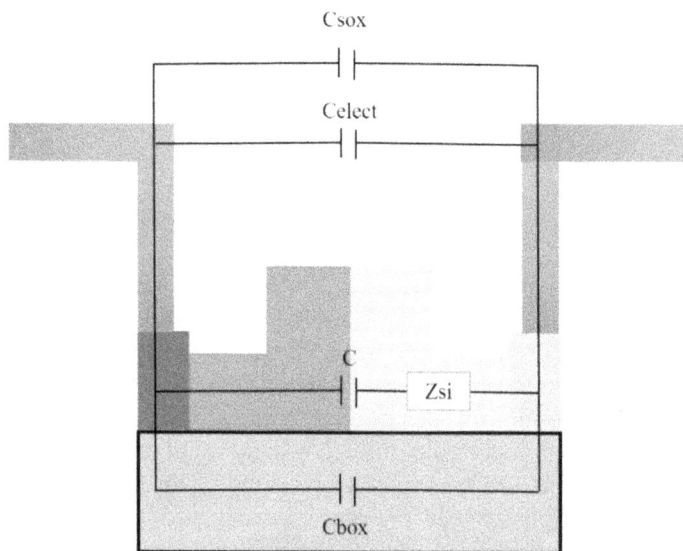

Figure 5.3. PN junction-equivalent circuit with device model.

tuned such that only a portion of it is doped or severely doped, while the rest remains intrinsic or minimally doped [45, 46]. As a consequence, the optical loss may be significantly decreased while modulation efficiency remains extremely high. Another advantage of this doping profile is that the depletion zone is contained inside the high concentration region. The device may achieve considerably better linearity since any excess depletion area over the active zone does not contribute to the effective index change [46]. When compared to the change in free electron concentration, the change in free hole concentration has a bigger influence on the change of refractive index and a smaller effect on the change of absorption coefficient. Therefore, p-type dopant is selected at a higher concentration than the n-type dopant. The waveguide is another essential trade-off in MZM design. The waveguide should be constructed to accommodate just the basic mode, hence the rib waveguide's size (thickness and width) should be kept to a minimum to avoid higher-order optical modes. In addition, a narrow core area results in a low parasitic capacitance. The narrow centre rib region, on the other hand, causes a lack of overlap between the carrier depletion zone and the optical mode, resulting in low modulation efficiency. It also causes a high level of series parasitic resistance, resulting in a significant loss of the electrical driving signal. Thus, the trade-off conditions of the doping pattern, concentration, waveguide dimensions, and the TWE have to be considered when designing a MZM for high-speed operation.

5.4 Ring modulator

The ring modulator is a frequently used silicon-based resonant optical modulator. The ring modulator may be used for two different types of intensity modulation. We could either modify the effective index of the waveguide generating the ring resonator, or we could change the coupling coefficients using index modulation on the ring-to-bus coupler. The first is known as an intracavity ring modulator (figure 5.4), while the second is known as a coupling-modulated ring resonator (figure 5.5).

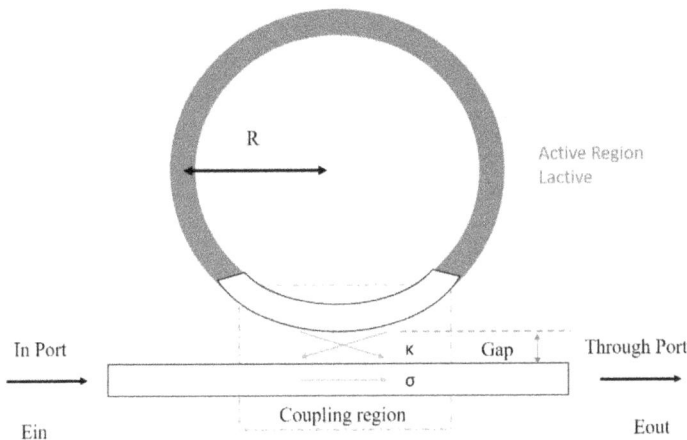

Figure 5.4. Intracavity RM schematic.

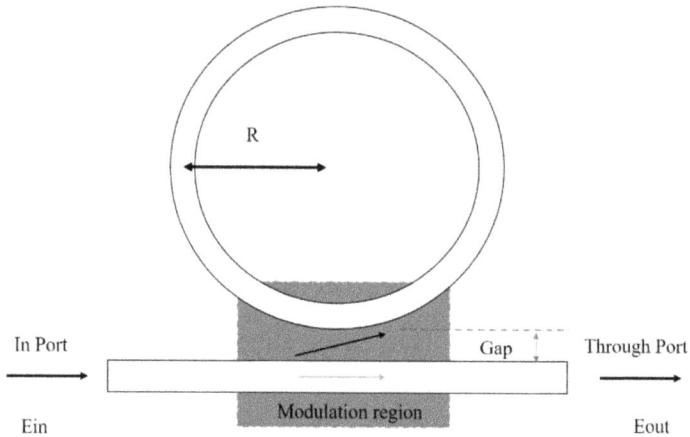

Figure 5.5. Coupling-modulated RM schematic.

The bulk of today's ring modulators is controlled by modulating the intracavity parameters. The ring resonator waveguide (except the coupling zone) is doped in the intracavity ring modulator, and round-trip characteristics such as phase accumulation and the field attenuation are regulated by the plasma dispersion effect. A modulating voltage V_m along the active length L_{active} can generate an electro-optical effect because an active junction is positioned inside the ring. When a round-phase trip's accumulation is an integer multiple of 2π, the circulating light constructively interacts with the incoming light coupled from the bus waveguide, causing light resonances within the cavity. The resonance wavelength shifts as the refractive index of the ring resonator waveguide is changed. The amount of light transmitted in the bus waveguide is determined by the resonance wavelength's position, which is regulated by an applied voltage representing input data.

In order for this form of intensity modulation to work, the light from the ring resonator must be depleted in the on-state and injected into the resonator in the off-state. As a result, the speed at which the cavity switches between on and off states are determined by the cavity's response time. When the cavity's quality factor (Q) is large, the cavity's lifetime is extended, and the rise and fall periods of the field amplitude take longer (Leading to signal distortion). As a result, the cavity linewidth limits the operating bandwidth. The larger the bandwidth, the smaller the photon lifetime and electrical time constant. However, because a shorter photon lifetime indicates a lower Q, the modulator must use a greater driving voltage to attain a given ER.

The modulator for the ring resonator is a lumped component. It is primarily constrained by two factors: an electrical contribution owing to junction construction, and an optical contribution due to photon lifetime τ_p. The efficiency of the PN junction in ring modulators is stated as pm V^{-1}, which reflects how much the resonance shifts in response to the applied voltage.

In the coupling-modulated (CM) ring resonator, the intracavity parameters are kept constant while the cross (k) and through (t) coupling coefficients are adjusted.

Figure 5.5 illustrates the CM-ring modulator's architecture. The ring modulator is brought in and out of critical coupling condition ($t = $ a, where a is the ring transmissivity) by varying the coupling coefficient between the ring and bus wave-guide, which causes the notch depth to vary at around resonance wavelength and thus creates intensity modulation.

Researchers are exploring photonic crystal-based ring resonators as a viable option for futuristic demands. Many photonic components like routers, splitters, filters, logical circuits, and multiplexers [47–51] are being designed using photonic crystals. This allows the integration of the components to obtain a PIC.

5.5 Modulator performance metrics

The total performance of the modulator is determined by design choices such as the passive structure utilised to house the modulator, the modulator length, and the driving voltage. Below is a list of the various performance measures for the overall modulator structure, along with their descriptions.

1. Modulator drive voltage/power consumption/Energy per bit transmission

    The device's power consumption is a crucial parameter to consider, especially in applications with several short-range channels. According to some, the modulator's maximum power consumption should be on the order of 10 fj/bit. Modulator's power consumption is estimated based on whether it is travelling wave electrodes or lumped element driven. A travelling wave electrode is generally utilised in MZM to enable high-speed operation across a long device. The RM is usually driven as a lumped element with no extra termination element. The power consumption is normally determined by the capacitance and driving voltage in this scenario. The power consumption/ energy per bit utilised for transmission is required to be as low as possible. The modulator is among the major contributors to the power budget of the transmission system.

2. Modulator loss

    The modulator's loss is determined by the PS length and loss per unit length. The loss from waveguide combiners and bends also contribute to the total loss. Losses from waveguide roughness, free carrier absorption loss, and any interactions with the electrode are all included in the PS loss. Modulator loss is required to be low in a modulator. Loss affects the eye-opening and leads to reduction in modulator performance for data centre application.

3. Modulation speed

    Modulation speed refers to the capacity to modulate data at certain speeds. Modulation speed/switching speed is usually measured in terms of rise time and fall time in an eye diagram. The switching speed has to be fast (time required to shift the levels needs to be small) for the modulator to be used for data centre application.

4. Modulation phase efficiency

    This is the amount of phase shift created for a particular phase shifter length and driving voltage. It is frequently expressed in V cm units (voltage required to produce π phase shift on a 1 cm long PS). The value of the modulation efficiency needs to be low (>V cm) to have high modulation efficiency. Voltage and length are the two parameters responsible for modulation efficiency.

5. Device footprint

It refers to the physical size of the modulator on the chip. Because length is the dominant dimension of Mach–Zehnder devices it is typically termed as such. In the case of RM, which is substantially more compact, the ring radius/diameter is commonly specified. The footprint of the device is expected to be small. Smaller footprint will help to improve the substrate real estate.

6. Modulation depth or extinction ratio

The extinction ratio, also known as modulation depth, is the difference in modulator's 1 and 0 optical output power levels. The degree of phase shift performed impacts the modulation depth and, as a result, the device's bit error rate performance. The modulation depth is expected to be high so that the differentiation between the levels is observed. Higher extinction ratio will also lead to high bit rate and distance transmission support.

7. Temperature sensitivity

It demonstrates how temperature affects the device's performance. Due to silicon's very strong thermo-optic coefficient (1.8×10^4 K^{-1}), ring resonators are highly sensitive to fluctuations in temperature. MZM is more temperature tolerant than RM. There are many techniques being researched to make RM more temperature tolerant.

8. Bandwidth

This is the wavelength range in which the device can work without needing to be tuned. Symmetrical Mach–Zehnder-based devices, on the other hand, have larger bandwidth when compared with ring resonator-based modulators whose bandwidth is generally around 1 nm. Bandwidth needs to be large so modulators can be used without tuning. MZM offers better bandwidth when compared with RM.

9. Jitter

Jitter is the time deviation from the ideal timing of a data-bit event and is one of the most important characteristics of a high-speed data transmission link. It is measured from the crossing point of the rising and falling edges in the eye diagram. The increase in jitter also reduces the pulse width, which leads to eye closure. The important contributor to the jitter is the carrier absorption loss. Jitter has to be reduced for the modulator to be used for long distance or speed transmission.

5.6 Data centre requirements

Artificial intelligence, the IoT, cloud computing, 5G, and other modern technologies have prepared the path for increased internet traffic [52]. The data centres that process these requests have been put under strain as a result of the growth in network

Table 5.1. MSA and IEEE ethernet standards and specifications for different optical transmission lengths.

Terminology	Data rate per lane (Gbps)	Number of wavelengths	Number of optical fibers	Optical fiber type	Transmission length
100GBASE-SR10	10	1	10	MMF at	Up to 500 m
100GBASE-SR4	25		4	0.85 μm	
100GBASE-PSM4				SMF at 1.3 μm	
100GBASE-10 × 10	10	10 ($\Delta\lambda = 7.2$ nm)	1	SMF at 1.55 μm	Up to 2 km
100GBASE-CWDM4	25	4 ($\Delta\lambda = 20$ nm)			
100GBASE–CLR4				SMF at 1.3 μm	
100GBASE-10 × 10	10	10 ($\Delta\lambda = 7.2$ nm)	1	SMF at 1.55 μm	Up to 10 km
100GBASE–LR4	25	4 ($\Delta\lambda = 20$ nm)		SMF at 1.3 μm	

traffic. This necessitates a very high data transmission rate between users and data centres, as well as between data centres and inside data centres. Optical communication is point-to-point communication ranging from back-to-back communication on a waveguide to between data centres of 10–20 km apart.

The data rate in optical transmission is dependent on the number of symbols per second, number of bits per second transmitted, number of optical fibres used, and number of operating wavelengths (wavelength division multiplexing (WDM)). The data centres are upgrading to 100 Gbps transceivers in order to meet the demands. The main standards provided by IEEE and Multi-Source Agreement groups for 100 Gbps are mentioned in table 5.1. All of the stated standards (binary pulse-amplitude modulation PAM-2) have one bit per symbol [53]. Depending on the reach, they are separated into three sections. The data rate is increased over short distances up to 500 m by increasing the number of optical fibres, either MMF or SMF. The data rate is then increased by multiplexing four or ten distinct wavelengths for greater reach.

Multilevel amplitude modulation, often known as M-ary pulse-amplitude modulation, is a pure amplitude modulation (PAM-M). It entails coding the bits on M separate intensity amplitude levels. Between the two commonly used modulations (i.e., PAM-2 and PAM-4) the number of symbols each period is doubled by two in the latter. As the number of levels of pulse-amplitude modulation increases, so does the amount of noise and complexity in the transceiver section. As a result, choosing PAM-4 versus complexity is a trade-off. Using a high-speed electrical amplifier to generate a high-speed electrical PAM4 signal is costly. Combining two OOK signals of various voltage levels and modulating with a single optical modulator overcomes this. For four-channel CWDM and PSM4 applications, PAM-4 modulation is appropriate [54].

5.7 Conclusion

The ever-growing network traffic demand has propelled the research to scale up and scale out the devices. The prerequisite for a high-speed operation is efficient modulation. Silicon modulators are the present choice to meet data centre demands. The data centre requirements will be based on the intra- and inter-data centre transmission. At present, Mach–Zehnder and ring modulators are preferred. The modulator's performance is governed by various parameters, and the phase shifter plays a vital role. Past years have seen a remarkable improvement in silicon modulators. The biggest difficulty, however, is to meet future needs for single-lane data speeds of 100 Gbaud and higher while being energy-efficient. Wavelength Division Multiplexing, or WDM, is the current approach to meet future network demands. Researchers are looking to incorporate materials with potent electro-optic effects into the SiPh modulators. The choice of such cost-effective and CMOS fabrication-friendly materials is a challenge. Though it may not replace or overthrow the existing plasma dispersion modulators, the selection will depend on the need of the application. Scaling out along with scaling up the optical devices will be the better choice for meeting futuristic network demands.

References

[1] *Internet of Things (IoT) connected devices installed base worldwide from 2015 to 2025 (in billions)* Statista https://statista.com/statistics/471264/iot-number-of-connected-devices-worldwide

[2] Boynton N, Cai H, Gehl M, Arterburn S, Dallo C, Pomerene A, Starbuck A *et al* 2020 A heterogeneously integrated silicon photonic/lithium niobate travelling wave electro-optic modulator *Opt. Express* **28** 1868–84

[3] Li Y, Lan T, Yang D, Bao J, Xiang M, Yang F and Wang Z 2023 High-performance Mach–Zehnder modulator based on thin-film lithium niobate with low voltage-length product *ACS Omega* **8** 9644–51

[4] Soref R and Bennett B 1987 Electrooptical effects in silicon *IEEE J. Quantum Electron.* **23** 123–9

[5] Reed G T, Mashanovich G Z, Gardes F Y, Nedeljkovic M, Hu Y, Thomson D J, Li K, Wilson P R, Chen S-W and Hsu S S 2014 Recent breakthroughs in carrier depletion based silicon optical modulators *Nanophotonics* **3** 229–45

[6] Akiyama S, Baba T, Imai M, Akagawa T, Takahashi M, Hirayama N, Takahashi H *et al* 2012 12.5-Gb/s operation with 0.29-V· cm V π L using silicon Mach–Zehnder modulator based-on forward-biased pin diode *Opt. Express* **20** 2911–23

[7] Xiao X, Xu H, Li X, Li Z, Chu T, Yu Y and Yu J 2013 High-speed, low-loss silicon Mach–Zehnder modulators with doping optimization *Opt. Express* **21** 4116–25

[8] Maegami Y, Cong G, Ohno M, Okano M, Itoh K, Nishiyama N, Arai S and Yamada K 2017 High-efficiency strip-loaded waveguide based silicon Mach–Zehnder modulator with vertical pn junction phase shifter *Opt. Express* **25** 31407–16

[9] Timurdogan E, Sorace-Agaskar C M, Sun J, Shah Hosseini E, Biberman A and Watts M R 2014 An ultralow power athermal silicon modulator *Nat. Commun.* **5** 4008

[10] Ding R, Liu Y, Li Q, Xuan Z, Ma Y, Yang Y, Lim A E-J *et al* 2014 A compact low-power 320-Gb/s WDM transmitter based on silicon microrings *IEEE Photonics J.* **6** 1–8

[11] Cai H, Fu S, Yu Y and Zhang X 2022 Lateral-zigzag PN junction enabled high-efficiency silicon micro-ring modulator working at 100 Gb/s *IEEE Photonics Technol. Lett.* **34** 525–8

[12] Hamdani M A, Qazi G, Naik S K, Sugesh R G J and Magray M A 2022 Highly efficient and novel lumped Michelson modulator using vertical PN junction based phase shifter *2022 Int. Conf. on Numerical Simulation of Optoelectronic Devices (NUSOD)* (Piscataway, NJ: IEEE) pp 1–2

[13] Irace A, Breglio G and Cutolo A 2003 All-silicon optoelectronic modulator with 1 GHz switching capability *Electron. Lett.* **39** 1

[14] Meister S, Rhee H, Al-Saadi A, Franke B A, Kupijai S, Theiss C, Eichler H J *et al* 2015 High-speed Fabry–Pérot optical modulator in silicon with 3-μm diode *J. Lightwave Technol.* **33** 878–81

[15] Hinakura Y, Arai H and Baba T 2019 64 Gbps Si photonic crystal slow light modulator by electro-optic phase matching *Opt. Express* **27** 14321–7

[16] Wu X, Dama B, Gothoskar P, Metz P, Shastri K, Sunder S, Van der Spiegel J, Wang Y, Webster M and Wilson W 2013 A 20Gb/s NRZ/PAM-4 1 V transmitter in 40 nm CMOS driving a Si-photonic modulator in 0.13 μm CMOS *2013 IEEE Int. Solid-State Circuits Conf. Digest of Technical Papers* (Piscataway, NJ: IEEE) pp 128–9

[17] Baba T, Akiyama S, Imai M and Usuki T 2015 25-Gb/s broadband silicon modulator with 0.31-V· cm VπL based on forward-biased PIN diodes embedded with passive equalizer *Opt. Express* **23** 32950–60

[18] Sciuto A, Libertino S, Alessandria A, Coffa S and Coppola G 2003 Design, fabrication, and testing of an integrated Si-based light modulator *J. Lightwave Technol.* **21** 228–35

[19] Akiyama S, Baba T, Imai M, Mori M and Usuki T 2014 High-performance silicon modulator for integrated transceivers fabricated on 300-mm wafer *2014 The European Conf. on Optical Communication (ECOC)* (Piscataway, NJ: IEEE) pp 1–3

[20] Xu Q, Schmidt B, Pradhan S and Lipson M 2005 Micrometre-scale silicon electro-optic modulator *Nature* **435** 325–7

[21] Akiyama S, Imai M, Baba T, Akagawa T, Hirayama N, Noguchi Y, Seki M *et al* 2013 Compact PIN-diode-based silicon modulator using side-wall-grating waveguide *IEEE J. Sel. Top. Quantum Electron.* **19** 74–84

[22] Sobu Y, Simoyama T, Tanaka S, Tanaka Y and Morito K 2019 70 Gbaud operation of all-silicon Mach–Zehnder modulator based on forward-biased PIN diodes and passive equalizer *2019 24th OptoElectronics and Communications Conf. (OECC) and 2019 Int. Conf. on Photonics in Switching and Computing (PSC)* (Piscataway, NJ: IEEE) pp 1–3

[23] Li K, Liu S, Thomson D J, Zhang W, Yan X, Meng F, Littlejohns C G *et al* 2020 Electronic–photonic convergence for silicon photonics transmitters beyond 100 Gbps on–off keying *Optica* **7** 1514–6

[24] Patel D, Ghosh S, Chagnon M, Samani A, Veerasubramanian V, Osman M and Plant D V 2015 Design, analysis, and transmission system performance of a 41 GHz silicon photonic modulator *Opt. Express* **23** 14263–87

[25] Baehr-Jones T, Ding R, Liu Y, Ayazi A, Pinguet T, Harris N C, Streshinsky M *et al* 2012 Ultralow drive voltage silicon traveling-wave modulator *Opt. Express* **20** 12014–20

[26] Thomson D J, Gardes F Y, Liu S, Porte H, Zimmermann L, Fedeli J-M, Youfang H *et al* 2013 High performance Mach–Zehnder-based silicon optical modulators *IEEE J. Sel. Top. Quantum Electron.* **19** 85–94

[27] Ziebell M, Marris-Morini D, Rasigade G, Fédéli J-M, Crozat P, Cassan E, Bouville D and Vivien L 2012 40 Gbit/s low-loss silicon optical modulator based on a pipin diode *Opt. Express* **20** 10591–6

[28] Jesuwanth Sugesh R G and Sivasubramanian A 2022 Silicon MZM with carrier depletion type PIPN phase shifter *2022 6th Int. Conf. on Devices, Circuits and Systems (ICDCS)* (Piscataway, NJ: IEEE) pp 144–7

[29] Li M, Wang L, Li X, Xiao X and Yu S 2018 Silicon intensity Mach–Zehnder modulator for single lane 100 Gb/s applications *Photonics Res.* **6** 109–16

[30] Jesuwanth Sugesh R G and Sivasubramanian. A 2022 Modelling and analysis of a corrugated PN junction phase shifter in silicon MZM *Silicon* **14** 2669–77

[31] Jain S, Rajput S, Kaushik V and Kumar M 2019 High speed optical modulator based on silicon slotted-rib waveguide *Opt. Commun.* **434** 49–53

[32] Sugesh R G, Jesuwanth and Sivasubramanian A 2021 High performance of core type phase shifter in silicon MZM *Optoelectron. Adv. Mater. Rapid Commun.* **15** 32–8

[33] Jesuwanth Sugesh R G and Sivasubramanian A 2022 High modulation efficient silicon MZM with core-based split PN junction phase shifter *Silicon* **14** 7033–41

[34] Jain S, Rajput S, Kaushik V and Kumar M 2020 Efficient optical modulation with high data-rate in silicon based laterally split vertical pn junction *IEEE J. Quantum Electron.* **56** 1–7

[35] Eltes F, Kroh M, Caimi D, Mai C, Popoff Y, Winzer G, Petousi D *et al* 2017 A novel 25 Gbps electro-optic Pockels modulator integrated on an advanced Si photonic platform *IEEE Int. Electron Devices Meeting (IEDM)* (Piscataway, NJ: IEEE) pp 24–5

[36] Verbist J, Verplaetse M, Lambrecht J, Srivinasan S A, De Heyn P, De Keulenaer T, Pierco R *et al* 2018 100 Gb/s DAC-less and DSP-free transmitters using GeSi EAMs for short-reach optical interconnects *Optical Fiber Communication Conf.* (Washington, DC: Optica Publishing Group) pp W4D–4

[37] Abel S, Eltes F, Ortmann J E, Messner A, Castera P, Wagner T, Urbonas D *et al* 2019 Large Pockels effect in micro-and nanostructured barium titanate integrated on silicon *Nat. Mater.* **18** 42–7

[38] Eltes F, Mai C, Caimi D, Kroh M, Popoff Y, Winzer G, Petousi D *et al* 2019 A BaTiO 3-based electro-optic Pockels modulator monolithically integrated on an advanced silicon photonics platform *J. Lightwave Technol.* **37** 1456–62

[39] Wolf S, Zwickel H, Hartmann W, Lauermann M, Kutuvantavida Y, Kieninger C, Altenhain L *et al* 2018 Silicon-organic hybrid (SOH) Mach–Zehnder modulators for 100 Gbit/s on-off keying *Sci. Rep.* **8** 1–13

[40] Kieninger C, Kutuvantavida Y, Elder D L, Wolf S, Zwickel H, Blaicher M, Kemal J N *et al* 2018 Ultra-high electro-optic activity demonstrated in a silicon-organic hybrid modulator *Optica* **5** 739–48

[41] Tamalampudi S R, Dushaq G, Villegas J E, Paredes B and Rasras M S 2023 A multi-layered GaGeTe electro-optic device integrated in silicon photonics *J. Lightwave Technol.* **41** 2785–91

[42] Li Q, Ho C P, Takagi S and Takenaka M 2020 Optical phase modulators based on reverse-biased III–V/Si hybrid metal-oxide-semiconductor capacitors *IEEE Photonics Technol. Lett.* **32** 345–8

[43] Giambra M A, Sorianello V, Miseikis V, Marconi S, Montanaro A, Galli P, Pezzini S, Coletti C and Romagnoli M 2019 High-speed double layer graphene electro-absorption modulator on SOI waveguide *Opt. Express* **27** 20145–55

[44] Datta I, Chae S H, Bhatt G R, Tadayon M A, Li B, Yu Y, Park C *et al* 2020 Low-loss composite photonic platform based on 2D semiconductor monolayers *Nat. Photonics* **14** 256–62

[45] Rasigade G, Marris-Morini D, Vivien L and Cassan. E 2010 Performance evolutions of carrier depletion silicon optical modulators: from PN to PIPIN diodes *IEEE J. Sel. Top. Quantum Electron.* **16** 179–84

[46] Yu H, Bogaerts W and Keersgieter A D 2010 Optimization of ion implantation condition for depletion-type silicon optical modulators *IEEE J. Quantum Electron.* **46** 1763–8

[47] Sridarshini T, Dhanabalan S S, Balaji V R, Manjula A, Indira Gandhi S and Sivanantha Raja A 2023 Photonic crystal based routers for all optical communication networks *Modeling and Optimization of Optical Communication Networks* (Hoboken, NJ: Wiley) pp 137–62

[48] Geerthana S, Sridarshini T, Balaji V R *et al* 2023 Ultra compact 2D-PhC based sharp bend splitters for terahertz applications *Opt. Quantum Electron.* **55** 778

[49] Geerthana S, Sridarshini T, Syedakbar S, Nithya S, Balaji V R, Thirumurugan A and Dhanabalan S S 2023 A novel 2D-PhC based ring resonator design with flexible structural defects for CWDM applications *Phys. Scr.* **98** 105975

[50] Sridarshini T, Geerthana S, Balaji V R, Thirumurugan A, Sitharthan R and Dhanabalan S S 2023 Ultra-compact all-optical logical circuits for photonic integrated circuits *Laser Phys.* **33** 076207

[51] Dhanabalan S S, Thirumurugan A, Raju R, Kamaraj S K and Thirumaran S (ed) 2023 *Photonic Crystal and Its Applications for Next Generation Systems* (London: Springer Nature)

[52] Cisco Annual Internet Report (2018–2023) White Paper, CISCO https://cisco.com/c/en/us/solutions/collateral/executive-perspectives/annual-internet-report/white-paper-c11-741490.html

[53] IEEE 2018 IEEE Standard for Ethernet https://ieeexplore.ieee.org/document/8457469

[54] Jesuwanth Sugesh R G, Sivasubramanian A and Balaji V R 2022 Step PN junction-based silicon microring modulator for high-speed application *Silicon* **14** 10651–60

IOP Publishing

Advances in All-optical Communication

Shanmuga Sundar Dhanabalan, Arun Thirumurugan and Sridarshini Thirumaran

Chapter 6

MIMO-FSO system for various weather conditions

C Palaniappan and S Robinson

The next generation of optical technologies will present several novel features, including ultra-high data rates, broadband multiple services, scalable bandwidth, and adaptable communications for various end users. Free-space optical (FSO) technology, one of the optical technologies, is essential for achieving free-space data transmission following the needs of future technologies since it is affordable, simple to implement, able to support high bandwidth, and highly secure. Massive multiple-input and multiple-output (mMIMO) is a key technology for delivering high-speed data transit, which is necessary. Using the traditional RF transmission technology, mMIMO system has already been implemented on a limited scale. However, the RF transmission technology has some drawbacks, such as licence constraints, fragmentation, and bandwidth depletion. A good substitute for this could be the FSO system. FSO can achieve high capacity with a large unlicensed optical spectrum and less operational costs. One of the biggest obstacles to FSO communication is the weather, specifically clean air, haze, and fog. This chapter provides a summary of current developments in FSO technology as well as the variables that will enable its widespread use. This chapter analyses the detailed survey of a mMIMO FSO system under atmospheric turbulence. In addition, the systems with MIMO FSO (8×8) and MIMO FSO (16×16) have been evaluated in a variety of meteorological conditions, including clear air, haze, and fog. The Q-factor and bit error rate (BER) of the MIMO-FSO system have been evaluated. Finally, the important applications of MIMO-based FSO systems will be presented.

6.1 Introduction to free-space optical communication

Multimedia applications of all kinds have grown significantly in recent years, producing a significant amount of mobile data and high-speed wireless connectivity. Future 5G technology will provide a variety of appealing services, including large system capacity, strong levels of security, incredibly low latency, very little power consumption, and fantastic quality of experience (QoE) [1]. Notably, compared to current wireless networks, 5G communication is anticipated with ultra-dense heterogeneous networks, enabling hundreds of times more wireless device connectivity [2]. As a result, high-capacity backhaul connectivity was needed for 5G and beyond networks to provide extremely dense fast access networks, low power consumption, and minimal end-to-end delays [3]. To ensure the end user's quality of service (QoS), for the end users, a robust technical solution is needed to handle the unprecedentedly high volume of information that 5G connectivity must handle. It is well known that radio frequency (RF) is extensively used in wireless communications, which are more confined due to the lack of spectrum resources [4]. The Internet of Things (IoT), envisioned for social, industrial, and economic applications, has the potential to enable real-time communication, sensing, monitoring, and resource sharing, while supporting extensive connectivity among smart devices. The number of physical smart devices connected to networks has greatly increased as a result of the Internet of Things (IoT) and Internet of Everything (IoE) technologies exponential expansion [5].

As a result, IoT devices produce a large amount of data. It is anticipated that the existing available electromagnetic frequency range will not be enough to meet the enormous needs of the IoT paradigm and the ever-increasing demand for 5G networks. This RF band, however, has a limited spectrum, restrictions on how it may be used due to regulations, and a high amount of interference from the nearby RF access points. Many times, mobile cellular operators, TV broadcasters, and end-to-end microwave communications are given complete access to RF sub-bands. Researchers are searching for an alternative method of wireless communication using millimetre and nanometer waves due to the evidence of RF-based wireless network disadvantages. Therefore, in the context of optical wireless communication (OWC), academia and industry are currently interested in license-free optical spectrum spanning 1 mm–10 nm as an emerging alternative to RF for future ultra-density and ultra-capacity networks [6].

Wireless communication employing optical spectrum is referred to as OWC. OWC technologies provide several outstanding features that address the high demand requirements of 5G/B5G communications, such as broad spectrum, ultra-high data rates, extremely low latency, cheap costs, and decreased power consumption. Compared to RF-enabled networks, OWC technology has certain enviable benefits, such as high data rate transmission capability for both indoor and outdoor applications at distances of many kilometres, ranging from nanometers. Additionally, because OWC technology uses a wide optical spectrum, it can provide remarkable communication attributes such as trustworthy security, electromagnetic interference-free transmission, and excellent system efficiency [7]. Reference [8] exhibits the capabilities of OWC technology providing energy-efficient connectivity and can reach 100 Gb s^{-1} under normal indoor lighting. The most crucial aspect of OWC technologies is that they do not need a comprehensive infrastructure, which lowers installation costs while still upholding the green objective for high-speed communication. OWC technology's visible light communication (VLC) and light fidelity (Li-Fi) approaches make use of the existing illumination structure to enable wireless data transfer [9]. OWC provides a better level of data protection due to the barriers' inability to let light waves through.

Typically, spectral ranges of visible light (VL), infrared (IR), or ultraviolet (UV) are used as propagation media. The most advanced wireless OWC technologies, including VLC, LiFi, optical camera communication (OCC), and free-space optical (FSO) communication, are being developed on the foundation of the three optical bands. In terms of communication protocol, propagation media, design, and applications, VLC, LiFi, OCC, and FSO technologies have some similarities and variations. However, a hybrid RF/OWC system that combines RF and OWC possibilities may be able to address the large upcoming customer demands.

6.1.1 Free-space optical communication

A type of OWC technology called FSO often uses the NIR spectrum, which has substantially lower attenuation levels, as the channel medium. FSO can be used to work in the VL and UV spectra without the need for light. High-speed communication networks use narrow-beam laser diode (LD) light sources rather than LEDs, with PDs acting as optical receivers. Due to the coherent nature of laser technology, the FSO signal may traverse a distance, allowing for point-to-point communication at a high data rate.

Optical power amplifiers can boost the modulated laser output's signal intensity. The laser light beam is then collected and refocused before transmission. Numerous modulation techniques and channel coding are utilised to generate a high optical power across a wide temperature range. The PD output must be converted into voltage using electronic circuits. A low-pass filter (LPF) reduces noise levels, and a demodulator completes the essential steps to transfer the original data to its destination. Video surveillance, campus connectivity, cellular network backhaul, LAN-to-LAN connectivity, chip-to-chip communication, communication in space and the ocean, fibre backup, etc., are all applications for FSO systems. Despite the many advantages of FSO systems across a wide range of applications, link reliability, high sensitivity to some limiting factors, such as atmospheric turbulences, physical obstructions, and outdoor weather conditions (e.g., heavy rain, fog, smoke, storms, deep clouds, snow, and scintillation), negatively affect the performance of FSO links. The PD output must be converted into voltage using electronic circuits. A LPF reduces noise levels, and a demodulator completes the essential steps to transfer the original data to its destination [10].

6.2 FSO communication principles

Infrared (IR) and visible light-based point-to-point wireless optical transmission is known as FSO across unguided propagation surfaces. In other words, FSO is a wireless optical data transmission line-of-sight (LOS) technique that uses eye-safe laser beams in free space. Generally, LOS communication in optical systems refers to a method of transmitting data where the transmitter and receiver must have a clear, unobstructed path between them. This is crucial for ensuring effective signal

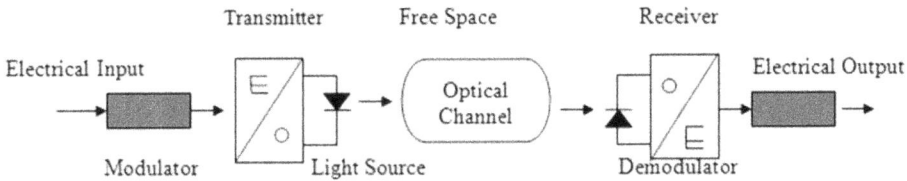

Figure 6.1. Block diagram of an optical wireless link.

propagation, as any obstacles can significantly degrade or block the optical signals. Telescopic lenses used in FSO receivers can gather light streams and transfer digital data at rates up to Gbps to their intended recipient. The capacity to manage large amounts of data is enhanced by the availability of a broad optical spectrum. As light travels more quickly through air than through glass, FSO offers a promising alternative to radio relay links. In general, VL and UV bands are utilised for outdoor FSO links, whereas for indoor and underwater FSO, near IR and VL bands are utilised [11]. In contrast to RF networks, which are very vulnerable to multipath fading, the photodetectors become resistant to multipath signal fluctuations when FSO is operating at exceptionally great frequencies (i.e., short wavelength). Internet data, voice, images/videos, computer files, and other digital signals are first converted into light signals using an optical transmitter, modulated with an appropriate technique, and then transmitted as multiplexed signals through the FSO channel in the optical domain. The incoming signal is then detected using an optical photodetector. After the necessary electronic switching, the demultiplexed signals are transmitted to their destination. Figure 6.1 shows the block diagram of an optical wireless link.

6.2.1 FSO transceiver

FSO communication can be roughly divided into three stages: an optical channel that travels through the atmosphere according to Beer–Lambert's law, a channel link where various noises and turbulences build up, and a receiver section with electronic equipment to process the signal that is received.

6.2.1.1 FSO transmitter

An optical amplifier (if used), a light source, a modulator, and a beamforming mechanism make up an optical transmitter. Before modulation, the incoming data bits may optionally be encoded. The optical pre-amplifier increases the optical intensity of the modulated light beams. Before transmission, the laser beams are then gathered and refracted using beamforming optics. Based on their applications, the most commonly used optical sources, such as LEDs (light emitting diodes) and LDs (LASER diodes), have different advantages and disadvantages for FSO systems.

6.2.1.2 FSO receiver

At the optical receiver, photons representing the emitted optical signal are optically collected, and a photodetector (PD) uses a trans-impedance circuit to convert the photons into electrical current. To retrieve the original transmitted data signals, the

FSO receiver manages the electrical current that has been detected through electronic components. To reduce background noise and thermal noise levels, the optical signal is transformed into electrical and then sent via a LPF. Avalanche photodetector (APD) and positive-intrinsic-negative (PiN) PD are frequently utilised in FSO systems. PiN PDs, which are typically used for low data rate and short-range FSO lines, are inexpensive, capable of operating in low bias, and tolerant of wide-range temperature variations. Thermal noise is a constraint on PiN PD's performance.

APDs are the favoured option for high data rate and long-distance FSO systems because they have higher multiplication gain than PIN PDs (owing to the impact ionisation process) and better SNR due to operating at high reverse bias. Despite having higher performance, APDs are expensive, have significant power consumption, and have temperature-sensitive avalanche gain.

Commercial FSO systems that operate at readily accessible wavelengths like 850 and 1550 nm frequently employ solid-state devices with good quantum efficiency. At about 1550 nm, InGaAs PDs provide the highest sensitivity in terms of extremely short transit times and rapid reaction detectors whereas Si PDs are appropriate for operation at lower wavelengths around 850 nm. The development of ultra-speed photodetectors over a broad range of wavelengths is made possible by recent advances in graphene materials, plasmonic nanomaterials, two-dimensional materials, and quantum dot nanoparticles [12].

To enhance connection performance, optical amplifiers (OAs) have been suggested for long-distance, multihop FSO communication. A better option than a semiconductor optical amplifier (SOA) operating at 1550 nm wavelength is an erbium-doped fibre amplifier (EDFA). The receiver performance in terms of possible SNR might, however, be negatively impacted by the amplified spontaneous emission (ASE) noise produced by the optical amplifier. However, in the case of weak turbulence, OAs can deal with electronic noise and lessen the scintillation impact [13].

Dark current noise, shot noise (sometimes referred to as quantum noise), and thermal noise are some of the types of associated noises at the receiver. Most of the time in practice it is possible to disregard the PD dark current noise. The instability of the intensity of the LDs results in another noise at the receiver that generates photocurrent variations known as laser relative intensity noise (RIN). RIN has a minor effect on the receiver's SNR performance, similar to PD dark current [10]. If the background light is deemed insignificant, the two primary noise sources that affect the receiver performance are shot noise and thermal noise. Thermal noise is often limited in a PIN PD. In contrast, the sensitivity of an APD-based receiver depends on the load resistance and either the noise or the impact ionisation gain. However, the electronic circuitry, which may be modelled as zero-mean Gaussian random noise, produces thermal noise, which is a function of load resistance. Shot noise, on the other hand, results from background radiation or electrical currents in PD that fluctuate randomly and are modelable by the Poisson process [14]. However, the shot noise can be roughly described as a Gaussian process if the number of absorbed photons is rather high. The shot noise distribution can be roughly approximated by a Gaussian process in the majority of FSO applications [15].

6.2.2 FSO classifications

The proliferation of wireless devices and the wide variety of multimedia services place a tremendous demand on the RF spectrum, where the majority of the sub-bands are licenced. As a result, the demand for RF bands is outpacing supply, which is fundamentally constrained in terms of cost and capacity. As a result, the most urgent issue is the paradigm of transferring the upper electromagnetic spectrum for cellular communications. FSO has only been used for inter-satellite links in space applications and short-range military applications over the past few decades. With the development of FSO communications devices and wireless optical terrestrial networks, a variety of services are now available to meet the continuously increasing need for faster data speeds [16–18]. For heterogeneous building communication networks, the development of effective wireless technology is crucial. FSO can be used in a variety of applications, from satellite linkages to outdoor inter-building connections.

FSO technology can be used in four various circumstances, including the atmosphere, indoors, space, and underwater, depending on the location and link distance. A space scenario is categorised as FSO space communication, and you can think of the air environment as extraterrestrial FSO communication. Home networking FSO also includes inside communications. Long-distance point-to-point outdoor optical communications through turbulent atmospheric channels are established by terrestrial FSO networks using a LOS mechanism. A promising future for wireless telecommunications, particularly for broadband internet access, is offered by this FSO communication. Terrestrial FSO is an emerging alternative for last-mile access. Despite the widespread use of fibre connectivity for fibre-to-the-home (FTTH) service, many end users are still not connected to FFTH service because of geographical limitations. With the use of this technology, remote users can connect with high bandwidth over long communication distances. In other words, without the use of optical fibre cable, terrestrial FSO technology creates a bridge for communications across communities or between buildings that are geographically apart.

The integration of terrestrial FSO with wireless RF networks is crucial because it improves QoS while minimising capacity and scalability issues. It should be noted that when the hop length and hop count are high, the spectral efficiency and fairness of RF wireless networks are substantially reduced.

FSO home networking (FSO-HN), on the other hand, is frequently used for indoor broadband wireless communications in a home, building, or business. Using FSO-HN, one may communicate over short distances of a few to tens of metres. The equipment typically facilitates eye safety issues and is portable, affordable, and lightweight. The structure is equipped with several tiny cell terminals linked by short-range optical wireless networks. Each small cell is contained within a single room because, unlike radio signals, optical light cannot pass through walls. All terminals are linked to a high-speed backbone infrastructure. As each terminal is free of mutual interference from the surrounding cells, the optical beams can be reused. Due to an improved beam steering mechanism and power budget, LOS FSO technology achieves extremely low BER and great throughput when compared to non-LOS links. To distribute light beams in non-LOS networks, a diffused light source

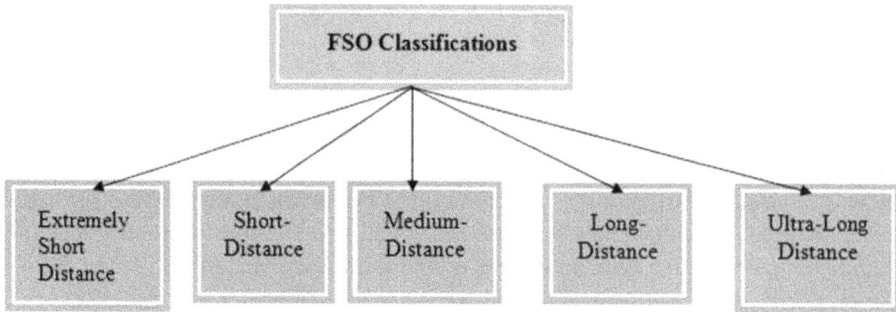

Figure 6.2. Classification of FSO.

is frequently employed in indoor environments. Diffused connections (or non-LOS links) perform better because they make the most of multipath propagations brought on by various objects including ceilings, furniture, walls, and floors [19].

A diffused link supports a lower bit rate compared to LOS lines, but there is a trade-off between link dependability and network capacity. Inter-orbital, inter-satellite, and deep space links are further categories for FSO space links. High-quality data services can be provided through satellite communications using FSO technology as a worldwide space backbone. The development of FSO technology in satellite-to-earth downlink space communications allows end users to experience a variety of advantages whether the receiver is fixed or in motion (such as an aeroplane, cruise ship, moving vehicle, etc.). Broadband FSO-based satellite communications offer complete coverage for remote locations where cellular network access is limited. The reliability of acoustic communications in underwater situations is currently being thoroughly researched while taking into account all limiting constraints. Classification of FSO is mentioned in figure 6.2.

The FSO classification can be encapsulated as follows:

- **Extremely short distance:** multichip packages with integrated chip-to-chip communication at nm distance.
- **Short-distance:** Wireless LANs and wireless communications in the sea.
- **Medium-distance:** Vehicle-to-vehicle (V2V), vehicle-to-everything (V2X), and indoor visible light communication technologies for WLANs.
- **Long-distance:** Communications between buildings.
- **Ultra-long distance:** Satellite links, such as satellite-to-ground and satellite-to-aircraft.

6.2.3 FSO applications

The FSO systems have been popular because they effectively bridge the distance between optical fibre infrastructure and destination customers, solving the last-mile problem's bottlenecks. To make the most of the current configuration, telecom operators are making huge investments in fibre backbone and FSO link expansion, along with significant growth at the network perimeter where end users may easily reach with high-speed systems. When buried optical fibre connectivity is expensive

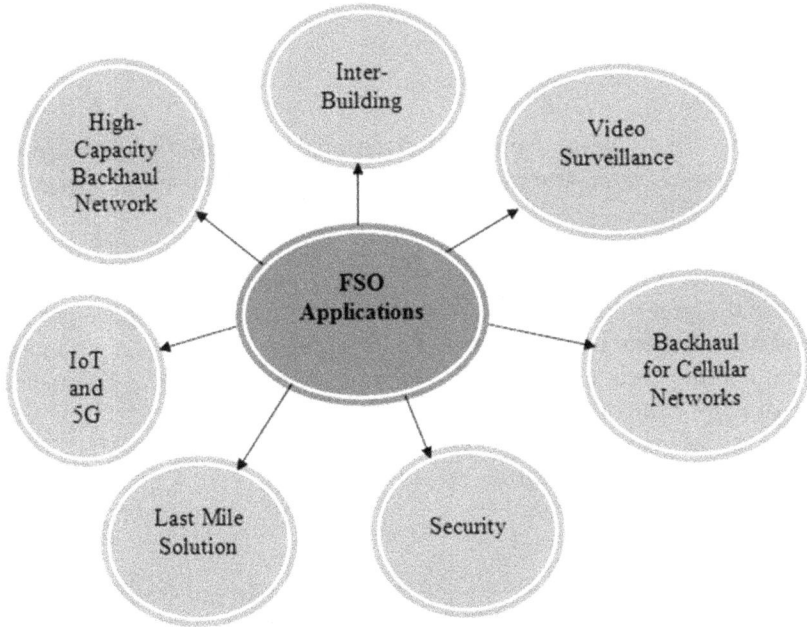

Figure 6.3. Applications of FSO.

or unworkable, FSO solutions can be a promising alternative. The primary applications of FSO are reported in figure 6.3.

6.3 Performance evaluation of the FSO system

Atmospheric attenuation, path loss, received optical power, modulation type, detector type, pointing loss factors, data rate, and BER are only a few of the variables influencing FSO system performance and connection budget. The log-normal and gamma–gamma distribution models are the two most widely used probability density functions (PDFs). The gamma–gamma distribution is valid in all turbulence situations, but the log-normal distribution is only valid in weak turbulence [20, 21].

6.3.1 Link budget

The Friis transmission format offers the pointing loss factors that are presented in [22].

$$P_R = P_T\, \eta_T\, \eta_R\, L_S\, G_T\, G_R\, L_T\, L_R \exp(-\alpha Z) \tag{6.1}$$

where
P_T is the transmitted power,
P_R is the received power,
η_T is the transmitter efficiency,
η_R is the receiver efficiency,
G_T is the transmitter effective antenna gain,

G_R is the receiver effective antenna gain,
L_T is the transmitter pointing loss factor,
L_R is the receiver pointing loss factor,
L_S is the space loss factor, and
α is the coefficient of atmospheric attenuation depends on the type of scattering, signal wavelength, size of the particles of the atmosphere, and the link visibility [23].

The beam divergence behaviour is illustrated in figure 6.4 [24], and it shows that space loss is the largest loss effect on signal strength through free-space propagation. The traversing optical beam experiences diffraction, which causes the beam to spread out and results in an increase in the received beam width to be much greater than the receiver telescope. Therefore, the receiver aperture only collects a fraction of the received beam.

The space loss factor L_S is given by [24]:

$$L_S = (\lambda/4\pi L)^2 \qquad (6.2)$$

where
λ is the signal wavelength
L is the link distance between the transmitter and receiver

Due to the wavelength dependence, the optical system loss in free space is much larger than in the RF system, so the Ls loss factor is much smaller.

The transmitter and receiver gain, G_T and G_R, can be estimated by [25]:

$$G_T = (\pi D_T/\lambda)^2 \qquad (6.3)$$

$$G_T = (\pi D_R/\lambda)^2 \qquad (6.4)$$

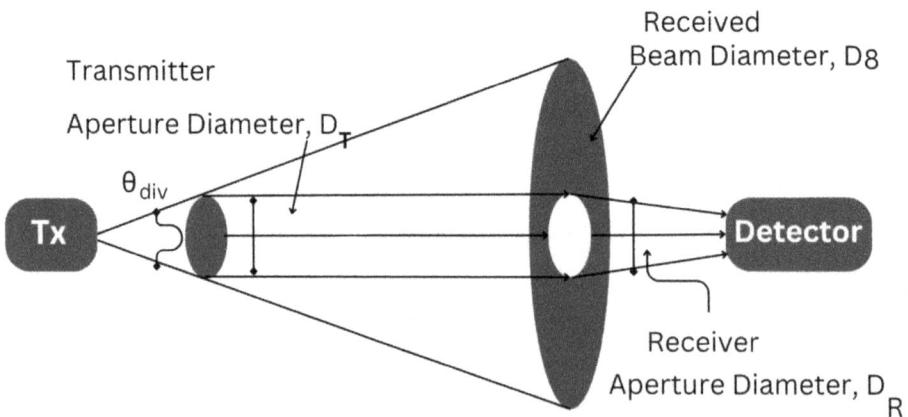

Traversing Beam Divergence

Figure 6.4. Traversing beam divergence.

where

D_T is transmitter aperture diameters

D_R is receiver aperture diameters

The transmitter and receiver pointing loss factors, L_T and L_R, can be estimated by [25]:

$$L_T = \exp\left(-G_T\left(\theta_T\right)^2\right) \tag{6.5}$$

$$L_R = \exp\left(-G_R\left(\theta_R\right)^2\right) \tag{6.6}$$

where

θ_T is the transmitter pointing error angle

θ_R is the receiver pointing error angle

In the FSO link, the power budget is given by [25]:

$$P_{RX} = P_{TX} - \alpha_{\text{sys}} - \alpha_{\text{atm}} \tag{6.7}$$

where

α_{sys} is system attenuation

α_{atm} is total atmospheric attenuation

6.3.2 FSO received optical power

Beer–Lambert's law governs how optical radiation moves through the atmosphere. The theoretical model of the received power, $P_R(\text{dBm})$, used for the FSO channel is given by [26–28]:

$$P_R = P_T \frac{D^2}{\theta^2_{div} L^2} 10^{-\gamma L/10} \eta_T \eta_R \tag{6.8}$$

where

P_T is the transmitted power (dBm)

D is the receiver diameter (m)

θ_{div} is the beam divergence angle (rad)

L is the optical link length (km)

γ the attenuation coefficient (dB km^{-1})

6.3.3 Data rate

The achievable data rate, R, can be estimated by [28]:

$$R = \frac{4}{\pi E_p N_b} P_r \tag{6.9}$$

where

N_b is the receiver sensitivity (photons/bits) in (dBm)

E_p is the photon energy given by:

$$E_p = \frac{hc}{\lambda} \tag{6.10}$$

where

    h is Planck's constant
    c is the free-space speed of light
    λ is the signal wavelength

6.3.4 Signal-to-noise ratio and BER

The optical signal's power is converted to electrical current by the photo detectors (APD and PIN). Shot noise, thermal noise, and background noise all contribute to currency fluctuation and imprecise bit recognition.

The BER is the primary component of performance evaluation in digital systems. It is the likelihood that the received bit stream will result in a non-precision choice. The BER for both APD and PIN is provided by [29, 30] where P (0/1) is the probability of deciding 0 when 1 is received and P (1/0) is the probability of deciding 1 when 0 is received.

$$\text{BER} = \frac{1}{2} \left[p(0/1) + p(1/0) \right] \tag{6.11}$$

For a Q-factor > 3, then

$$\text{BER} = \frac{1}{2} \text{erfc} \left(\frac{1}{2\sqrt{2}} \sqrt{\text{SNR}} \right) \tag{6.12}$$

with

$$Q = \frac{I_1 + I_0}{\sigma_1 + \sigma_0}$$

where σ_1^2 and σ_0^2 are the variances of equal value and I_1 and I_0 are the average currents for bits of 1 and 0, respectively.

The SNR for a PIN photodiode is given by [30]:

$$\text{SNR}_{\text{PIN}} = \frac{I_P^2}{2q_e B(I_P + I_D) + 4k_B T_{\text{PIN}} BF_n / R_L} = \frac{(P_r P_{\text{PIN}})^2}{2q_e B(P_r P_{\text{PIN}} + I_D) + \frac{4k_B T_{\text{PIN}} BF_n}{R_L}} \tag{6.13}$$

where

    q_e is electron charge
    B is bandwidth
    I_D is the dark current
    T_{PIN} is the absolute photodiode temperature
    F_n is the photodiode noise figure (equal to 1 for PIN photodiode)
    R_L is the PIN load resistance

According to [31], the SNR for an APD photodiode is provided.

$$\text{SNR}_{\text{APD}} = \frac{(R_0 P_{\text{rec}} M)^2}{2q_e B(R_0 P_{\text{rec}} + I_D)M^{x+2} + 2qI_L B + 4k_B T_{\text{APD}} B F_T/R_{\text{eq}}} \tag{6.14}$$

where

R_0 is the sensitivity of the APD

M is the APD gain

I_L is the surface leakage current

x is the excess noise factor

B is the equivalent noise bandwidth

T_{APD} is the system temperature

F_T is the APD photodiode noise figure

R_{eq} is the equivalent circuit resistance

6.3.5 Channel models

The FSO channel can be described by a number of statistical models, including the Gamma–Gamma distribution model, the Negative exponential model, the Kolmogorov spectrum model, and the Log-normal model, taking atmospheric turbulence into account. The Gamma–Gamma model is selected for this investigation among them because it performs better in strong, moderate, and light atmospheric turbulences. Here, the Gamma–Gamma model is expressed using the probability density function, as shown in equation (6.14):

$$f(I) = \frac{2(\alpha\beta)^{\frac{\alpha+\beta}{2}}}{\Gamma(\alpha)\Gamma(\beta)} I^{\frac{\alpha+\beta}{2}-1} K_{\alpha-\beta}(2\sqrt{\alpha\beta I})2, \ I > 0 \tag{6.15}$$

where

'I' represents the intensity of the received signal, while the parameters 'α' and 'β' represent small-scale fading and large-scale fading, respectively. These are caused by atmospheric turbulence, as expressed in equations (6.15) and (6.16).

$$\alpha = \left(\exp\left(\frac{0.49\sigma^2 R}{\left(1 + 1.11\sigma^{\frac{12}{5}} R\right)^{\frac{7}{6}}}\right) - 1\right)^{-1} \tag{6.16}$$

$$\beta = \left(\exp\left(\frac{0.51\sigma^2 R}{\left(1 + 0.69\sigma^{\frac{12}{5}} R\right)^{\frac{5}{6}}}\right) - 1\right)^{-1} \tag{6.17}$$

The Rytov variance can be expressed as

$$\sigma^2 R = 1.23 C_n^2 K^{\frac{7}{6}} L^{\frac{11}{6}} \tag{6.18}$$

where

L = length of propagation path,

K = optical wave number, and

C_n^2 is a measure of the variations of refractive index, which indicates the turbulence strength. Values of span from 10^{-17} m$^{-2/3}$ to 10^{-13} m$^{-2/3}$ for weak to strong turbulences [64].

6.4 Introduction to MIMO system

Every year, the number of mobile users drastically rises. Users expect immediate access to multimedia services and fast internet speeds. Furthermore, the advancement of smart cities has led to the distribution of numerous interconnected devices throughout urban areas, generating exabytes of data that require efficient transfer [32]. This necessitates higher data speeds, more network capacity, better mobility, and higher spectral and energy efficiency [33]. Researchers have therefore suggested the use of 5G networks to address the resulting problems, and effective technologies like device-to-device (D2D) communication, ultra-dense networks (UDNs), spectrum sharing, centimetre- or millimetre-wave technology, the Internet of Things (IoT), and mMIMO will be implemented in 5G networks [34]. Since the third generation (3G) of wireless networks, MIMO has been a crucial technology for improving the performance of wireless transceivers. To improve spectral efficiency, range, and/or link reliability, several antennas are used in the transmitter and receiver.

However, the MIMO receiver is supposed to use a detection technique to distinguish the symbols that are distorted by interference and noise owing to the many interfering signals being delivered from different antennas. Over the past 50 years, there has been a lot of interest in the MIMO detector. With just one or a few antennas operating in the same frequency band, the mMIMO base station (BS) can service a sizable number of user terminals. The number of BS antennas is more than the total number of antennas in the user equipment inside the cell or service area. This is the main characteristic of the traditional mMIMO system operating below 6 GHz carrier frequency. Due to the difficulties in channel estimation brought on by pilot contamination, multiuser interference eventually averages out to appear as increasing additive noise. Optical networks integrate transmitters, receivers, amplifiers, and routers for data transmission via light through fiber optic cables. Photonic crystal-based waveguides and fibers, coupled with MIMO-enabled antennas, enhance capacity and reliability. These components, alongside logic gates [65, 66], adder [67], modulators and filters [68], multiplexers and demultiplexers [69], splitters [70], circulators [71], and encoders and decoders [72, 73], revolutionise high-speed, long-distance data communication in modern networks.

6.4.1 From SISO to MIMO to mMIMO

Unlike MIMO systems, which use multiple antennas at both the transmitter and the receiver (i.e., UEs), SISO systems only use a single antenna at the transmitter (i.e., BSs). MIMO systems provide more capacity and reliability than SISO systems because of the channels' significant multiplexing, diversity, and array gain benefits over SISO channels [35]. While the maximum multiplexing gain is the lesser of the number of antennas at the transmitter and receiver units, the diversity gains of MIMO scale with the number of independent channels between the transmitter and receiver. However, because of a fundamental trade-off between the two, maximal multiplexing and diversity gains cannot be obtained concurrently from MIMO systems. It is impossible to make both of the best gains at once [36]. Contrarily, mMIMO employs a significantly higher number of antennas than traditional MIMO systems.

The single-cell downlink wireless communication system model is considered when assessing the performance of cellular systems with various antenna configurations, ranging from massive MIMO to SISO. The specific scenarios examined include:

SISO (single-input–single-output)
SU-MIMO (single user MIMO)
MU-MIMO (multi user MIMO)
Massive MIMO.

These configurations represent different MIMO technologies and their impacts on system performance. The system consists of a transmitter (Tx) with N transmit antennas, k users, and receivers (Rx) with M receive antennas for each user.

6.4.1.1 SISO [N = 1, M = 1 and K = 1]

A BS (Tx) and a UE (Rx), each with a single antenna, make up SISO. Given that the general model for the received signal y is

$$y = Hx + n \qquad (6.19)$$

where

H is the transmit signal vector
x is the channel matrix
n is the for noise vector

It is assumed that n represents additive white Gaussian noise (AWGN), which follows a complex normal distribution $\mathscr{CN}(0, \sigma)$ with a mean of zero and σ standard deviation. The channel matrix in (1) is reduced to a one-dimensional scalar for the SISO case.

The signal that is received is reduced to

$$y = hx + n \qquad (6.20)$$

Thus, the single link's possible capacity (in bits/s/Hz) can be written as

$$C_{\text{SISO}} = \log_2(1 + \gamma) = \log_2\left(1 + h^2 \frac{P_t}{\sigma_n^2}\right) \qquad (6.21)$$

where

γ is the signal-to-noise ratio (SNR)
P_t is the transmit (signal) power'
σ_n^2 is the noise power
h is the channel coefficient

6.4.1.2 Single-user MIMO [N > 1, M > 1 and K = 1]

In MIMO antenna array systems, both the transmitters and receivers are equipped with multiple antennas. According to the Shannon capacity theorem, this setup leads to a substantial increase in data rates and average spectral efficiency in wireless systems without requiring an increase in SNR or bandwidth. Furthermore, spatial multiplexing—enabled by broadcasting multiple streams from several antennas—provides an additional boost to system capacity [37, 38].

The channel capacity of SISO systems scales with the smaller number of transmit or receive antennae when the channel is full rank, but the channel capacity of MIMO systems scales linearly with an increasing number of antennas. When the channel is not at full rank, the profits may be restricted (i.e., not exactly a linear increase) [39]. However, this improvement comes at the trade-off of higher signal processing complexity, space limitations (especially for mobile terminals), and higher deployment costs for many antennas [40].

The received signal vector , $y_m \in C^{M \times 1}$ for the SU-MIMO system with multi-antenna transmitter and receivers, but where only one active user is served or planned at a transmission time interval (TTI), can be written as

$$y_m = \sqrt{p} H_{m,n} x_n + n_m \tag{6.22}$$

The notation n representing transmit antennas and $m = \{1, 2, ..., M\}$ representing receive antennas is standard in the context of MIMO systems. This notation effectively conveys that there are N transmit antennas and M receive antennas involved in the communication system.

- Transmit antennas: Let $n = \{1, 2, ..., N\}$ denote the set of transmit antennas at the base station (BS), where N denote the set of receive antennas at the user terminals, where M is the total number of receive antennas.
- Receive antennas: Let $m = \{1, 2, ..., M\}$ refers to receive antennas.

$x_n \in C^{N \times 1}$ is the transmit signal vector

$n_m \in C^{M \times 1}$ is the noise and interference vector

$H_{m,n} \in C^{M \times N}$ is the assumed narrow-band time-invariant channel with deterministic and constant channel matrix

ρ is a scalar representing the normalized transmit power (i.e., the total power of the transmit signal sum to unity, $E\{\|x_n\|^2\} = 1$)

The instantaneous attainable rate (bits/s/Hz) is given by the following equation, assuming independent and identically distributed Gaussian transmit signals, zero-mean circularly symmetric complex Gaussian noise, and perfect channel state information (CSI) at the receiver.

$$C_{\text{SU-MIMO}} = \log_2 \left| \left(I + \frac{\rho}{N} HH^* \right) \right| \tag{6.23}$$

where m, n channel subscripts have been eliminated and N is the total number of transmit antennas. The expression is bounded by

$$\log_2(1 + \rho M) \leqslant C_{\text{SU−MIMO}} \tag{6.24}$$

$$\leqslant \min(N, M) \log_2\left(1 + \frac{\rho \max(N, M)}{N}\right)$$

6.4.1.3 Multiuser MIMO [N > 1, M > 1, Mk = 1 and K > 1]

SU-MIMO and MU-MIMO are the two fundamental configurations for MIMO systems [41]. Greater benefits of MU-MIMO include the following:

- Through the use of transmit diversity, spatial multiplexing, and beamforming techniques, SU-MIMO transmissions allocate all time-frequency resources to a single terminal. The advantage of MU-MIMO over SU-MIMO, especially when the channels are spatially correlated, is that it takes the use of multiuser diversity in the spatial domain by assigning a time-frequency resource to many users [41].
- In MU-MIMO many users can be served simultaneously by BS antennas because user terminals can use relatively inexpensive single-antenna devices and the BS only needs expensive equipment, which lowers costs.
- Rich scattering is typically not needed since MU-MIMO systems are less sensitive to the propagation environment than SU-MIMO systems are.

In this case, the BS uses a single antenna to simultaneously communicate with numerous users. The vector of the received signal, $y_k \in C^{K \times 1}$, can be written as

$$y_k = \sqrt{\rho}\, H_{k,n}\, x_n + n_k \tag{6.25}$$

$x_n \in C^{N \times 1}$ is the transmit signal vector, $n_k \in C^{K \times 1}$ is the noise and interference vector, and $H_{k,n} \in C^{K \times N}$ is the channel matrix. The specified attainable capacity (bits/s/Hz) is provided as

$$C_{\text{MU−MIMO}} = \max_P \log_2 |(1 + \rho\, \mathbf{HPH}^*)| \tag{6.26}$$

P is a positive diagonal matrix with power allocations $\mathbf{P} = \{p_1, p_2, \dots, p_K\}$, which maximizes the sum transmission rate. Here, the k,n channel subscripts have been dropped.

Due to its advantages, MU-MIMO has been used by numerous antenna systems in recent years and is a candidate for multiple wireless standards [42]. mMIMO is being developed to scale up the advantages of MIMO greatly, even though MU-MIMO has been identified as a non-scalable technology, despite its enormous advantages over SISO and SU-MIMO antenna systems. mMIMO includes more service antennas than active terminals, which can be used for upgrades like beam forming to increase throughput and energy efficiency, in contrast to MU-MIMO, which essentially has the same number of terminals and service antennas. Therefore, its potential to scale up is increased by its extra antennae [43].

MIMO is an intelligent technique that, in general, aims to increase the effectiveness of wireless communication lines. Studies and commercial deployment scenarios in various wireless standards, including the IEEE 802.11 (WiFi), IEEE 802.16

(WiMAX), the third generation (3G) Universal Mobile Telecommunications System (UMTS), the High-Speed Packet Access (HSPA) family series, as well as the LTE, have demonstrated that multiple antenna systems offer significant improvements in the performance of cellular systems, concerning both capacity and reliability.

6.4.1.4 mMIMO [N ≫ M, N → ∞ or M ≫ N, M → ∞]
Large-Scale Antenna Systems (LSAS), Full Dimension MIMO (FDMIMO), Very Large MIMO, and Hyper MIMO are other names for mMIMO. A few hundred BS antennas are used in an antenna array system to serve many tens of user terminals at once in the same time-frequency resource.

In massive MIMO systems, this term refers to the number of independent signal paths that can be utilized for transmitting data. The concept of degrees of freedom (DoF) is crucial because it directly impacts the system's ability to support multiple users simultaneously and enhance overall spectral efficiency. Typically, DoF refers to the number of independent signal paths available in a communication system, allowing for the simultaneous transmission of multiple data streams, thereby enhancing capacity and spectral efficiency [43, 44].

The achievable rate for MIMO decreases as the number of antennas rises to $N \gg M$ and $N \to \infty$,

$$C_{\mathrm{mMIMO}} \approx M \log_2\left(1 + \frac{\rho M}{N}\right) \tag{6.27}$$

Equations (6.22) and (6.23) show the benefits of mMIMO, where the capacity increases linearly with the number of deployed antenna at the BS or the UE, as the case may be. They assume that the row or column vectors of channel H are asymptotically orthogonal. However, it should be emphasised that the aforementioned equations (6.15)–(6.23) only provide a brief review of SISO, MIMO, and huge MIMO systems.

6.4.2 mMIMO

mMIMO is a scaled-up version of the conventional small-scale MIMO systems [45]. As shown in figure 6.5, a mMIMO system is a multiuser communications solution that employs a large number (practically some dozens or hundreds, theoretically up to thousands) of antenna elements to serve simultaneously multiple users with the flexibility to opt what users to schedule for reception at any given time. The most common mMIMO concept assumes that user terminals have only a single antenna, while the number of antennas at the BS is significantly larger than the number of users being served.

The introduction of mMIMO has had a tremendous impact on the research and development community during the past decade. As a result, many next-generation communication technologies such as 5G below 6 GHz adopted mMIMO as their key technology. Most of the mMIMO literature focuses on mobile broadband type high rate problems with large data packets such that channel estimation and training make clear sense. The other application of interest is massive machine-type

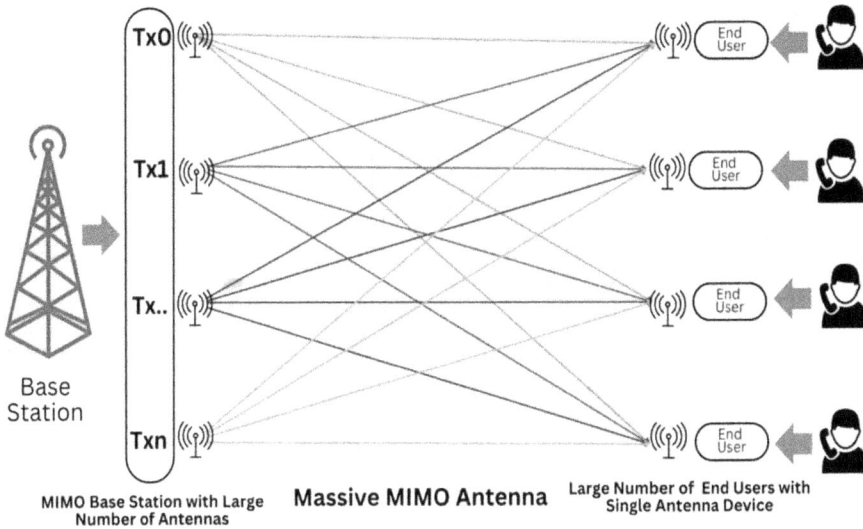

Figure 6.5. mMIMO architecture.

communications (mMTC) wherein a large number of connected devices are only sporadically active [46, 47].

6.4.3 Benefits of mMIMO

mMIMO technology allows users to handle signals from all of the antennas with extremely simple processing by enhancing the diversity and spatial multiplexing gains by adding more antennas at the base station [48, 49]. Here is a summary of the possible advantages of mMIMO:

- **Capacity and link reliability:** Due to its resistance to fading, mMIMO boosts diversity gain and thus offers network robustness [50]. When using multicell minimum mean square error (MMSE) precoding/combining and spatial channel correlation, it has been shown that capacity increases indefinitely as the number of antennas grows, even in the presence of pilot contamination [51].
- **Spectral efficiency:** mMIMO increases the cellular network's spectral efficiency (SE) by spatially multiplexing several user devices per cell [52]. High spectrum efficiency is thus achieved by using several antennas, which further increase throughput, multiplexing gain, and spatial data streams [53]. It is demonstrated that, when tens of users are provided concurrently in the same time-frequency resources, the total spectral efficiency in the mMIMO can be ten times higher than in the conventional MIMO.
- **Energy efficiency:** The transmitted power is inversely proportional to the number of transmit antennas because of coherent combining. The transmit power will decrease dramatically as the number of transmit antennas rises. μ 1 nt, where nt is the number of antennas, should be the power per antenna. Additionally, by adding more transmit antennas without raising the transmit power, the throughput might be raised [54]. Milliwatts, or incredibly low power, is used by each antenna [55]. As a result, system reliability rises along with energy efficiency.

- **Security enhancement and robustness improvement:** In today's wireless communication networks, purposeful jamming and man-made interference are major challenges. In order to cancel the signals from deliberate jammers, a large number of antenna terminals results in a large number of degrees of freedom. Furthermore, beamforming makes big MIMO systems intrinsically resistant to passive eavesdropping attempts. Nevertheless, the eavesdropper can deploy countermeasures by taking advantage of the user's close proximity to a high channel correlation or the channel estimation's weakness [56].
- **Cost efficiency:** mMIMO reduces the cost of system deployment by doing away with bulky items like coaxial cables, which are needed to connect the BS components. Furthermore, rather than utilizing multiple expensive, high-power amplifiers, mMIMO use inexpensive milliwatt amplifiers [57]. Furthermore, it can significantly increase data rates while simultaneously reducing radiated power by a factor of 1000.
- **Signal processing:** Signal processing is made simpler when there are several antennas because they reduce the impacts of thermal noise, uncorrelated noise, fast fading, and interference [58, 59]. Furthermore, when the base station's and user terminals' channel responses diverge (i.e., when they are mutually orthogonal, or when the inner products are zero), a favorable propagation environment is created. Non-orthogonal channel vectors, on the other hand, require sophisticated signal processing to reduce interference.

Channel hardening is one of the most important characteristics of large-scale beam steering antenna arrays. It describes the phenomenon wherein, as the number of antennas approaches infinity, the mMIMO channel matrix approaches its values. Stated otherwise, the off-diagonal components of the Gramian matrix get weaker relative to diagonal terms as the size of the channel gain matrix rises, and the effective channel approaches are deterministic. The detection method and channel estimation can both take advantage of this feature. In this kind of situation, the simple matched filter (MF) gets close to optimality [60]. This is only valid, though, in the case of rich scattering and genuinely huge antenna arrays. As a result, in correlated fading channels and real-world propagation circumstances, sophisticated detection algorithms are valuable.

6.4.4 mMIMO FSO system

Uninterrupted data transfer at a high transmission rate is essential in this era of wireless communication. Both the transmitting and receiving sides of the mMIMO technologies employ a wide array of antennas. The additional antennas significantly enhance the system's transmission rate. mMIMO builds upon the already established MIMO system [61, 62]. Figure 6.6 depicts the mMIMO FSO system.

Three fundamental components make up a FSO communication system: a transmitter, the propagation channel, and a receiver. The light carrier in this concept has been a continuous wave (CW) laser source. A Pseudo-Random Bit Sequence (PRBS) generator is first used to create the random bit sequence. Then, a

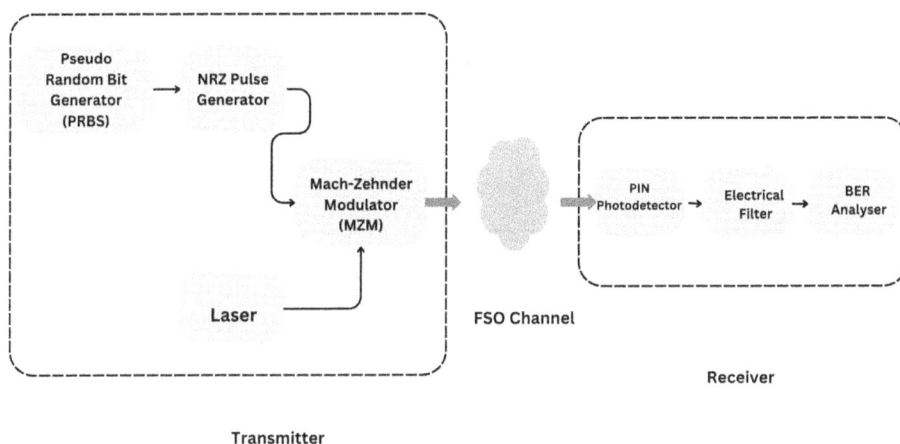

Figure 6.6. Illustration of a mMIMO FSO system.

non-return-to-zero (NRZ) pulse generator converts this data stream into voltage levels. The baseband signal is modulated onto the optical carrier using a Mach–Zehnder (MZ) modulator. The CW lasers at the transmitting end transmit this modulated optical signal across the FSO channel. This signal is received at the receiving end by an APD detector, and the original data is then recreated using a LPF and a 3R Regenerator. The received signal has been examined using a BER analyzer.

This chapter's goal is to evaluate a huge MIMO FSO system under various turbulence-prone weather circumstances, see how it performs in terms of BER, power consumption, transmission range, etc., and determine the system's desirable parameters. Concerning cost-effectiveness and ease of use, the huge MIMO FSO system of diversity order has been chosen for the analysis. The performance outcomes of this system are further contrasted with an improved system.

6.5 Result analysis

Different weather scenarios were examined when analysing the two suggested system [63, 64]. Sixteen were chosen as the diversity order for the simulation. As the optical source, a CW laser was employed. In the simulation, four different weather conditions were used. Clear skies, hazy, light rain, and heavy rain are among them. From this simulation, the FSO channels were built with similar attenuation values in each weather scenario.

Initially, the performance of the 16×16 FSO system was examined at a transmission rate of 10 Gbps, and it was found, as seen in table 6.1 [63], that the model produces excellent results at this data rate. During heavy rain, the minimum BER recorded at a distance of 1 km was $3.19 \times 10-119$ when using a 0 dBm source power.

For a distance of two kilometers, the minimum BER at a transmission rate of 10 Gbps under various weather conditions is noted. There was a discernible increase in

Table 6.1. Min. log of bit error rate versus source power (dBm).

Weather condition	Data rate (Gbps)	16 × 16 mMIMO FSO system		
		Source power (dBm)	Distance (km)	Minimum BER
Heavy rain	10	0	1	$3.19 \times 10-119$
		10	2	$3.14 \times 10-14$

Table 6.2. Distance vs. data rate of mMIMO FSO system for a BER of 10^{-10}.

Weather condition	Data rate (Gbps)	Attenuation (dB km^{-1})	BER	Optimum distance (8 × 8 mMIMO FSO system)	Optimum distance (16 × 16 mMIMO FSO system)
Clear weather	40	0.43	10^{-10}	6.32	8.2
Haze		4.2		2.53	2.97
Moderate rain		5.8		2.11	2.43
Heavy rain		9.2		1.6	1.79

the BER at this distance. Under heavy rain, the BER is $3.14 \times 10-14$ when 10 dBm of source power is used.

This chapter demonstrated that the 16 × 16 system performs admirably at 10 Gbps in all weather situations. As a result, this technology can transmit data at a higher pace across a greater distance. Therefore, additional analysis was done to determine the ideal mMIMO FSO system parameters.

The optimal parameters of the proposed mMIMO FSO system were determined through additional analysis. With a constant BER of 10^{-10}, simulations were run to determine the maximum data rate for the ideal distance for a 16 × 16 system. As seen in table 6.2 [63], a data rate of 40 Gbps was found to be optimal based on the investigation. It was determined that the ideal distance was significantly greater during clear weather compared to other weather conditions. The maximum distance at the optimal data rate of 40 Gbps in clear conditions was 8.2 km. A comparable analysis was also conducted for the 8 × 8 system. Under clear weather, the distance in this instance was 6.32 km.

The transmission range drastically decreased when severe weather was considered. Even so, the distance remained noteworthy given the conditions. From table 6.4 [63], in haze, the 16 × 16 system's range was found to be 2.97 km, while the 8 × 8 system's range was 2.53 km. The transmission ranges recorded for the 8 × 8 system were 2.11 km in moderate rain and 1.6 km in heavy rain. For the 16 × 16 system, these values increased to 2.43 km in moderate rain and 1.79 km in heavy rain. As a result, compared to the 8 × 8 system, the transmission range of the 16 × 16 system is typically enhanced by 200–300 m during periods of high wind.

A MIMO system's power penalty is displayed in table 6.3 [63] for each of the four weather scenarios. This table presents the values for an ideal data rate of 40 Gbps and a distance of 1.79 km. According to the table, clear weather is required for an 8 × 8 mMIMO FSO system to achieve a BER of 10^{-10}. This means that the source power needs to be −3.5 dBm. The need for power rises to 13 dBm during periods of heavy rain.

Table 6.3 [63] illustrates the power penalty of the 16 × 16 FSO system. It is evident that in clear weather, a BER of 10–10 requires a source power of −6.5 dBm. About 0 dBm is needed for haze, while 10 dBm is needed for heavy rain. Thus, we can observe that doubling the number of antennas reduces the power required by 3 dBm under the same weather conditions. In other words, the amount of source power needed decreases by half when the number of antennas doubles.

Here, a constant source power of 10 dBm is used to run the simulation. Table 6.4 [63, 64] lists the ideal system parameters that were discovered from this investigation.

The above comparison table considers the maximum attenuation range as 9.2 (dB km^{-1}) for heavy rain and two different ranges of 1 and 2 km. The same data rate was considered as 10 Gbps. The performance of two different power sources is analyzed at a distance of 1 km: one with an output of –1 dBm and the other with an output of 0 dBm, compared to a third power source with an output of 10 dBm at a distance of 2 km. The quality factor for system 1 and system 2 for 1 km is 16.3994 and 6.687 45 respectively. Another quality factor for system 1 and system 2 for 2 km is 7.195 49 and 4.7589. Based on the comparison of the two systems, the better quality factor for system 1 is 6.687 45. The minimum bit error rate for system

Table 6.3. Power penalty mMIMO FSO system in different weather conditions at a distance of 1.79 km.

Weather condition	Data rate (Gbps)	Distance (km)	BER	8 × 8 mMIMO FSO system	16 × 16 mMIMO FSO system
Clear weather	40	1.79	10^{-10}	−3.5	−6.5
Haze				0	0
Heavy rain				13	10

Table 6.4. Comparison of optimum system parameters.

Attenuation (dB km^{-1})	Range (km)	Power (dBm) System 1	Power (dBm) System 2	Data rate (Gbps)	Quality factor System 1	Quality factor System 2	BER System 1	BER System 2
(Heavy rain) 9.2	1	−1	0	10	16.3994	6.687 45	$9.13005e^{-061}$	$3.19e^{-119}$
9.2	1	10	10	10	7.195 49	4.7589	$3.06994e^{-013}$	$3.19e^{-14}$

1 and system 2 for 1 km is $9.13005e^{-061}$ and $3.19e^{-119}$. The minimum bit error rate for system 1 and system 2 for 1 km is $9.13005e^{-061}$ and $3.19e^{-14}$.

The analysis encompasses the mMIMO system with various configurations, such as 8×8 and 16×16, evaluated across different atmospheric conditions, distances, source powers, and data rates. mMIMO emerges as a transformative solution, addressing FSO limitations arising from atmospheric turbulence. Its deployment within 5G systems below 6 GHz and extension to higher carrier frequencies marks a revolution in wireless communication. Leveraging extensive antenna arrays, mMIMO effectively addresses interference in various atmospheric conditions, particularly in scattering-prone settings, thereby enhancing system performance with improved BER, higher data rates, and optimal source power utilization. This technology optimizes spectral efficiency, throughput, and reliability by concurrently serving multiple users through spatial multiplexing. Despite the complex antenna infrastructure, advanced signal processing techniques facilitate the efficient management of user signals, minimizing interference. This proficiency in interference handling distinguishes mMIMO, paving the way for higher data rates and improved standards in wireless communication networks. Its pivotal role in navigating challenging environments, especially at higher carrier frequencies like cmWave or mmWave bands, solidifies mMIMO as an indispensable cornerstone propelling the evolution of cellular and wireless communication.

6.6 Conclusions

The chapter investigates the performance of a mMIMO FSO system under various turbulent weather conditions, determining optimal parameters for each scenario. The system achieves an ideal transmission speed of 40 Gbps using a 16x16 antenna array in simulations. While the results indicate that performance could be enhanced with a greater number of antennas, the choice of 16 antennas for both the transmitter and receiver was made to balance performance with cost efficiency.

To model atmaospheric turbulence, the Gamma–Gamma distribution was employed, allowing for a detailed examination of key performance metrics such as BER, quality factor, source power, and receiver sensitivity. The findings suggest that by effectively mitigating the challenges posed by atmospheric turbulence, the system can achieve a satisfactory transmission range at elevated data rates. However, it is noted that data transmission over extended distances remains constrained by insufficient source power. Therefore, further research is warranted to develop a mMIMO FSO system utilizing visible light sources to significantly enhance performance.

References

[1] Jaber M, Imran M A, Tafazolli R and Tukmanov A 2016 5G backhaul challenges and emerging research directions: a survey *IEEE Access* **4** 1743–66

[2] Chowdhury M Z, Hossan M T, Islam A and Jang Y M 2018 A comparative survey of optical wireless technologies: architectures and applications *IEEE Access* **6** 9819–40

[3] Siddique U, Tabassum H, Hossain E and Kim D I 2015 Wireless backhauling of 5G small cells: challenges and solution approaches *IEEE Wirel. Commun.* **22** 22–31

[4] Ghassemlooy Z, Arnon S, Uysal M, Xu Z and Cheng J 2015 Emerging optical wireless communications-advances and challenges *IEEE J. Sel. Areas Commun.* **33** 1738–49

[5] Hassan W A, Jo H-S and Tharek A R 2017 The feasibility of coexistence between 5G and existing services in the IMT-2020 candidate bands in Malaysia *IEEE Access* **5** 14 867–88

[6] Koonen T 2018 Indoor optical wireless systems: technology, trends, and applications *J. Lightwave Technol.* **36** 1459–67

[7] Grobe L, Paraskevopoulos A, Hilt J, Schulz D, Lassak F, Hartlieb F, Kottke C, Jungnickel V and Langer K-D 2013 High-speed visible light communication systems *IEEE Commun. Mag.* **51** 60–6

[8] Tsonev D, Videv S and Haas H 2015 Towards a 100 Gb/s visible light wireless access network *Opt. Express* **23** 1627–37

[9] Wang Q, Wang Z, Dai L and Quan J 2017 Dimmable visible light communications based on multilayer ACO-OFDM *IEEE Photonics J.* **8** 1–11

[10] Malik A and Singh P 2015 Free space optics: current applications and future challenges *Int. J. Opt.* **2015** 945483

[11] Rabinovich W *et al* 2010 Free space optical communications research at the US Naval research laboratory *Free-Space Laser Communication Technologies XXII* vol **7587** (Bellingham, WA: International Society for Optics and Photonics) p 758702

[12] Komine T and Nakagawa M 2004 Fundamental analysis for a visible-light communication system using LED lights *IEEE Trans. Consum. Electron.* **50** 100–7

[13] Grubor J, Randel S, Langer K-D and Walewski J W 2008 Broadband information broadcasting using LED-based interior lighting *J. Lightwave Technol.* **26** 3883–92

[14] Elgala H, Mesleh R and Haas H 2011 Indoor optical wireless communication: potential and state-of-the-art *IEEE Commun. Mag.* **49** 56–62

[15] Borah D K, Boucouvalas A C, Davis C C, Hranilovic S and Yiannopoulos K 2012 A review of communication oriented optical wireless systems *EURASIP J. Wirel. Commun. Netw.* **2012** 91

[16] Son I K and Mao S 2017 A survey of free space optical networks *Digit. Commun. Netw.* **3** 67–77

[17] Gong S, Shen H, Zhao K, Wang R, Zhang X, de Cola T and Fraire J A 2020 Network availability maximization for free-space optical satellite communications *IEEE Wirel. Commun. Lett.* **9** 411–5

[18] Hamza A S, Deogun J S and Alexander D R 2019 Classification framework for free space optical communication links and systems *IEEE Commun. Surv. Tutor.* **21** 1346–82

[19] Liu C-H, Chang Y-C, Norris T B and Zhong Z 2014 Graphene photodetectors with ultra-broadband and high responsivity at room temperature *Nat. Nanotechnol.* **9** 273–8

[20] Morsy M A and Alsayyari A S 2020 Performance analysis of coherent BPSK-OCDMA wireless communication system *Wirel. Netw.* **26** 4491–505

[21] Morsy M A and Abdulaziz S A 2019 Multi-rate OCDMA system BER performance evaluations for different MLcode sequences *Opt. Quantum Electron.* **51** 1–18

[22] El-Wakeel A S, Mohammed N A and Aly M H 2016 Free space optical communications system performance under atmospheric scattering and turbulence for 850 and 1550 nm operation *Appl. Opt.* **55** 7276–86

[23] Mohammed N A, El-Wakeel A S, Malek Mohammadi A and Aly M H 2012 Performance evaluation of FSO link under NRZ-RZ line codes, different weather conditions and receiver types in the presence of pointing errors *Open Electr. Electron. Eng. J.* **6** 28–35

[24] Mohammed N A, El-Wakeel A S and Aly M H 2012 Pointing error in FSO link under different weather conditions *Int. J. Video Image Process. Netw. Secur. IJVIPNS-IJENS* **15** 84–904

[25] Sri I K and Srinivasulu A 2019 Performance analysis of inter-satellite optical wireless communication using 12 and 24 transponders *Proc. of 2nd Int. Conf. on Microelectronics, Computing and Communication Systems (MCCS 2017)* (Singapore: Springer) pp 811–21

[26] Singh S and Soni G 2013 Pointing error evaluation in FSO link *5th Int. Conf. on Advances in Recent Technologies in Communication and Computing (ARTCom 2013) (Bangalore)*

[27] Noor N H M, Naji A W and Al-Khateeb W 2012 Performance analysis of a free space optics link with multiple transmitters/receivers *Int. Island Univ. Malaysia (IIUM) Eng. J.* **13** 445–7

[28] Ghoname S, Fayed H A, Abd El Aziz A and Aly M H 2016 Performance analysis of FSO communication system: effects of fog, rain and humidity *2016 6th IEE Int. Conf. on Digital Information Processing and Communications (ICDIPC) (Beirut, Lebanon)* pp 151–5

[29] Mikołajczyk J, Bielecki Z, Bugajski M, Piotrowski J, Wojtas J, Gawron W and Prokopiuk A 2017 Analysis of free-space optics development *Metrol. Meas. Syst.* **24** 653–74

[30] El Mashade M B, Toeima A H and Aly M H 2016 Receiver optimization of FSO system with MIMO technique over log-normal channels *Optoelectron. Adv. Mater. Rapid Commun.* **10** 497–502

[31] Ghoname S, Fayed H A, Abd El Aziz A and Aly M H 2017 FSO system performance enhancement: receiver impact *J. Adv. Res. Appl. Mech.* **37** 1–8

[32] Adjoudani A *et al* 2003 Prototype experience for MIMO BLAST over third-generation wireless system *IEEE J. Sel. Areas Commun.* **21** 440–51

[33] Han L Q and You Y H 2016 Performance of multi-input and multi-output free space optical communication under atmospheric atmospheric attenuation and atmospheric turbulence *Chin. J. Lasers* **7** 223–30

[34] Wang X, Wang H Q and Cao M H 2017 Channel capacity of correlated wireless optical MIMO system *J. Electron. Meas. Instrum.* **31** 663–8

[35] Lu L, Li G Y, Swindlehurst A L, Ashikhmin A and Zhang R 2014 An overview of mMIMO: benefits and challenges *IEEE J. Sel. Top. Signal Process.* **8** 742–58

[36] Ordóñez L G, Palomar D P and Fonollosa J R 2011 Fundamental diversity, multiplexing, and array gain tradeoff under different MIMO channel models *Proc. IEEE Int. Conf. Acoust. Speech Signal Process. (ICASSP)* (Prague, Czech Republic) pp 3252–5

[37] Vahid S, Tafazolli R and Filo M 2015 Small cells for 5G mobile networks *Fundamentals of 5G Mobile Networks*; J Rodriguez (Chichester: Wiley) pp 63–104

[38] Mattigiri S and Warty C 2013 A study of fundamental limitations of small antennas: MIMO approach *Proc. IEEE Aerosp. Conf. (Big Sky, MT)* pp 1–8

[39] Khan F 2009 *LTE for 4G Mobile Broadband: Air Interface Technologies and Performance* (New York: Cambridge University Press)

[40] Goldsmith A 2005 *Wireless Communications* (Cambridge: Cambridge Univeristy Press) pp 321–50

[41] Li Q *et al* 2010 MIMO techniques in WiMAX and LTE: a feature overview *IEEE Commun. Mag.* **48** 86–92

[42] Larsson E G, Edfors O, Tufvesson F and Marzetta T L 2014 Massive MIMO for next generation wireless systems *IEEE Commun. Mag.* **52** 186–95

[43] Chin W H, Fan Z and Haines R 2014 Emerging technologies and research challenges for 5G wireless networks *IEEE Wirel. Commun* **21** 106–12

[44] Wang S, Xin Y, Chen S, Zhang W and Wang C 2014 Enhancing spectral efficiency for LTE-Advanced and beyond cellular networks [guest editorial] *IEEE Wirel. Commun* **21** 8–9

[45] Pappa M, Ramesh C and Kumar M N 2017 Performance comparison of mMIMO and conventional MIMO using channel parameters *Proc. Int. Conf. on Wireless Commun., Signal Process. and Networking* pp 1808–12

[46] Senel K and Larsson E G 2018 Grant-free massive MTC-enabled mMIMO: a compressive sensing approach *IEEE Trans. Commun.* **66** 6164–75

[47] Dawy Z, Saad W, Ghosh A, Andrews J G and Yaacoub E 2017 Toward massive machine type cellular communications *IEEE Trans. Wirel. Commun.* **24** 120–8

[48] Basar E 2016 Index modulation techniques for 5G wireless networks *IEEE Commun. Mag.* **54** 168–75

[49] Luo X 2016 Multiuser mMIMO performance with calibration errors *IEEE Trans. Wirel. Commun.* **15** 4521–34

[50] Basnayaka D A and Haas H 2015 Spatial modulation for mMIMO *Proc. IEEE Int. Conf. Commun.* pp 1945–50

[51] Björnson E, Hoydis J and Sanguinetti L 2018 Massive MIMO has unlimited capacity *IEEE Trans. Wirel. Commun.* **17** 574–90

[52] Marzetta T L 2010 Noncooperative cellular wireless with unlimited numbers of base station antennas *IEEE Trans. Wirel. Commun.* **9** 3590–600

[53] Björnson E, Larsson E G and Debbah M 2016 Massive MIMO for maximal spectral efficiency: How many users and pilots should be allocated? *IEEE Trans. Wirel. Commun.* **15** 1293–308

[54] Chen J, Chen H, Zhang H and Zhao F 2016 Spectral-energy efficiency tradeoff in relay-aided mMIMO cellular networks with pilot contamination *IEEE Access* **4** 5234–42

[55] Rusek F, Persson D, Lau B K, Larsson E G, Marzetta T L, Edfors O and Tufvesson F 2013 Scaling up MIMO: Opportunities and challenges with very large arrays *IEEE Signal Process. Mag.* **30** 40–60

[56] Do T T, Björnson E, Larsson E G and Razavizadeh S M 2017 Jamming-resistant receivers for the mMIMO uplink *IEEE Trans. Inf. Forensics Secur.* **13** 210–23

[57] Alshamary H 2017 Coherent and non-coherent data detection algorithms in mMIMO *Master's Thesis* (Iowa City, IA: University of Iowa)

[58] Björnson E, Larsson E G and Marzetta T L 2016 Massive MIMO: ten myths and one critical question *IEEE Commun. Mag.* **54** 114–23

[59] Jin S, Wang X, Li Z, Wong K K, Huang Y and Tang X 2016 On mMIMO zero-forcing transceiver using time-shifted pilots *IEEE Trans. Veh. Technol.* **65** 59–74

[60] Narasimhan T L and Chockalingam A 2014 Channel hardeningexploiting message passing (CHEMP) receiver in large-scale MIMO systems *IEEE J. Sel. Top. Signal Process.* **8** 847–60

[61] Albreem M A, Juntti M and Shahabuddin S 2019 Massive MIMO detection techniques: a survey *IEEE Commun. Surv. Tutor.* **21** 3109–32

[62] Mansour A, Mesleh R and Abaza M 2017 New challenges in wireless and free space optical communications *Opt. Lasers Eng.* **89** 95–108

[63] Ahmed S, Syed M, Hasan A and Nazrul Islam A K M 2022 Design of an optimum mMIMO FSO system and analysis of its performance in different weather conditions *MIJST* **10** 43–52

[64] Md S, Hasan A, Ahmed S and Nazrul Islam A K M 2021 Simulation of a mMIMO FSO system under atmospheric turbulence *2021 5th Int. Conf. on Electrical Engineering and Information & Communication Technology (ICEEICT), UTC from IEEE* pp 1–6

[65] Sridarshini T, Geerthana S, Balaji V R, Arun T, Sitharthan R, Sivanantha Raja A and Shanmuga Sundar D 2023 Ultra-compact all-optical logical circuits for photonic integrated circuits *J. Laser Phys.* **33** 076207

[66] Geerthana S, Syedakbar S, Sridarshini T, Balaji V R, Sitharthan R and Shanmuga Sundar D 2022 2D-PhC based all optical AND, OR and EX-OR logic gates with high contrast ratio operating at C band *J. Laser Phys.* **32** 106201

[67] Sivaranjani R, Shanmuga Sundar D, Sridarshini T, Sitharthan R, Karthikeyan M, Sivanantha Raja A and Marcos Flores C 2020 Photonic crystal based all-optical half adder: a brief analysis *J. Laser Phys.* **30** 116205

[68] Sridharshini T, Preethi C, Geerthana S, Balaji V R, Arun T, Sitharthan, Kartikeyan M and Shanmuga Sundar D 2022 Current and future horizon of optics and photonics in environmental sustainability *J. Sustain. Comput.* **36** 100815

[69] Kavitha V, Balaji R, Shanmuga Sundar D, Sridarshini T, Robinson S, Massoudi R, Gopalkrishna H and Jesuwanth Sugesh R G 2023 Design and performance analysis of eight channel demultiplexer using 2D photonic crystal with trapezium cavity *J. Opt.* **25** 065102

[70] Arunkumar R, Jayson K J and Robinson S 2019 Design and analysis of optical Y-splitters based on two-dimensional photonic crystal ring resonator *J. Optoelectron. Adv. Mater.* **21** 435–42

[71] Arunkumar R and Robinson S 2021 Investigation on ultra-compact 2D-PC based optical circulator for photonic integrated circuits *J. Optoelectron. Adv. Mater.* **23** 112–8

[72] Arunkumar R, Kavitha V and Rama Prabha K 2022 Investigation on ultra-compact, high contrast ratio 2D-photonic crystal based all optical 4×2 encoder *Opt. Quantum Electron.* **54** 110

[73] Arunkumar R and Robinson S 2023 Realization of an all-optical 2×4 decoder based on a two-dimensional photonic crystal *Opt. Quantum Electron.* **55** 570

Chapter 7

AI in optics and photonics

Vineeth Palliyembil and E G Anagha

Light-based technologies play a crucial role in improving the quality of the various aspects of our daily lives. Advances in the fields of optics and photonics continue to respond to the needs of the humankind by providing breakthroughs in various fields such as medicine, telecommunication, astronomy, life sciences, energy, and manu-facturing. The opportunities for research in optics and photonics will continue to arise in the upcoming decades with greater benefits on combining with Artificial Intelligence-based technologies. Artificial Intelligence (AI) has the potential to revolutionize the field of optics by enabling the development of new algorithms and systems that can analyse, interpret, and make decisions based on complex datasets. Intelligent optics is an upcoming field that interfaces Machine Learning (ML) with optics for optical data analysis by providing efficient, accurate, and rapid analysis of optical spectra and images for various applications such as medical diagnostics, information technology, agriculture, and so on. Photonics and opto-electronics utilize ML-based algorithms for optimization, forward modelling, and inverse design of photonic structures and optical materials, and also for enhancing the prediction accuracy of photonic sensors. AI algorithms can be used to control photonics devices in real-time, adjusting their parameters based on feedback from sensors or other data sources. Another important application is the realization of AI-based photonic processors along with Optical Neural Networks (ONNs) with the capability of efficient and ultra-high speed computing capabilities for application in cloud computing, communication systems, and data centres. Some of the key research applications of AI in optical communication include Optical Performance Monitoring (OPM) for bit error rate optical signal-to-noise ratio estimation; Modulation Format Identification (MFI) including analysis of eye diagram, constellation diagram, and scatterplots; signal processing with nonlinear equalization, dispersion compensation, Autoencoder, and soft-demapping; and short-reach communication involving intelligent Visible Light Communication (VLC) and indoor optical wireless communication. The integration of AI and photonics has the potential to enable new applications and to improve the performance of existing photonics-based technologies. Although the prospects of AI with light wave technology seem bright, there are several challenges that still need addressing. Computational complexity, long training times, and generalizability are the main challenges for ML-based intelligent VLC systems. In optical communication systems, there is a lack of sufficient experimental and practical data from network operators due to difficulty in collecting both image and sequential data. Insufficient data and diversity pose difficulties in realizing robust and generalized AI-based models for analysing and optimizing optical communication systems. In order to address the pressing need for high-quality datasets for algorithm training, more effective approaches are required for data augmentation. Overall, the use of AI in photonics and optics has the potential to enable the development of more advanced and efficient photonics systems, which can lead to numerous benefits in a wide range of fields.

7.1 Introduction

7.1.1 Introduction to optics and photonics

Light plays a significant and fundamental role in shaping the lives of human beings and societies. Light is an electromagnetic wave that travels through space at a constant speed and can be described by its wavelength, frequency, and amplitude. Light has fundamentally changed the way human beings use energy and has brought us into the information era. The two branches dealing with the science and technology of light are termed optics and photonics because of the dual, powerful nature of light. Optics studies the nature and characteristics of light, while photonics is the science of generating, manipulating, and detecting photons (particles of light) for various applications. Optics and photonics are closely related and together form a multidisciplinary field that spans across physics, engineering, and materials science. Both optics and photonics have revolutionized our understanding of light and its properties. They have also led to numerous technological advancements that have transformed our daily lives. The future looks bright for these fields as they continue to push boundaries in science and technology.

Some of the earliest means of communication used by people were based on light by making use of fire signals for sending alarms, as a call for help, or for making announcements. Most experts believe that the invention of the first laser in 1960s marked the modern era of optics followed by the introduction of optical fibers and semiconductor optoelectronics [1]. Over a period of 10 years, the technical developments in optical fiber communication have brought about a 100-fold increase in the amount of information that can be transmitted from one place to another, thus allowing the internet to transform society [2]. The burgeoning demands on communication networks for various services such as remote education, on-demand video, telemedicine, video conferencing, home shopping, and so on fuelled by the development of personal computers with high storage and processing capabilities require a high-speed fiber-optic backbone [3, 4].

7.1.2 Introduction to AI

Basically, AI is the science and engineering of making machines to demonstrate intelligence especially speech recognition, visual perception, decision making, and human language processing. In other words, any task performed by a program or machine that otherwise requires application of human intelligence to accomplish is AI. It includes reasoning, learning, planning, problem solving, self-correction, manipulation, motion, knowledge representation, perception, and creativity. The motive of AI is to acquire the scarce resource of intelligence, then promote and utilize it through the spread of computer technology. AI refers to the myriad of technologies that enable machines to simulate human-like intelligence. This includes the ability to perceive their environment, understand and interpret information, perform tasks, and learn from experiences to improve their performance over time [5].

AI is starting to play an increasingly important role in our society and in our daily lives. Machine Learning (ML) and Deep Learning (DL) are two exciting areas of AI as depicted in figure 7.1. ML involves the use of algorithms to automatically make sense of data and generate insights. This technology makes use of the idea that learning is a dynamic process made possible through examples and experiences; in other words, the idea that a machine can retain information and become smarter over time. This revolutionary technology is fascinating because a machine is less likely than a human to suffer from sleep deprivation, diversions, information overload, and short-term memory loss. There are three main classifications of ML techniques: supervised learning, unsupervised learning, and reinforcement learning. ML is inherently related to data analysis and statistics as the success of the learning algorithm depends on the data used. ML algorithms are proficient at classifying various impairments in optical communication systems using images of eye diagrams. However, they currently only identify the type of distortion, not the quantity. This limitation can be addressed by integrating a nearest neighbors' technique after the ML algorithm.

The main benefits of AI over humans are its scalability, longevity, and capacity for ongoing improvement. This can significantly boost production, decrease costs,

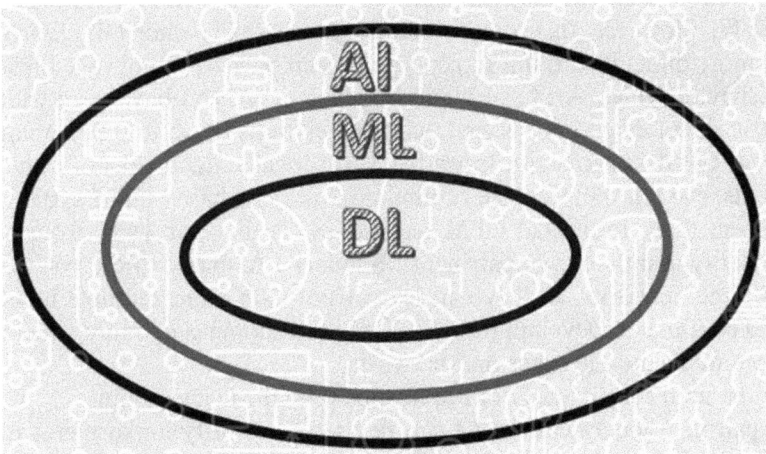

Figure 7.1. AI, ML, and DL.

and reduce human error. Even though it's a relatively new technology, AI has the potential to revolutionize the way we do business by improving productivity and ultimately turbo boosting economic prosperity.

7.2 Intersection of AI/ML in optics

Optical communication and networks are leveraging various AI techniques, ranging from early ML to more current DL. AI-based techniques have penetrated the field of optics covering various domains from quantum communication to nano-photonics, optical communication, and optical networks. There has been fascinating progress in the past few years in the combined areas of AI and optics ranging from photonic neural networks, AI-enabled optical computing and intelligent VLC systems, AI-enabled Optical Performance Monitoring (OPM), and Modulation Format Identification (MFI) to name a few. Here we briefly explore the various advances in the combined areas of AI and optics and their future implications.

7.2.1 Major applications

7.2.1.1 AI for optical performance monitoring

OPM is a critical function in optical communication systems and networks. It is used to measure and monitor the performance of optical links and components in order to ensure that they are operating within acceptable limits. OPM deals with overseeing the performance metrics such as symbol error rate, Q-factor, Polarization Mode Dispersion (PMD), etc., for an optical communication system [6]. Traditional OPM techniques are based on analytical models of the optical channel. However, these models can be inaccurate, especially in the presence of nonlinearities and other impairments.

During the performance monitoring, curves such as eye diagrams, phase portraits, and amplitude histograms are analysed. Several features from these plots are extracted and sometimes new features added to conduct the investigations. Manual definition of features is difficult and subsequently it is impossible to distinguish patterns among the plots if differences are slight. DL techniques are handy in these scenarios as they can learn on their own the various features from the input data [7]. However, this comes with more complexity since DL algorithms are generally more difficult to train. In general, AI can be used to improve the accuracy and efficiency of OPM. Artificial Neural Networks (ANN) have been widely used for OPM in direct detection systems, with correlations of up to 0.997 obtained [8].

AI-based OPM systems can learn from historical data to identify and predict impairments, even in the presence of nonlinearities. AI-based OPM systems can be more efficient than traditional OPM techniques, as they can learn from historical data to identify and predict impairments. Similarly, AI-based OPM systems can be scaled to large networks, as they can learn from data from multiple links. As AI techniques continue to develop, they are likely to become even more widely used in optical communication systems and networks.

AI can be used for a variety of applications in OPM, including:
- Impairment identification: AI can be used to identify impairments in optical links, such as chromatic dispersion, PMD, and OSNR.

- Impairment prediction: AI can be used to predict the performance of optical links under different conditions. This can be used to prevent outages and ensure that the links are operating within acceptable limits.
- Link optimization: AI can be used to optimize the performance of optical links. This can be done by adjusting the parameters of the links, such as the modulation format and the transmit power.
- Fault management: AI can be used to identify and troubleshoot faults in optical networks. This can help to reduce the time to repair faults and improve the availability of the networks.

Many techniques used for OPM are dependent on the signal type and assume that the monitoring unit already has knowledge of it. Multi-impairment monitoring of different signal types may require training multiple ANNs for each signal type. Some methods have been proposed that are transparent to the bit rate and modulation format, but they require significant training data. Other works have utilized multi-task learning and DL to identify the signal type and impairments at the same time. However, they also require large training datasets. ANNs are shown to be best suited for coherent detection systems compared to other ML algorithms. Deep learning has been the most common method of choice for MFI and bit rate identification, with most works achieving very high accuracy. Photonic reservoir computing is a promising technology for OPM and modulation format recognition since it reduces the training complexity and allows for high improved performance operation [9].

ML algorithms can be used to classify different impairments in optical communication systems using images of eye diagrams. However, it was able to only identify the type of distortion but not the quantity. This could be overcome by using an additional nearest neighbours' technique after the ML algorithm.

7.2.1.2 AI for visible light communication (VLC)

VLC is a wireless communication technology that uses optical signal as the carrier of information. It has the potential to offer high-speed, secure, and energy-efficient data transmission [10]. However, VLC systems are susceptible to several challenges, such as channel fading, nonlinearity, and multipath interference. VLC systems also suffer from the nonlinear effects that exist in almost every part of the system, such as the Intensity Modulation and Direct Detection (IM/DD) scheme, the LED nonlinearity, the channel noise and interference, and the receiver distortion [11].

AI has emerged as a promising solution to address these challenges. In recent years, there has been growing interest in the use of AI for VLC systems. Many research papers have proposed AI-based techniques for improving the performance of VLC systems and also for allocation of resources in VLC systems. A DL-based detection scheme for VLC systems uses a deep neural network to detect received signals and extract information bits efficiently. A physical-layer anti-eavesdropping framework for VLC links enhances confidentiality using smart beamforming over the Multiple Input Single Output (MISO) VLC wiretap channel. This is achieved through reinforcement learning and deep RL-based VLC beamforming control schemes [12]. Federated learning is another AI technique that can provide secrecy performance. Here, the cost of transferring raw data is lower and more secrecy is achieved by running AI locally client side [13].

AI-based techniques can be used to improve the performance of VLC systems in several ways, including:

- Channel estimation: AI can be used to estimate the VLC channel, which is essential for accurate data transmission.
- Equalization: AI can be used to equalize the VLC channel, which can help to mitigate the effects of channel fading and multipath interference.
- Signal detection: AI can be used to detect VLC signals in the presence of noise and interference.
- Resource allocation: AI can be used to allocate resources in VLC systems, such as bandwidth and power, in an efficient manner.

VLC-based indoor positioning has advantages such as low cost, high durability, and environmental friendliness. The block diagram for an intelligent VLC system is shown in figure 7.2. Ml techniques such as KNN, ANN, clustering, and fusion of multiple classifiers are used for improving indoor positioning, decreasing error rates, reducing flicker, and controlling dimming [15]. In vehicle-to-vehicle communication, AI-based VLC techniques such as those used in driverless cars provide sublime experience and safety. However, the development of smart vehicles depends on the progress of communication as well as embedded computing technologies.

7.2.1.3 AI for short reach optical communication
Short Reach Optical Communication (SROC) is a type of optical communication that is used to transmit data over distances of up to 100 km. SROC systems are typically used for data centre interconnects, enterprise networks, and other applications where high bandwidth and low latency are required over short distances. SROC systems have many challenges such as the increasing demand for high-speed,

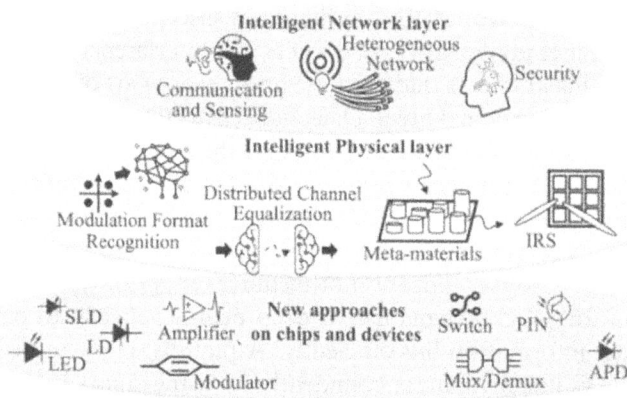

Figure 7.2. An intelligent visible light communication (IVLC) system. Reproduced from [14]. CC BY 4.0.

low-cost, and low-power data transmission in data centres, metro networks, and access networks. SROC systems are generally characterized by direct detection-based receiver configurations and other simple and inexpensive components. However, these components may induce transmission issues such as LED non-linearity, channel noise, interference, and distortions.

AI has the potential to revolutionize SROC by enabling new capabilities and improving the performance of existing systems. AI can be used to improve different SROC tasks, such as signal modulation and demodulation, channel modeling and estimation, nonlinear compensation, data augmentation, and system optimization. DL algorithms can be used to process different types of data collected from SROC systems, such as image data (e.g., constellation diagrams, eye diagrams, spectrograms) and sequential data (e.g., time series, bit streams). The advantages of DL algorithms over conventional ML algorithms for SROC tasks include their ability to perform end-to-end learning, feature extraction, complex analysis, and self-configuration [16].

AI-based techniques can be used to address a wide range of challenges in SROC, including [17]:

- **OPM**: AI can be used to develop more accurate and efficient OPM algorithms for detecting and identifying impairments in SROC systems. This can help to ensure the reliable operation of SROC systems and prevent service outages.
- **MFI**: AI can be used to develop more accurate and efficient MFI algorithms for identifying the modulation format of a transmitted signal. This can be useful for applications such as optical network management and security.
- **Signal processing:** AI can be used to develop more efficient and effective signal processing algorithms for SROC systems. This can improve the performance of SROC systems in terms of data rate, reach, and power consumption.
- **Indoor optical wireless communication (OWC)**: AI can be used to develop more robust and reliable OWC systems. This can be useful for applications such as home networking and industrial automation.

In addition to these specific applications, AI can also be used to improve the overall performance of SROC systems by:

- **Learning from historical data:** AI can be used to learn from historical data about the performance of SROC systems. This can be used to develop predictive models that can help to identify and prevent problems before they occur.
- **Optimizing system parameters:** AI can be used to optimize the parameters of SROC systems. This can help to improve the performance of the systems in terms of data rate, reach, and power consumption.
- **Self-healing:** AI can be used to develop self-healing capabilities for SROC systems. This means the systems can automatically detect and repair problems without human intervention

7.2.1.4 AI for modulation format identification

MFI is an important task in optical communication systems. It allows the receiver to determine the modulation format of the incoming signal, which is essential for decoding the signal correctly. Traditionally, MFI has been performed using statistical methods, such as peak-to-average power ratio (PAPR) analysis or histogram-based methods. However, these methods are often not robust to noise and other impairments.

In recent years, AI techniques have been shown to be very effective for MFI. AI techniques can learn the statistical characteristics of different modulation formats, even in the presence of noise and other impairments. This makes them ideal for use in optical communication systems, where the signal quality can be degraded by a variety of factors [18].

The performance of AI techniques for MFI has been shown to be very promising. In a recent study, SVMs and DNNs were shown to achieve an accuracy of up to 99.9% for MFI in optical communication systems. This is significantly higher than the accuracy of traditional MFI techniques [6]. Autonomous MFI schemes have been developed using various ML algorithms, such as k-means and ANN. Some works take advantage of the Stokes space signal representation, which is immune to offset issues in phase and frequency [19].

7.2.1.5 AI for optical sensing

One area where AI has shown remarkable advancements is in optical sensing. Optical sensing refers to the use of light-based technologies, such as cameras and sensors, to capture and interpret information from the surrounding environment. By combining AI with optical sensing, we can extract meaningful insights and make informed decisions in a wide range of applications.

AI plays a crucial role in enhancing optical sensing capabilities by providing sophisticated algorithms for data analysis, pattern recognition, and object detection. These algorithms enable optical sensors to process vast amounts of visual information in real-time, surpassing human accuracy and efficiency. Moreover, AI-powered optical sensing systems have the ability to learn and improve their performance over time through ML techniques [20].

Some of the domains where optical sensing is commonly used include:

Environmental monitoring: Optical sensors are used to measure various environmental parameters like air and water quality, pollution levels, and greenhouse gases.

Medical and healthcare: Optical sensing is employed in medical imaging (e.g., endoscopy, optical coherence tomography), glucose monitoring, pulse oximetry, and various other diagnostic and therapeutic applications.

Industrial and manufacturing: Optical sensors are utilized in industrial automation, quality control, and inspection processes for measuring parameters such as distance, displacement, colour, and temperature.

Aerospace and defence: Optical sensors are used for target detection, tracking, and missile guidance systems in the defence sector.

Biotechnology and life sciences: Optical sensing plays a significant role in DNA sequencing, fluorescence-based assays, and label-free biomolecule detection.

Robotics and autonomous systems: Optical sensors are integrated into robotics for object detection, localization, and navigation.

Telecommunications: Fiber-optic communication systems rely on optical sensing to transmit data as light pulses through optical fibers.

Automotive: Optical sensors are used for adaptive cruise control, collision avoidance, and vehicle navigation systems.

Structural health monitoring: Optical sensing helps assess the integrity of structures like bridges, buildings, and pipelines by measuring strain, deformation, and vibrations.

Agriculture: Optical sensors aid in precision farming by monitoring crop health, soil properties, and vegetation.

Consumer electronics: Optical sensing is found in devices like cameras, optical fingerprint scanners, and ambient light sensors in smartphones and other gadgets.

Security and surveillance: Optical sensors are used for perimeter security, motion detection, and facial recognition systems.

Some of the recent works on AI-enabled optical sensing are discussed briefly. A single embedded Biber Bragg Grating (FBG) sensor aided by ML models was utilized to create a smart helmet capable of real-time detection of blunt-force impact events on helmets. Implementation of ML-FBG smart helmet systems can act as an early-stage intervention approach both during and immediately after a concussive event [21]. A regularized polynomial regression-based supervised ML algorithm was utilized to realize an optical distance sensor for industrial scenario distance measuring and attained a performance improvement by a factor of 4 over sensor designs without ML. A method to improve the accuracy of rainfall rate assessment of optical satellite sensors was achieved by using a Random Forests (RF) model. The combination of specific characteristics in RF models makes them highly suitable for their application in precipitation remote sensing. Using the proposed model, it was possible to accurately assign rainfall rates, even on an hourly basis, with considerable precision. Moreover, these rainfall rates could be determined throughout the day, including day, night, and twilight conditions, thereby facilitating the estimation of 24 h rainfall rates [22]. AI-enabled optical sensing techniques based on Laser-Induced Fluorescence (LIF) equipped with drones were used to monitor seawater and detect oil pollution. Optical gas sensors are widely employed for air quality control and environmental protection due to their various characteristics such as refractive index and optical absorption. Various ML techniques such as regression, classification, and clustering were incorporated with these optical sensors to improve their data analysis and enhance their sensitivity.

AI-driven optical sensing is extensively used in the agricultural industry, offering innovative techniques that contribute to sustainable practices. While optical sensors offer a simpler data collection process, the data they generate can be complex and voluminous, particularly in the case of hyperspectral imaging (i.e. imaging across a wide range of wavelengths). As a result, handling such data necessitates the application of sophisticated data processing and statistical methods. Owing to the recent advancements in ML algorithms, images and spectra can now be analysed and classified in a completely automated and reproducible manner thereby reducing the requirement for complex image or spectrum analysis methods. The most common ML methods used in agriculture are ANNs and Support Vector Machines (SVMs). DL-based algorithms have been utilized to estimate disease severity from apple rot images to obtain 90.4% accuracy [23]. Similarly, SVMs were used on RGB images from smartphones to detect stripe rust in wheat. Various fungal diseases such as downy mildew and powdery mildew in cucumber were analysed based on deep CNN and RF models with 92.2% and 84.8% accuracy, respectively. Currently, most of the reported works are specific to some particular variety of plants and therefore the development of a more generalized approach is required for the future of modern agriculture. Another challenge is the discrimination of various plant stresses, which would require further improvements in the sensitivity of optical devices and advanced ML techniques to distinguish the targeted characteristic from noise.

In conclusion, the main limitations of optical sensors are (1) cross-sensitivity, (2) massive data generation, (3) slow data processing, (4) loss of signal-to-noise ratio across fibre length, and (5) total cost of sensor and interrogator systems. By developing robust data analytics engines made possible by recent advancements in ML and AI, these difficulties can be solved. Advanced regression techniques, such as ANNs and SVMs, have demonstrated their superiority over nonlinear curve fitting in terms of data processing speed and improved denoising capabilities. AI models used in computer vision, audio recognition, and machine translation have exhibited remarkable capabilities in addressing classification problems that involve hundreds of classes. The success of these models can be attributed to the extensive data collection, curation, and sharing efforts within these research communities.

7.2.2 Major challenges

The use of AI for Optics is still in its early stages, but there is a great deal of potential for this technology to revolutionize the field. As AI techniques continue to develop, we could witness further revolutionary and powerful applications of AI for Optics in the years to come. However, currently there are issues such as the lack of large-scale and high-quality experimental datasets for training and testing AI models and the trade-off between accuracy and complexity of AI models. Similarly, there are challenges in the integration of AI models with existing optical communication system architectures and protocols as well as the security and privacy issues of AI-based solutions.

7.3 Intersection of AI in photonics

Photonics combines principles from physics, optics, electronics, and materials science to develop innovative devices and systems that harness light for a wide range of applications. By using photons instead of electrons, photonics enables longer transmission distances, greater bandwidth capacity, and faster data rates. Photonics has huge potential for future improvements as the study and technical developments continue. It is an important field that is advancing in a variety of industries, from communication and healthcare to energy and beyond, thanks to its multidisciplinary character and versatility in light manipulation. AI has become an increasingly powerful tool in the field of photonics, revolutionizing the way light-based technologies are developed, optimized, and utilized. By integrating the capabilities of AI with photonics, researchers and engineers are able to tackle complex problems, enhance device performance, and unlock new possibilities. Some of the major applications of photonics integrated with AI are explored in the following sections.

7.3.1 Major applications

7.3.1.1 Inverse design of photonic structures

One area where AI has made significant contributions is in the design and optimization of photonic devices. Traditional design approaches often rely on trial and error or manual optimization, which can be time-consuming and resource-intensive. With AI, researchers can employ ML algorithms to analyse vast datasets and explore a wide range of design parameters. This enables the discovery of novel device architectures, such as photonic circuits, waveguides, and sensors, that exhibit improved efficiency, sensitivity, and functionality.

In general, device modelling in photonics can be generally categorized into forward and inverse modelling. To obtain a desired targeted output, initial designs are made based on geometric parameters that are verified using simulations. The initial designs are less likely to give the desired results and hence some adjustments to some of the design parameters and repeated revaluations by simulations are required. Although this forward modeling scheme has attained notable success, the trial-and-error mechanism becomes time-inefficient and computationally costly owing to the continuously increasing complexity of the nanophotonic devices. AI methods in forward modeling typically utilize discriminative neural networks. AI-based forward modeling techniques have been successfully utilized in the design of beam combiners, photonics topological insulators waveguides, and photonic crystal-based filters [24].

In photonics, inverse modelling is the process of figuring out the characteristics or parameters of a photonic system based on data that has been observed or the desired results. It entails resolving the inverse problem in which the parameters or internal properties that govern a system are unknown but its input and output are known. In photonics, this often entails identifying the optical characteristics of structures, devices, or materials. Applications for inverse modelling in photonics involve developing photonic devices with desired functionality, optimising the performance of existing devices, and identifying unknown or complicated photonic structures. It is especially helpful when doing analytical computations or direct measurements of the system's properties, which can otherwise be difficult or time-consuming. The various approaches to inverse design of photonic structures are shown in figure 7.3.

A novel method of inverse design using ML was used to model a Raman Amplifier (RA) by selecting pump powers and wavelengths to obtain the targeted gain profile [25]. This work is highly significant since next-generation optical communication systems are targeted to operate in all the five bands (O, E, S, C,

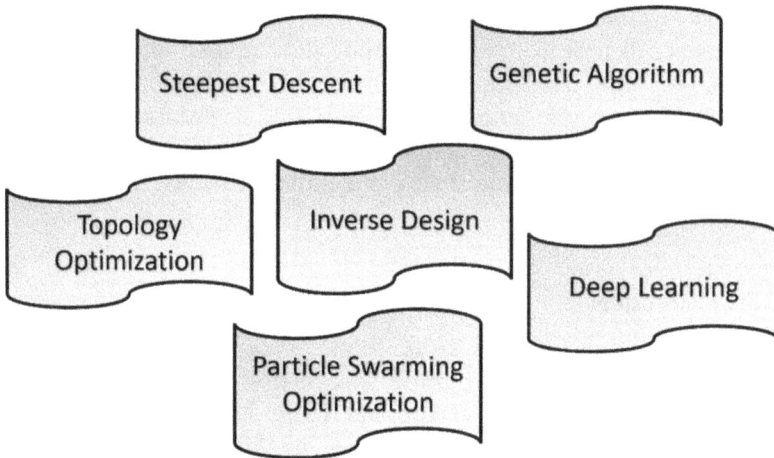

Figure 7.3. Various approaches to inverse design of photonic structures.

and L bands). Although the Erbium Doped Fiber Amplifiers (EDFA) are well established, RAs provide great flexibility in the gain profile design in addition to low noise properties and availability of gain across a broad range of wavelengths making them an attractive solution for future ultra-wideband optical networks. To learn the mapping between the gain profile and pump powers, a multilayer neural network is utilized. As soon as the model has been learned, the prediction of pump powers and wavelengths are obtained by a simple forward propagation through the multilayer neural network. The proposed work attained high accuracy, low complexity, and ultra-fast operation compared to state-of-the-art techniques for which each gain profile is a complex optimization problem that needs to be solved from scratch.

Earlier methods to design RAs using inverse design utilized genetic algorithms in combination with integration of propagation equations that was highly time-consuming and cumbersome. Further designs used a neural network in combination with genetic algorithms to achieve optimization. However, this technique needed to be run every single time a new gain profile was required resulting in added delays. Although powerful, genetic algorithms rely on random optimisation, which has a non-negligible convergence time, especially if initial conditions are not well established [26]. In another approach, a neural network was utilized to first learn the forward model and then the gradient descent was employed in combination with the forward model to only optimize pump powers. However, if initial conditions are not near to the solutions, gradient descent optimisation tends to be time-consuming and vulnerable to local minima.

The typical ML methods used in the inverse design approach are either gradient-based approaches or evolutionary approaches. Basically, these inverse design algorithms are rule-based approaches consisting of iterative steps often depending on numerical simulations at intermediate steps. However, these are limited by their random search nature and become inefficient for complex designs. Deep Learning (DL), on the other hand, enables a computational model consisting of multiple layers of processing units to learn multiple levels of abstraction in given data [27]. DL-based various architectures such as multilayer perceptron, convolutional network, recurrent network, generative model, or hybrid model have been utilized in the forward as well as inverse design of various photonic and plasmonic devices. The design of the target photonic device can be directly mapped to its optical properties, and vice versa, using well-trained DL models. Deep learning can also work in conjunction with traditional optimisation techniques to boost algorithm performance. The distinctive advantages of DL lie in its data-driven technique that enables the model to discover useful information automatically from a massive amount of data, which is in sharp contrast to traditional physics or rule-based approaches. The photonics research community has gained significant benefits from recent advances in DL such as Convolutional Neural Networks (CNNs), Recurrent Neural Networks (RNNs), Generative Adversarial Networks (GANs), and Variational Autoencoders (VAEs).

Some of the major model architectures utilized in the inverse design of photonic structures are discussed next. The basic Multilayer Perceptron (MLP) models and their variants have been used to design topological photonic structures and integrated photonic devices such as silicon-on-insulator-based 1×2 power splitters [28]. Although the MLP model has a simple approach it poses some disadvantages when the intrinsic structure, target response, and design space are either multimodal or hard to parameterize. To overcome this, several application-specific advanced models have been proposed to address photonic design problems. CNN-based techniques perform feature extraction by applying convolutional operations at each layer. Since a CNN's convolution operation is translationally symmetric, periodic photonic structures like photonic crystals or metamaterials can be accurately modelled. RNNs are ideal for modelling optical signals or spectra in the time domain with a particular line shape originating from different resonance modes for photonic design. The RNN-based framework has been utilized to analyse and fully capture strong variations, in particular the divergences associated with the band edge behaviour of a photonic reservoir. RNN-based framework opens the door to developing robust and efficient solutions to designing, controlling and quantifying the relative complexity of various quantum systems [29].

Autoencoders are a type of neural network architecture that can learn an efficient representation of input data by encoding it into a lower-dimensional latent space and then decoding it back to the original input space. The success of using an autoencoder for inverse photonic design depends on the complexity of the problem and the availability of a suitable dataset. In some cases, it might be required to combine the autoencoder approach with other optimization or search algorithms to find the best designs within a larger design space. A Variational Autoencoder (VAE) is a type of autoencoder that incorporates ideas from variational inference to learn a probabilistic latent space representation. Unlike traditional autoencoders, which map input data to a fixed, deterministic latent space, VAEs map the data to a probability distribution in the latent space. This probabilistic representation allows for more flexibility and robustness in generating new data samples and makes the VAE suitable for tasks like data generation and manifold learning. Nanopatterned power splitters with high transmission ratios and broadband spectral response for photonic integrated circuits have been designed using a novel Conditional Variational Eutoencoder (CVAE) model [30]. A Generative Adversarial Network (GAN) consists of two neural networks: a generator and a discriminator. GANs are designed to generate new data that is similar to a training dataset. Using GANs for inverse design in photonics allows for the generation of novel photonic structures that possess desired optical properties. The adversarial training process helps the generator learn the underlying structure–property relationship, enabling the generation of structures with specific functionalities. GANs provide a powerful framework to explore the design space and generate diverse and optimized photonic structures.

ML and DL are powerful tools that can be utilized to find complex relations between structures and their optical characteristics. Currently, most of the research works are related to analysis of optical properties related to transmission, reflection, scattering, and so on under linear and circular polarization. It is required to enhance the capability of ML and DL algorithms to work on more degrees of freedom, such as the phase, nonlinearity, angular momentum, topology, and near-field distributions. Some more modifications or preprocessing steps may be included to model the input-output relations for specific applications. Transfer learning and reinforcement learning techniques may be explored further to expand the photonic design capability to obtain complete control of light. Since huge amounts of data are required to train deep neural networks, data collection is a huge burden when it comes to photonic design. It is not practical to facilitate collection of a massive dataset as most of the data can be obtained only from simulations of physical experiments. So far, photonic optimization has been used in restricted design spaces that are largely constrained to the geometrical aspects of the structure. This in turn limits the feedback from other critical layers such as effect of electromagnetic properties with time or fabrication constraints. This creates a fundamental challenge preventing globally optimized solutions to photonic platforms.

7.3.1.2 Photonics for computing

Typically, the modelling, design, and analysis of photonic devices require hard computing methods to solve the complex Maxwell's equations, which tend to be computationally exhaustive, expensive, and laborious. The complexity of these computing methods depends directly on the computational domain size and precision requirements, which imposes several limitations in fully exploring the parameter space. Soft computing methods based on ML algorithms can provide a suitable solution to this problem, thus enabling enhanced design and implementation of photonic solutions to the electromagnetic problems. Use of ML-based algorithms can provide several advantages such as higher execution speeds and inclusion of uncertainties in fabrications, material parameters, and manufacturing thereby reducing fabrication tape-out and increasing the yield of photonics manufacturing. The process of optical solving wherein the fundamental properties of optical waveguides are analysed plays a crucial rule in photonic-integrated circuit design. Quick and precise mode solving is necessary for the efficient design of complex photonic structures such as resonators, modulators, arrayed waveguide gratings, and so on. Typically, such numerical problems are solved using techniques such as Finite Element Methods (FEM) or Finite Difference Time Domain (FDTD) methods. Even though these techniques are well established, they consume a larger number of computational resources when performing larger computations along with lower execution speeds. ML-based soft computing optical solvers are much more beneficial than typical physics-based solvers, especially for parameter sweeps and optimizations.

Recently, an RNN was used to predict the field patterns of optical waveguides by taking geometrical values of the waveguide as input and field values as outputs as shown in figure 7.4 [31]. DL models were also used to solve effective refractive indices in silicon nitride channel waveguides [32]. Recently photonic accelerators have come to replace their electronic counterparts due to advantages such as increased speed, energy-efficient operations, and parallelism [33]. Photonic computing involves using photons, or light particles, instead of traditional electronic signals for processing information as shown in figure 7.4. Photonic accelerators aim to enhance the speed and efficiency of computations by leveraging the unique properties of light.

High-performance computational methods have become indispensable in various scientific fields, including photonics, enabling the description, design, interpretation, and prediction of optical system behaviour. The widespread accessibility of high-performance photonic components stands as evidence of computing's significant impact on advancing the field of photonics. Additionally, photonic architectures present intriguing opportunities for scaling computations beyond current hardware capabilities, establishing a remarkably reciprocal relationship between photonics and computing.

Despite decades of research, the interest in photonics for computing and computing for photonics is now surging dramatically, driven by the increasing demand for high-speed computing capabilities. As traditional digital computers reach a performance plateau, novel concepts like Neural Networks (NNs) and combinatorial optimization are pioneering new frontiers in information processing, with significant commercial relevance already being explored. These emerging paradigms deviate substantially from conventional computing methods, driving a rapid quest for more suitable types of computing hardware, where photonics holds

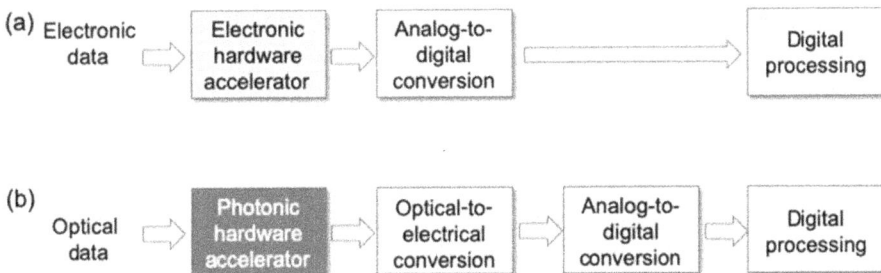

Figure 7.4. General schematic of electronic and photonic accelerators [33]. CC BY 4.0.

exceptional promise. Simultaneously, readily available high-performance computers can now intricately model and design increasingly complex photonic devices and systems with remarkable precision and accuracy. These advancements have given rise to a unique scenario: photonics emerges as a promising technology for the next generation of computing hardware, while recent progress in digital computers enables the design, modeling, and development of a new class of photonic devices and systems with unprecedented complexities.

7.3.1.3 All-optical machine learning

All-optical ML refers to the concept of performing ML tasks using purely optical components and techniques, without the need for any electronic processing. It involves using light-based systems and principles to perform various stages of the ML process, including data encoding, processing, training, and inference.

The traditional approach to ML involves using electronic components, such as CPUs and GPUs, to process data and perform mathematical operations required for training and inference. However, researchers have been exploring the potential of all-optical solutions to overcome certain limitations associated with electronic systems, such as speed, power consumption, and scalability [34].

In all-optical ML, data is typically encoded into optical signals, which are then manipulated using optical elements like waveguides, beam splitters, and photo-detectors. Light beams interact with each other, and the interference patterns or intensity changes carry out the required mathematical operations for ML tasks.

One of the key advantages of all-optical ML is the potential for extremely high processing speeds, as photons can travel at the speed of light. Additionally, all-optical systems can potentially be more energy-efficient for certain types of tasks, and they have the potential to handle massive parallelism due to the inherent nature of light. However, building practical all-optical ML systems comes with various challenges, such as signal loss, noise, and the lack of mature optical components for certain tasks.

In a recent work, researchers presented an all-optical DL framework where the NN is constructed using multiple layers of diffractive surfaces. These surfaces collaborate to perform optical operations, enabling the network to learn and execute arbitrary functions statistically. The physical network handles the inference and prediction processes entirely with optical mechanisms, while the learning phase responsible for designing the network is carried out using a computer [35]. In another approach, a silicon-on-insulator platform was utilized to propose an on-chip Diffractive Optical Neural Network (DONN) capable of performing ML tasks with remarkable integration and low power consumption features. Compared to conventional ANNs, on-chip DONNs were found to execute complex functions at higher speeds, lower latency, and with reduced power consumption levels [36].

Some of the limitations regarding all-optical ML are:

Signal loss and noise: Implementing all-optical components can introduce signal loss and noise, which can affect the accuracy and reliability of the results.

Increased complexity: Designing and building practical all-optical ML systems can be complex and require specialized expertise in optics and photonics.

Limited components: The range of available optical components for specific ML tasks may be limited, making it challenging to find suitable solutions for certain applications.

Scalability: Scaling up all-optical systems to handle large-scale ML tasks is a significant challenge, especially in comparison to the mature electronic counterparts.

Training: While the inference part can be done all-optically, the training process often relies on electronic computation, which introduces a hybrid approach and requires synchronization.

Data representation: Converting electronic data to optical signals and vice versa can be challenging and may lead to inefficiencies.

Advancements in all-optical ML could contribute to the development of optical computing, potentially revolutionizing various computational tasks beyond ML. All-optical systems can inherently handle parallelism, making them well-suited for large-scale parallel processing tasks, such as big data analysis and AI training. In addition, optical systems have the potential to be more energy-efficient than electronic counterparts, offering a greener and more sustainable approach to ML. Overall, the future of all-optical ML holds great promise, but it requires continued research and development to address existing challenges and fully realize its potential in various domains.

7.3.1.4 Deep learning-enabled photonics

Traditional inverse design approaches such as topology optimization, genetic algorithms, steep descent, and particle swarming optimization demand extensive computational resources and prolonged durations to discover the ideal local structure. In contrast, DL, embraced as a subset of ML, has garnered global interest due to its ability to swiftly and effectively handle substantial datasets for processing and analysis. DL-enabled inverse design of nanophotonic structures provides remarkable flexibility, which is difficult to achieve with conventional optimization techniques. DL-based approaches can consider several inter-linked parameters such as material types and geometrical parameters for effective inverse design as compared to conventional methods that can handle only one or two parameters at a time [37] (figure 7.5).

However, DL-enabled photonic design has some challenges that need to be overcome including:

Data requirements: DL models require substantial amounts of labelled data for training, and obtaining such data in photonics can sometimes be challenging due to limited datasets.

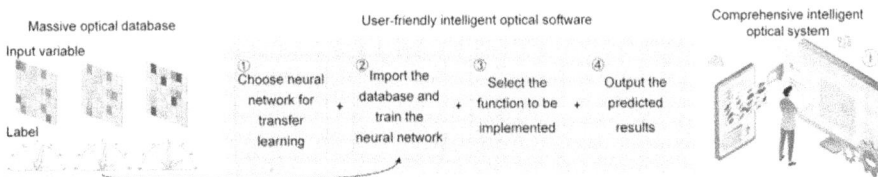

Figure 7.5. A user-friendly software system for photonic design [37]. CC BY 4.0.

Interpretability: Understanding how DL models arrive at certain conclusions or decisions in the context of photonics applications can be complex. Interpretability is crucial, especially in critical applications like medical imaging.

Computational resources: Training DL models in photonics might demand significant computational power and resources, making it inaccessible for some researchers or institutions.

Generalization: Ensuring that DL models trained on specific photonics datasets can generalize well to new and diverse data is a persistent challenge.

Physical constraints: Integrating DL models with actual photonics hardware can face constraints due to the physics of light propagation, noise, and real-world limitations, requiring careful consideration during model design and implementation.

7.3.2 Major challenges

The integration of AI in photonics brings forth exciting possibilities, but it also comes with significant challenges researchers and engineers must address to fully realize its potential. Some of the major challenges of AI in photonics are:

Data quality and quantity: AI algorithms, particularly DL models, often require large amounts of high-quality training data. Acquiring and curating such data in photonics applications, especially in specialized fields, can be challenging and time-consuming.

Computational resources: AI algorithms, especially DL, are computationally intensive, requiring significant processing power and memory. Implementing AI solutions in photonics systems while maintaining real-time performance can be a technical challenge.

Transferability and generalization: AI models trained for specific photonics tasks may struggle to generalize well to new and unseen situations or environments. Ensuring the transferability of AI models across different conditions is essential for real-world applications.

Lack of benchmark datasets: Photonics datasets for AI training may be scarce or not publicly available, limiting the ability to compare and benchmark different AI approaches effectively.

Integration with existing systems: Integrating AI capabilities into existing photonics systems or workflows may pose challenges in terms of compatibility, adaptability, and optimization.

Robustness to noise and variability: Photonics systems are often subject to noise and environmental variability. Ensuring the robustness of AI algorithms to handle such conditions is essential for reliable performance.

The synergy between AI and photonics has opened up new avenues for research, development, and application of light-based technologies. By harnessing the power of AI, photonics researchers and engineers can accelerate innovation, optimize performance, and unlock the full potential of photonics in areas ranging from telecommunications and healthcare to energy and beyond (table 7.1).

Table 7.1. Various ML/DL algorithms and their applications in optics and photonics.

S. No.	ML/DL algorithm	Application in optics/photonics	Remarks
1	Multilayer perceptron	3D vectorial holography, nanoparticle simulation	• Difficult-to-handle high-dimensional data • High reliability
2	Convolutional neural networks (CNNs)	Optical image processing, spectra analysis, data storage, optical communications, optical sensing	• High-dimensional data processing
3	Recurrent neural networks (RNNs)	Optical character recognition, transient electromagnetic modelling	• Sequential information processing
4	Generative model	Inverse design of nanophotonic devices	• Unsuitable for discrete data
5	Support vector machine	Modulation format identification, Optical sensing	• Memory efficient for large datasets • Computationally intensive
6	Random forests and decision trees	Optical sensing	• Vulnerable to small variations • Slower training time
7	K-nearest neighbours (KNNs)	VLC for indoor positioning, spectral classification, optical pattern matching	• Memory intensive • Computationally expensive • Inefficient with large datasets
8	Artificial neural networks (ANNs)	Intelligent VLC systems, optical performance monitoring	• Lack of robustness to adversarial inputs

7.4 Conclusion and future scope of AI/ML in optics and photonics

The future scope of AI/ML in Optics and Photonics is vast and holds tremendous potential for transformative advancements in various fields. The integration of AI and ML with optics and photonics technologies opens up new avenues for innovation, efficiency, and automation. Some key areas where AI/ML is expected to have a significant impact include optical design and simulation, adaptive optics, bio-photonics and medical applications, material discovery, quantum photonics, and so on. As AI/ML techniques continue to advance, their integration with optics and photonics will lead to groundbreaking discoveries, cost-effective solutions, and more efficient processes across various industries. The synergy between AI/ML and optics will foster innovations that have the potential to revolutionize technology, science, and society as a whole. To unlock this potential, interdisciplinary collaboration between AI/ML researchers and experts in optics and photonics will be vital in exploring and realizing the numerous opportunities that lie ahead.

References

[1] Sivaranjani R *et al* 2020 Photonic crystal based all-optical half adder: a brief analysis *Laser Phys.* **30** 116205

[2] Sridarshini T *et al* 2022 Current and future horizon of optics and photonics in environmental sustainability *Sustain. Comput.: Inform. Syst.* **36** 100815

[3] Geerthana S *et al* 2022 2D-PhC based all optical AND, OR and EX-OR logic gates with high contrast ratio operating at C band *Laser Phys.* **32** 106201

[4] Dhanabalan S S, Thirumurugan A, Raju R, Kamaraj S-K and Thirumaran S 2023 *Photonic Crystal and Its Applications for Next Generation Systems* Springer Tracts in Electrical and Electronics Engineering (Singapore: Springer)

[5] Zeng X and Long L 2022 Introduction to artificial intelligence *Beginning Deep Learning with TensorFlow* (Berkeley, CA: Apress) pp 1–45

[6] Tizikara D K, Serugunda J and Katumba A 2022 Machine learning-aided optical performance monitoring techniques: a review *Front. Commun. Netw.* **2** 756513

[7] Musumeci F *et al* 2019 An overview on application of machine learning techniques in optical networks *IEEE Commun. Surv. Tutor.* **21** 1383–408

[8] Wang D and Zhang M 2021 Artificial intelligence in optical communications: from machine learning to deep learning *Front. Commun. Netw.* **2** 656786

[9] Li S and Pachnicke S 2020 Photonic reservoir computing in optical transmission systems *2020 IEEE Photonics Society Summer Topicals Meeting Series (SUM) (Cabo San Lucas, Mexico)* pp 1–2

[10] Ulkar M G, Baykas T and Pusane A E 2020 VLCnet: deep learning based end-to-end visible light communication system *J. Lightwave Technol.* **38** 5937–48

[11] Haigh P A *et al* 2014 Visible light communications: 170 Mb/s using an artificial neural network equalizer in a low bandwidth white light configuration *J. Lightwave Technol.* **32** 1807–13

[12] Naser S *et al* 2022 Toward federated-learning-enabled visible light communication in 6G systems *IEEE Wirel. Commun.* **29** 48–56

[13] Ma S *et al* 2019 Signal demodulation with machine learning methods for physical layer visible light communications: prototype platform, open dataset, and algorithms *IEEE Access* **7** 30588–98

[14] Shi J *et al* 2022 AI-enabled intelligent visible light communications: challenges, progress, and future *Photonics* **9** 529

[15] Xiao L *et al* 2019 Deep reinforcement learning-enabled secure visible light communication against eavesdropping *IEEE Trans. Commun.* **67** 6994–7005

[16] Xie Y, Wang Y, Kandeepan S and Wang K 2022 Machine learning applications for short reach optical communication *Photonics* **9** 30

[17] Karar A S, Falou A R E, Barakat J M H, Gürkan Z N and Zhong K 2023 Recent advances in coherent optical communications for short-reach: phase retrieval methods *Photonics* **10** 308

[18] Khan F N *et al* 2016 Modulation format identification in coherent receivers using deep machine learning *IEEE Photonics Technol. Lett.* **28** 1886–9

[19] Mata J *et al* 2018 Artificial intelligence (AI) methods in optical networks: a comprehensive survey *Opt. Switching Networking* **28** 43–57

[20] Zhu C, Alsalman O and Naku W 2023 Machine learning for a Vernier-effect-based optical fiber sensor *Opt. Lett.* **48** 2488–91

[21] Zhuang Y *et al* 2021 Fiber optic sensor embedded smart helmet for real-time impact sensing and analysis through machine learning *J. Neurosci. Methods* **351** 109073

[22] Kühnlein M *et al* 2014 Improving the accuracy of rainfall rates from optical satellite sensors with machine learning—a random forests-based approach applied to MSG SEVIRI *Remote Sens. Environ.* **141** 129–43

[23] Wang G, Sun Y and Wang J 2017 Automatic image-based plant disease severity estimation using deep learning *Comput. Intell. Neurosci.* **2017** 1–8

[24] Khan Y *et al* 2018 Mathematical modeling of photonic crystal based optical filters using machine learning in *2018 Int. Conf. on Computing, Electronic and Electrical Engineering (ICE Cube) (Quetta, Pakistan)* pp 1–5

[25] Zibar D *et al* 2020 Inverse system design using machine learning: the Raman amplifier case *J. Lightwave Technol.* **38** 736–53

[26] Chen J and Jiang H 2018 Optimal design of gain-flattened Raman fiber amplifiers using a hybrid approach combining randomized neural networks and differential evolution algorithm *IEEE Photonics J.* **10** 1–15

[27] Ma W *et al* 2021 Deep learning for the design of photonic structures *Nat. Photonics* **15** 77–90

[28] Sridarshini T *et al* 2023 Ultra-compact all-optical logical circuits for photonic integrated circuits *Laser Phys.* **33** 076201

[29] Burgess A and Florescu M 2021 Modelling non-Markovian dynamics in photonic crystals with recurrent neural networks *Opt. Mater. Express* **11** 2037–48

[30] Tang Y *et al* 2020 Generative deep learning model for inverse design of integrated nanophotonic devices *Laser Photonics Rev.* **14** 2000287

[31] Alagappan G and Png C E 2021 Prediction of electromagnetic field patterns of optical waveguide using neural network *Neural Comput. Appl.* **33** 2195–206

[32] Alagappan G and Png C E 2019 Deep learning models for effective refractive indices in silicon nitride waveguides *J. Opt.* **21** 035801

[33] Kitayama K-I *et al* 2019 Novel frontier of photonics for data processing—photonic accelerator *APL Photon.* **4** 090901

[34] Lin X *et al* 2018 All-optical machine learning using diffractive deep neural networks *Science* **361** 1004–8

[35] Fu T *et al* 2023 Photonic machine learning with on-chip diffractive optics *Nat. Commun.* **14** 70

[36] Huang L, Xu L and Miroshnichenko A E 2020 Deep learning enabled nanophotonics *Advances and Applications in Deep Learning* (London: IntechOpen)

[37] Duan B *et al* 2022 Deep learning for photonic design and analysis: principles and applications *Front. Mater.* **8** 791296

IOP Publishing

Advances in All-optical Communication

Shanmuga Sundar Dhanabalan, Arun Thirumurugan and Sridarshini Thirumaran

Chapter 8

Blood components detection in octagonal-cored photonic crystal fiber biosensor for healthcare applications

Abdul Mu'iz Maidi, Nianyu Zou and Feroza Begum

This study aims to develop a blood component sensor based on photonic crystal fibre (PCF) with the ability to detect red blood cells, haemoglobin, white blood cells, plasma, and water. The sensor design includes an octagonal core hole surrounded by cladding air holes of octagonal and circular geometries, enhancing sensing performance. Rigorous numerical simulations using the finite element method (FEM) in COMSOL Multiphysics evaluate the optical and sensing properties. Parameters such as effective refractive index, power fraction, relative sensitivity, confinement loss, chromatic dispersion, effective area, nonlinear coefficient, and V-parameter were investigated. Remarkable outcomes are achieved at the optimum wavelength of 2.0 μm, showcasing impressive relative sensitivities: 95.27% for red blood cells, 94.02% for haemoglobin, 92.49% for white blood cells, 91.59% for plasma, and 89.47% for water. These findings affirm the feasibility and effectiveness of the sensor design, providing strong support for its future fabrication and deployment in blood sensing applications. These advancements hold significant potential for furthering progress in the medical industry.

8.1 Introduction

The field of optical fibres has traditionally been dominated by its primary applications in telecommunication. However, with advancements in fabrication techniques, optical fibres have demonstrated their versatility in various sensing applications. These advancements include modifications to the shape and size of the core and cladding holes, pitch distance, and other parameters, enabling PCFs to achieve values that conventional optical fibres cannot reach. In contrast to conventional fibres, PCFs exhibit exceptional resilience in harsh environmental conditions and offer a wide range of potential applications, encompassing optical

doi:10.1088/978-0-7503-5623-7ch8
8-1

communication [1–3], imaging [4], filters [5], optical computing [6–8], as well as sensing applications [9–11]. The development of photonic crystal-based all-optical logic gats and half adders has paved the way for potential applications in all-optical computing technology [8]. Furthermore, the role of optics and photonics in environmental sustainability has been highlighted as having a significant impact on global environmental applications and the potential for green future development [12].

Among the diverse array of sensing applications, considerable attention has been drawn to hollow core PCFs due to their unique capability to enable strong light-analyte interactions. Hollow core PCF provide an excellent platform for efficient interaction between light and analytes, enabling a wide range of sensing methodologies such as temperature [13], curvature [14], torsion [15], vibration [16], pressure [17], refractive index [18], electric field [19], magnetic field [20], gas [21], pH [22], chemicals [23], and even the detection of blood components [24–28]. Blood components play pivotal roles in the realm of medical diagnosis and treatment, underscoring the critical need for accurate and reliable detection methodologies. In this context, the development of PCF-based sensors presents a promising approach to addressing the challenges associated with blood component detection, leveraging the distinctive properties of PCFs. The detection and analysis of various components in human blood, such as red blood cells (RBCs), white blood cells (WBCs), haemoglobin (HB), plasma, and water, are crucial for diagnosing conditions like thalassemia, haemophilia, anaemia, myeloma, and more [29]. Current methods for their detection include ion chromatography [30] and the use of gold nanoparticles [31], as well as piezoelectric quartz crystals [32]. However, the new method employing biosensors based on PCFs has been introduced for the identification and analysis of these particular constituents in blood.

PCF-based biosensors have been developed to enable the detection of various blood elements. Researchers [24] presented an innovative blood component sensor that operated within the optical wavelength range. The PCF design featured a core with a hollow ring and four layers of circular air holes with different diameters (d_1 and d_2). Specifically, the first and fourth layer had the largest diameter ($d_1 = 1.75$ µm), while second and third rings employed the smallest diameter ($d_2 = 0.8$ µm). The sensor demonstrated relative sensitivities of 56.05%, 53.72%. 66.46%. 54.04%, and 55.09% for RBCs, WBCs, HB, plasma, and water, respectively, at the set optimum wavelength. Following their previous work, the same researchers [25] made further enhancements to the PCF design. Notably, the dimensions of the core and cladding holes has been increased, where the core sensing ring is 0.6 µm, diameter $d_1 = 3.6$ µm, and diameter $d_2 = 1.8$ µm. However, these modifications did not yield any significant changes in the relative sensitivity results. The determined relative sensitivities for RBCs, HB, WBCs, plasma, and water remained at 55.83%. 58.05%, 62.72%, 65.05%, and 66.47%, respectively.

In a distinct research conducted by Islam *et al* [26], a PCF-based sensor designed for the detection of plasma was introduced. The sensor boasted a sophisticated core design, featuring a unique hexagonal benzene-shaped configuration consisting of nineteen core holes. The cladding region incorporated five layers of circular air holes in a circular arrangement. The sensor was rigorously evaluated across a range of parameters, including relative sensitivity, confinement loss, effective area, and numerical aperture. Specifically focusing on the relative sensitivity result, at the optimised wavelength of 1.33 μm, the sensor demonstrated a sensitivity of 77.84% in detecting plasma. Moreover, Ahmed *et al* [27] introduced an intricately designed PCF core, consisting of a circular arrangement comprising seven non-overlapping core holes surrounded by an additional six holes, resulting in a total of 49 core holes. The cladding air holes were similarly arranged in this specific pattern. This innovated configuration, while complex in nature, yielded remarkable improvements in relative sensitivity: 80.93% for RBCs, 80.56% for WBCs, 80.13% for HB, 79.91% for plasma, and 79.39% for water. The primary focus of this design was to emphasise intricate structural features with enhancing the overall performance of the sensor. In addition, a group of researchers [28] introduced a rectangular core PCF sensor for the precise detection of blood components. This sensor design employed a configuration consisting of an analyte-infiltrated rectangular core, characterised for specific width W and length L dimensions. Additionally, the cladding region incorporated a total of six rectangular holes, each with different sizes denoted as $R1$, $R2$, $R3$, $R4$, and $R5$. These deliberate design choices led to significant improvements in relative sensitivity outcomes. By optimising key parameters, the sensor achieved remarkable relative sensitivities of 93.50%, 92.15%. 92.41%, 90.48%, and 89.14% for RBCs, WBCs, HB, plasma, and water, respectively. These advancements in PCF-based sensor design demonstrate their potential for efficient and accurate detection of blood components.

Accurate detection of blood components is crucial in preventive health measures and medical diagnosis. However, these current PCF sensor designs [24–28] have limitations in feasible fabrication and sensitivity, necessitating the development and improvements of PCF-based sensors. This research explores a novel blood component sensor based on PCF with exceptional relative sensitivity. The sensor exhibits a streamlined design characterised by one analyte-infiltrated octagonal core and two layers of cladding air holes. It demonstrates remarkable sensing capabilities in the optical wavelength range of 1.6–4.0 μm, as confirmed through numerical simulations. The results indicate relative sensitivities exceeding 90% for all blood components, while confinement losses remain impressively low, below 10^{-7} dB m^{-1} at the optimal wavelength. Furthermore, the sensor propagates only one mode, as confirmed through the evaluation of V-parameter, which then results in favourable effects in chromatic dispersion, effective area, and nonlinear coefficient.

This novel PCF-based blood component sensor holds promise for advancing blood component detection in medical applications. The enhanced sensitivity and minimised confinement losses provide a reliable foundation for accurate analysis. The sensor's single-mode functionality further enhances its potential for precise and efficient detection.

8.2 Design

Figure 8.1 illustrates the cross-sectional design of the proposed blood component sensor, featuring a total diameter of 35.7 μm. The sensor architecture encompasses a

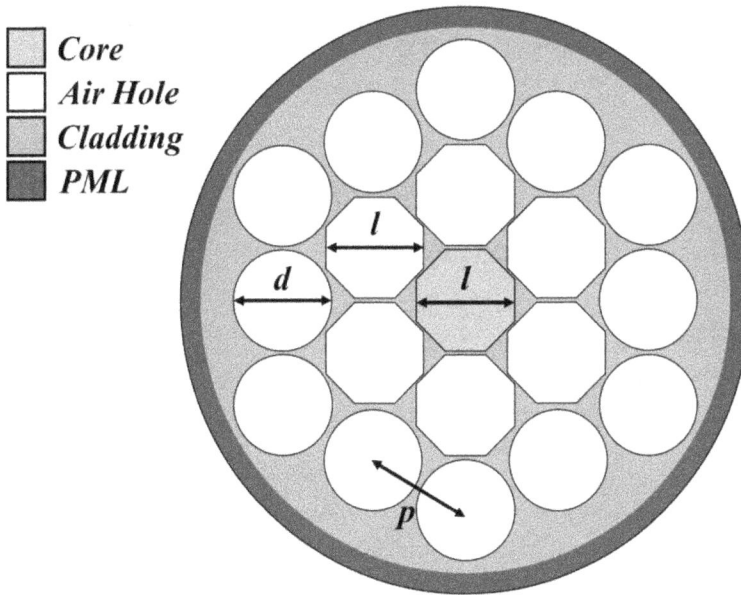

Figure 8.1. Cross-sectional design of the proposed PCF sensor for blood component detection.

core, cladding, and a perfectly matched layer (PML) to ensure optimal functionality. The cladding consists of two layers of air holes, with the first ring comprised of octagonal hollow holes encircling a solitary core of an octagonal shape, while the second ring incorporates circular holes. Utilising an octagonal core enhances specific sensing capabilities by providing increased surface area for light-analyte interaction due to its asymmetrical design, ultimately improving relative sensitivity in detecting various blood components. The core, with length l of 6.6 µm, is deliberately injected with the sensed analyte to allow light-analyte interaction to enhance its sensing capabilities. The octagonal-shape cladding holes align precisely with the core length (l of 6.6 µm), while the circular air holes in the second ring possess a diameter d of 6.65 µm.

Additionally, the pitch distance p between neighbouring cladding air holes is precisely set at 8.0 µm. Thus, all these parameters play a critical role in determining the overall structure and performance of the PCF sensor. Moreover, the air-filling fraction (AFF), defined as the ratio of the cladding air hole diameter to the pitch distance, assumes significant importance in the sensor design. The AFF has been deduced to 0.95, thereby influencing the presence of air within the cladding region and exerting an impact on the optical and sensing properties. The fabrication techniques for PCFs have evolved significantly, embracing methods such as stack-and-draw [33], extrusion [34], and innovative approaches like 3D printing [35–37]. These advancements in fabrication methods have streamlined the production of unconventional PCF geometries, ensuring their relevance and applicability across various fields.

To accommodate for fibre loss characteristics, the PML is established at 10% of the overall fibre diameter. The blood components comprise various constituents, including water, plasma, WBCs, HB, and RBCs, with corresponding refractive indices of 1.33, 1.35, 1.36, 1.38, and 1.40, respectively [38]. The chosen substrate is fused silica, and its refractive index is calculated using the Sellmeier equation [39, 40]:

$$n^2 = 1 + \frac{0.69617\lambda^2}{\lambda^2 - 0.0684^2} + \frac{0.40794\lambda^2}{\lambda^2 - 0.11624^2} + \frac{0.89748\lambda^2}{\lambda^2 - 9.89616^2} \tag{8.1}$$

where λ is the operating wavelength.

8.3 Methodology

In this investigation, the finite element method (FEM) is employed, incorporating a PML boundary imposed on the fibre. This numerical approach is widely recognised for its efficiency in engineering design simulations. By employing PML boundary conditions, the propagation characteristics of leaky modes in PCFs can be effectively eliminated. The operation wavelength range considered in this study spans from 1.6 to 4.0 μm, encompassing the visible and infrared (IR) wavelength region. To evaluate the performance of the blood component sensor, the primary metric of measurement is the relative sensitivity. However, to gain a comprehensive under-standing of the sensor's overall capabilities, other essential optical properties are computed, including confinement loss, chromatic dispersion, effective area, non-linear coefficient, and V-parameter.

Power fraction P represents the ratio of the optical power transmitted through a specific region to the total are of the fibre. It is defined as [41–45]:

$$P = \frac{(\text{sample}) \int \text{Re}\,(E_x H_y - E_y H_x) dx dy}{(\text{total}) \int \text{Re}\,(E_x H_y - E_y H_x) dx dy} \times 100 \tag{8.2}$$

where E and H are the electric fields and magnetic fields, respectively, and the x and y subscripts represent the polarisation of these fields in the x and y directions.

Relative sensitivity S quantifies the effectiveness of the sensor in detecting variations in the refractive index of the analyte and its surrounding medium, and it can be expressed as [41–45]:

$$S = \frac{n_r}{\text{Re}(n_{\text{eff}})} \times P \tag{8.3}$$

where n_r is the refractive index of the test analyte, $\text{Re}(n_{\text{eff}})$ is the real part of the effective refractive index, and P is the power fraction.

Confinement loss L refers to the light signal lost because of its leakage from the core region into the cladding region. It can be quantified by [46–51]:

$$L = \frac{40\pi}{\ln\,(10)\lambda}\text{Im}(n_{\text{eff}}) \times 10^6 \tag{8.4}$$

where $\text{Im}(n_{\text{eff}})$ is the imaginary part of the effective refractive index and λ is the operating wavelength.

Chromatic dispersion D is the temporal spreading of light signal due to the phenomenon of different wavelengths propagating through a medium at different velocities, and it can be determined by [42, 52–54]:

$$D = -\frac{\lambda}{c}\frac{d^2}{d\lambda^2}\,\mathrm{Re}(n_{\mathrm{eff}}) \qquad (8.5)$$

where c is the speed of light and $\mathrm{Re}[n_{\mathrm{eff}}]$ is the real part of the effective refractive index.

Effective area A_{eff} is the parameter that quantitively describes the transverse coverage of the fundamental mode in an optical fibre, and it can be expressed as [21, 42, 47, 53, 55, 56]:

$$A_{\mathrm{eff}} = \frac{\left(\int_{-\infty}^{\infty}\int_{-\infty}^{\infty}|E|^2\,dxdy\right)^2}{\int_{-\infty}^{\infty}\int_{-\infty}^{\infty}|E|^4\,dxdy} \qquad (8.6)$$

Nonlinear coefficient γ quantifies the ability of the fibre to confine and propagate high intensity light, reflecting the extent of refractive index changes in response to the intensity of the propagating light and contributing to various nonlinear optical phenomena, and it is found by [53, 54, 57]:

$$\gamma = \left(\frac{2\pi}{\lambda}\right)\left(\frac{n_2}{A_{\mathrm{eff}}}\right) \qquad (8.7)$$

where n_2 is the nonlinear refractive index, and it is the change in the refractive index, which is dependent upon the intensity of light in the optical fibre.

V parameter V is a dimensionless quantity that characterises the guidance properties of the fibre and determines the number of modes supported by the waveguide by [58, 59]:

$$V = \frac{2\pi}{\lambda}R\sqrt{n_{co}^2 - n_{cl}^2} \qquad (8.8)$$

where R is the radius of the core and n_{co} and n_{cl} are the refractive index of core and cladding, respectively.

8.4 Results and discussion

In the evaluation of the proposed PCF as a sensor for blood components (RBCs, HB, WBCs, plasma, and water), numerous optical parameters have been taken into account. These parameters encompass effective refractive index, power fraction, relative sensitivity, confinement loss, chromatic dispersion, effective area, nonlinear coefficient, and V parameter. Figure 8.2 showcases the profile of mode distribution within the fibre core for the tested analytes at an operating wavelength of 2.0 μm.

The COMSOL software was utilised to evaluate the effective refractive index, and figure 8.3 displays the outcomes of the tested analytes. The effective refractive indices of the examined blood components exhibit a strong correlation with their

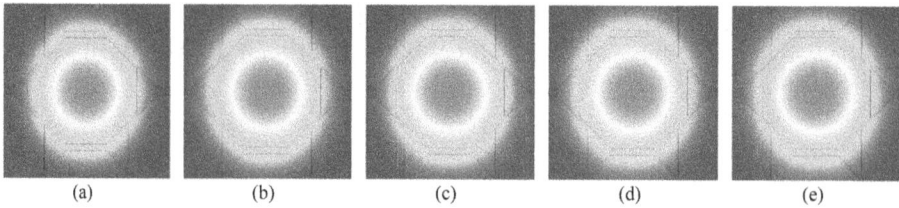

Figure 8.2. Mode distribution profile in the core at the operating wavelength of 2.0 μm for (a) RBCs, (b) HB (c), WBCs (d), plasma, and (e) water.

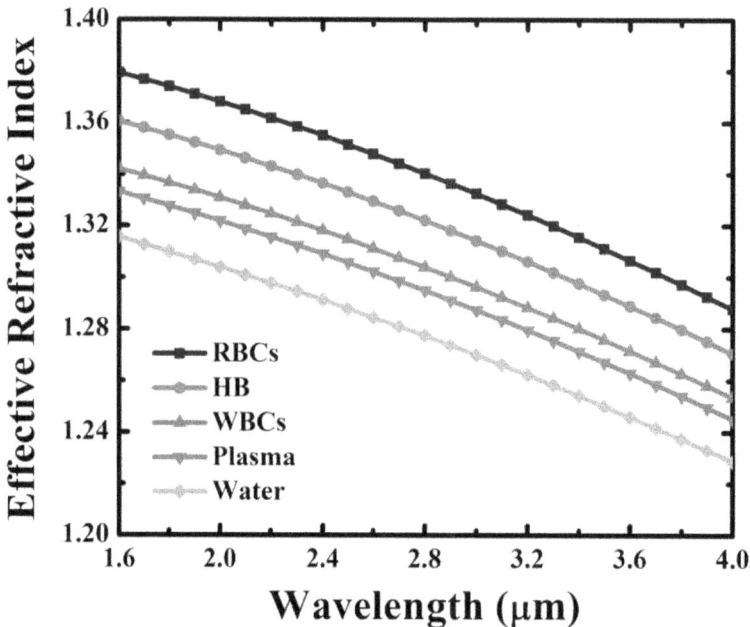

Figure 8.3. Variation of effective refractive index with operating wavelength for the test analytes: RBCs ($n = 1.40$), HB ($n = 1.38$), WBCs ($n = 1.36$), plasma ($n = 1.35$), and water ($n = 1.33$) in the proposed PCF sensor.

individual refractive indices. Among these components, the highest refractive index is observed for RBCs, followed by HB, WBCs, plasma, and water. From the depicted figure, it is evident that the effective refractive index of the analytes diminished with the increment of wavelength within the range of 1.6–4.0 μm.

Figure 8.4 exhibits the power fractions for the sensing analytes within the core of the proposed PCF. The power fractions of the blood constituents (RBCs, HB, WBCs, plasma, and water) exhibit a comparable pattern. Initially, the power fractions rise to a peak value that is specific to each analyte and subsequently decline as wavelength increases. These findings, with the effective refractive index, are subsequently employed to calculate the relative sensitivities.

Figure 8.5 showcases the results of relative sensitivities, which gauges the interaction between the optical light signal and the sensing analytes. The measured

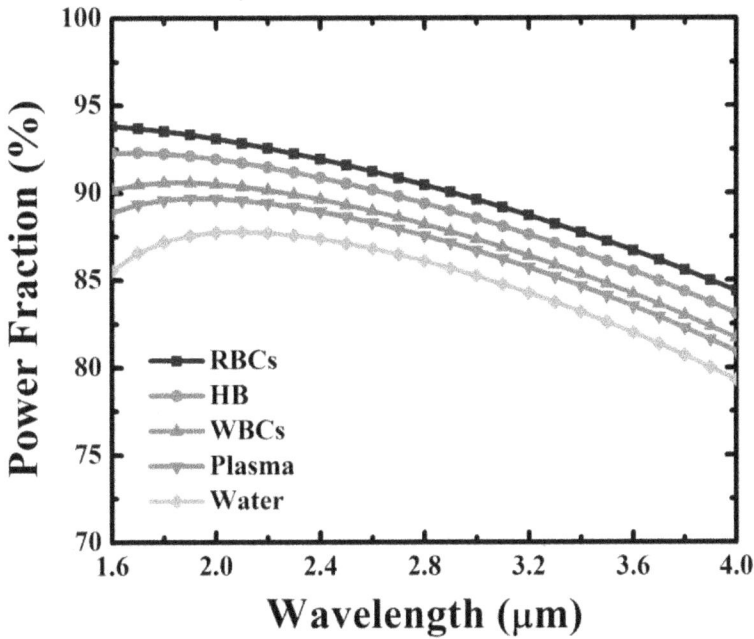

Figure 8.4. Variation of power fraction with operating wavelength for the test analytes: RBCs ($n = 1.40$), HB ($n = 1.38$), WBCs ($n = 1.36$), plasma ($n = 1.35$), and water ($n = 1.33$) in the proposed PCF sensor.

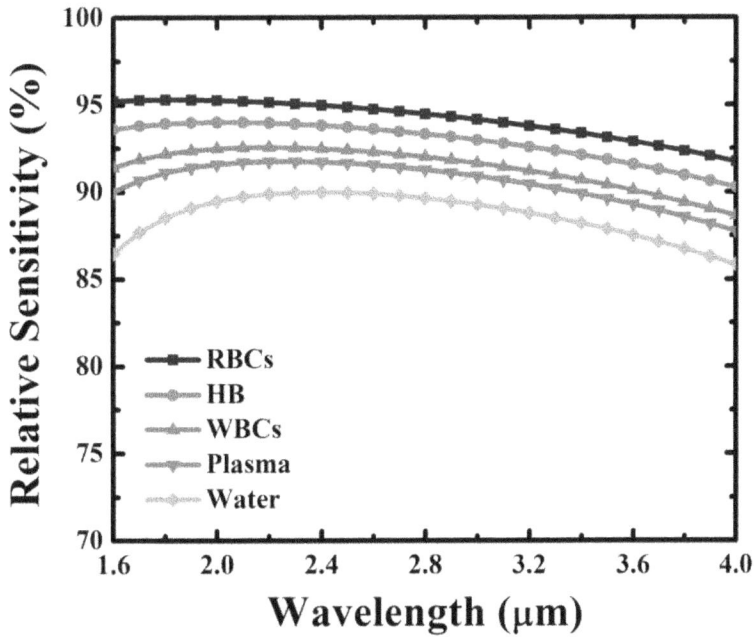

Figure 8.5. Variation of relative sensitivity with operating wavelength for the test analytes: RBCs ($n = 1.40$), HB ($n = 1.38$), WBCs ($n = 1.36$), plasma ($n = 1.35$), and water ($n = 1.33$) in the proposed PCF sensor.

values pertain to RBCs, HB, WBCs, plasma, and water. In accordance with equation (8.3), relative sensitivity aligns with the results derived from the effective refractive index and power fraction, thereby exhibiting a relative pattern on the graph. Within the lower range of wavelengths, the relative sensitivity of the blood analytes experiences an initial increase, followed by a subsequent decrease as the wavelength further increases. The notable relative sensitivity values obtained can be attributed to the large core hole dimensions. The enlarged core holes size promotes large amount of analyte penetration into the fibre, thus enhancing the interaction between light and analytes, and resulting in increased relative sensitivity. The optimal operating wavelength for this study is set at $\lambda = 2.0$ μm, as the maximum relative sensitivities are achieved at this specific wavelength. Consequently, at the optimum operating wavelength, the corresponding relative sensitivity values are recorded as 95.27% for RBCs, 94.02% for HB, 92.49% for WBCs, 91.59% for plasma, and 89.47% for water.

The phenomenon of confinement loss, which refers to the decrease in light intensity as it escapes from the fibre core, is influenced by the wavelength. This behaviour is demonstrated in figure 8.6, which presents the confinement loss results for the analysed constituents aligned with the operating wavelength. As the wavelength increases, the confinement losses also increase. Moreover, a noticeable correlation exists between the refractive index of the examined test analytes and the confinement loss. This can be explained by the difference in refractive index between the core and cladding, leading to enhanced confinement of light within the core of the

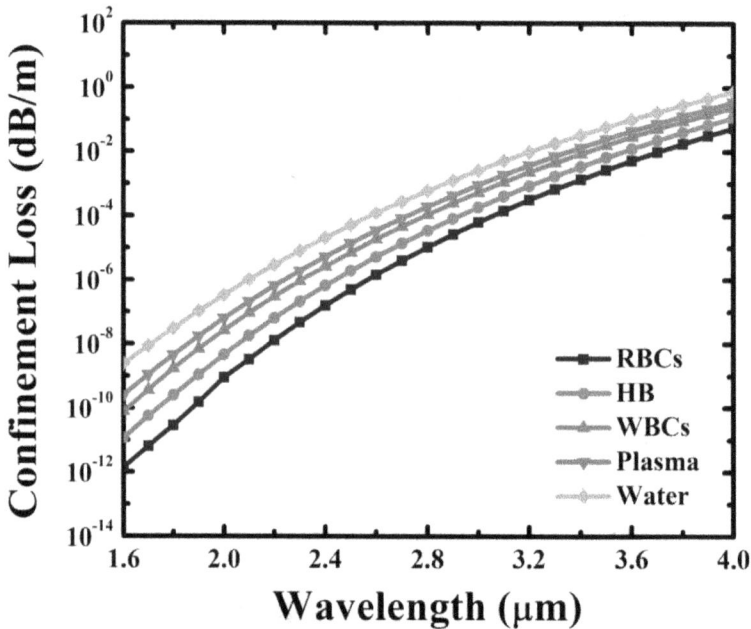

Figure 8.6. Variation of confinement loss with operating wavelength for the test analytes: RBCs ($n = 1.40$), HB ($n = 1.38$), WBCs ($n = 1.36$), plasma ($n = 1.35$), and water ($n = 1.33$) in the proposed PCF sensor.

fibre. At the optimal wavelength of 2.0 μm, the respective confinement loss values are 9.04×10^{-10} dB m^{-1} for RBCs, 4.57×10^{-9} dB m^{-1} for HB, 2.64×10^{-8} dB m^{-1} for WBCs, 6.23×10^{-8} dB m^{-1} for plasma, and 3.33×10^{-7} dB m^{-1} for water.

Figure 8.7 presents the chromatic dispersions with respect to the wavelength for the test analytes, illustrating a linear decrease in chromatic dispersion as the wavelength decreases. This behaviour arises due to the higher frequencies associated with shorter wavelengths, leading to increased interaction between the propagating light and the material properties of the test analytes. Importantly, the chromatic dispersion values exhibit variation among the different blood components, indicative of their unique refractive indices. A higher refractive index of the analyte corresponds to a higher chromatic dispersion value. This phenomenon arises due to the intensified interaction between light and the medium, resulting in a more pronounced dispersion effect. At the operating wavelength of 2.0 μm, the chromatic dispersions for the blood components are as follows: -0.0032 ps km^{-1} nm^{-1} for RBCs, -0.0031 ps km^{-1} nm^{-1} for HB, -0.0030 ps km^{-1} nm^{-1} for WBCs, -0.0029 ps km^{-1} nm^{-1} for plasma, and -0.0028 ps km^{-1} nm^{-1} for water. The encouraging findings related to confinement loss and chromatic dispersion in this research emphasise the potential use of these results in optical communication applications.

Figure 8.8 presents the variations in effective area for RBCs, HB, WBCs, plasma, and water as a function of operating wavelength, spanning the range from 1.6 to

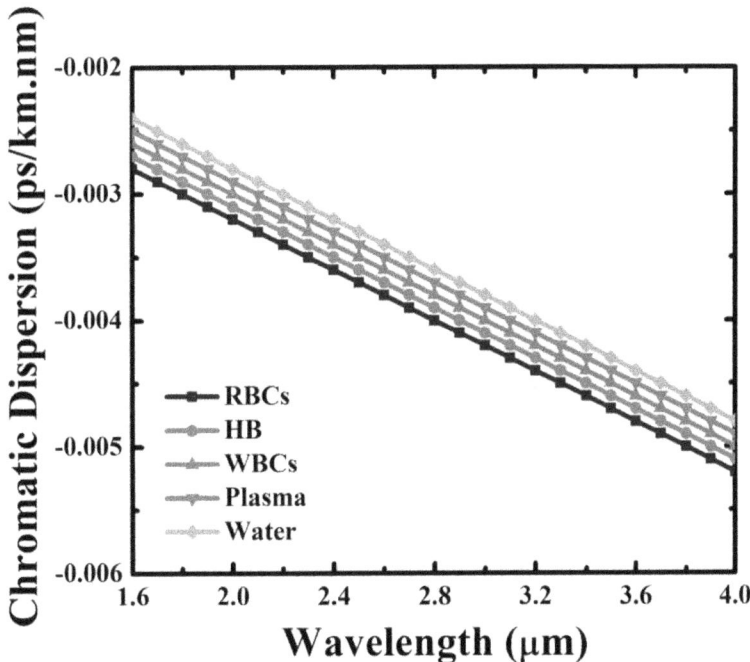

Figure 8.7. Variation of chromatic dispersion with operating wavelength for the test analytes: RBCs ($n = 1.40$), HB ($n = 1.38$), WBCs ($n = 1.36$), plasma ($n = 1.35$), and water ($n = 1.33$) in the proposed PCF sensor.

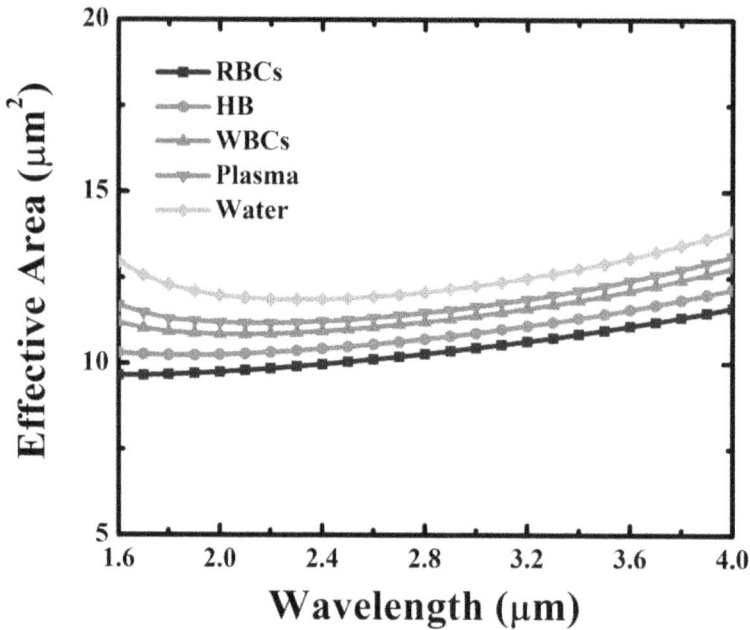

Figure 8.8. Variation of effective area with operating wavelength for the test analytes: RBCs ($n = 1.40$), HB ($n = 1.38$), WBCs ($n = 1.36$), plasma ($n = 1.35$), and water ($n = 1.33$) in the proposed PCF sensor.

4.0 μm. The obtained results unveil intriguing trends that can be attributed to the intricate interplay between the refractive indices of the analytes and the operating wavelength. At the earlier wavelengths, a slight decrease in the effective area is observed across all analytes, with a little more reduction observed for water, plasma, and WBCs, as indicated in figure 8.8. This reduction in effective area can be comprehended by examining the interaction between the incident light and the PCF structure. At shorter wavelengths, the higher refractive index of the analytes intensifies their interaction with the PCF, leading to enhanced light confinement within the core and consequently resulting in a decrease in the effective area. Conversely, as the operating wavelength increases, the influence of the refractive index difference between the core and cladding diminishes. This facilitates improved light confinement, but confinement loss increases as wavelength increases and thus manifests as an increase in the effective area. The corresponding values for effective areas are denoted as 9.75, 10.3, 10.8, 11.2, and 12.0 μm² for RBCs, HB, WBCs, plasma, and water, respectively.

Figure 8.9 illustrates a comprehensive portrayal of the results pertaining to the nonlinear coefficient as a function of wavelength for the various test analytes. The nonlinear coefficient serves as a pivotal parameter in characterising the nonlinear optical properties inherent to the optical fibre and exhibits an inverse relationship with the effective area, as explicitly stated in equation (8.7). Notably, all observed outcomes of the nonlinear coefficient consistently demonstrate a decline as the wavelength increases. Specifically, at shorter wavelength, the light energy fosters a more pronounced interaction thereby yielding amplified nonlinear effects.

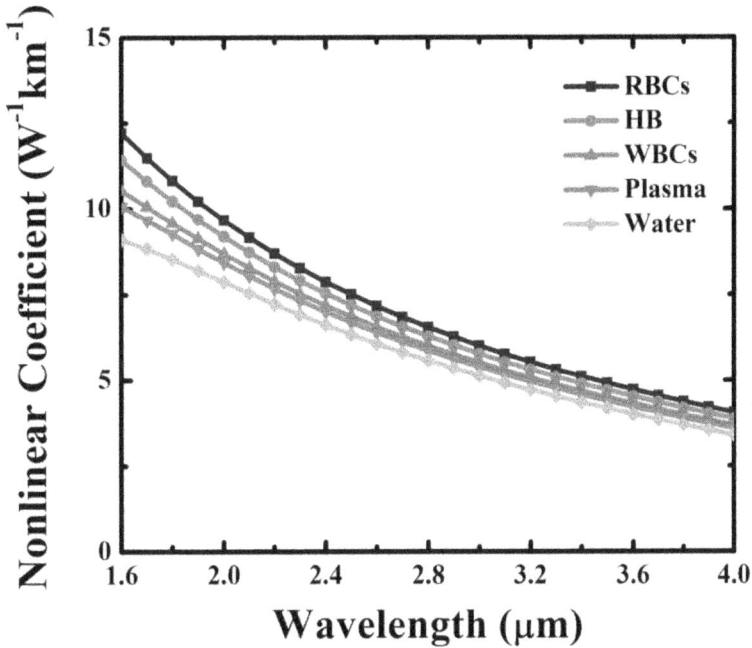

Figure 8.9. Variation of nonlinear coefficient with operating wavelength for the test analytes: RBCs ($n = 1.40$), HB ($n = 1.38$), WBCs ($n = 1.36$), plasma ($n = 1.35$), and water ($n = 1.33$) in the proposed PCF sensor.

Consequently, the effective area diminishes due to heightened confinement of light within the core. Conversely, as the wavelength increases, the light energy progressively diminishes, leading to a decrease in the intensity of nonlinear effects, subsequently resulting in a decline in the nonlinear coefficient. At the set optimum wavelength, the reported nonlinear coefficients for RBCs, HB, WBCs, plasma, and water are respectively denoted as 9.67, 9.20, 8.69, 8.43, and 7.87 W^{-1} km^{-1}. These findings effectively represent its applicability in nonlinear optics. The favourable results obtained concerning the nonlinear coefficient in this investigation indicate the suitability of applying them in optical communication systems and nonlinear optics.

The V parameter is a crucial factor in determining the operational mode of the proposed fibre, distinguishing between single-mode and multi-mode operation. Figure 8.10 showcases the results of the V-parameter analysis for the blood component analytes infiltrated at different optical wavelengths. As the wavelength increases, the V parameter gradually decreases. This behaviour aligns with the reduction in the refractive index difference between the core and cladding, which occurs with increasing wavelength. According to the criterion that V_{eff} should be less than or equal to 2.405 for single-mode fibre operation [58, 59], it can be conducted that the proposed fibre design functions as a single-mode fibre across the entire range of operating wavelengths.

Table 8.1 compares the relative sensitivity values for detecting various blood components using the proposed PCF sensor and previous PCF designs [24–28]. The results emphasise the influence of design complexity on the performance of the

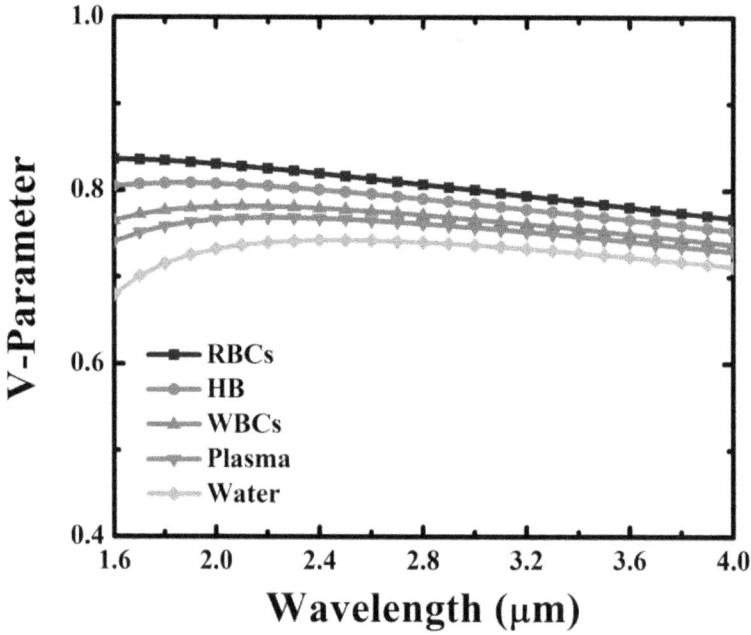

Figure 8.10. Variation of V-parameter with operating wavelength for the test analytes: RBCs ($n = 1.40$), HB ($n = 1.38$), WBCs ($n = 1.36$), plasma ($n = 1.35$), and water ($n = 1.33$) in the proposed PCF sensor.

Table 8.1. Comparison of relative sensitivity values for detecting various blood components using the proposed PCF sensor and prior PCF designs.

References	Design	Relative sensitivity (%)				
		RBCs	HB	WBC	Plasma	Water
[24]	Single sensing core ring and 4 layers of circular cladding holes	56.05	66.47	53.72	54.04	55.09
[25]	Single sensing core ring and 4 layers of circular cladding holes	55.83	58.06	62.72	65.05	66.47
[26]	Hexagonal benzene-shaped core and 5 rings of circular cladding air holes	—	—	—	77.84	—
[27]	Circular lattice cladding air holes and 49 circular core holes	80.93	80.56	80.13	79.91	79.39
[28]	Six cladding segment holes and a rectangular core hole	93.50	92.41	92.25	90.48	89.14
Proposed PCF	Two layers of octagonal and circular cladding air holes and one octagonal core hole	95.27	94.02	92.49	91.59	89.47

sensor. The PCF sensor proposed in this study utilizes a streamlined design that incorporates a solitary core hole and two cladding air holes with distinct geometries. Despite its simplicity, this design yields significantly higher relative sensitivity values compared to the more elaborate previous PCF designs. Impressively, the proposed PCF sensor achieves remarkable relative sensitivities of 95.27% for RBCs, 94.02% for HB, 92.49% for WBCs, 91.59% for plasma, and 89.47% for water. These findings highlight the remarkable capacity of the sensor to precisely identify and distinguish different blood components. Moreover, the findings suggest that the streamlined design of the proposed sensor does not compromise its sensing performance but rather enhances its ability to analyse blood components. As a result, the suggested PCF sensor exhibits significant potential for real-world applications in blood analysis and medical diagnostics.

8.5 Conclusion

An innovative PCF-based sensor has been developed with the purpose of detecting a wide range of blood constituents, such as red blood cells, haemoglobin, white blood cells, plasma, and water. The sensor design incorporates a unique cladding configuration with two layers. The first layer consists of octagonal holes arranged in a ring, while the second layer features circular holes. Additionally, the sensor includes a single octagonal core hole at its centre. The design underwent comprehensive numerical analysis using the FEM simulation on the COMSOL Multiphysics platform.

The PCF sensor operates as a single-mode fibre and demonstrates exceptional performance across multiple parameters. Notably, it achieved remarkable results in terms of relative sensitivity, confinement loss, chromatic dispersion effective area, nonlinear coefficient, propagation constant, and V parameter. At the optimal wavelength of 2.0 μm, the sensor exhibits impressive relative sensitivities: 95.27% for red blood cells, 94.02% for haemoglobin, 92.49% for white blood cells, 91.59% for plasma, and 89.47% for water. These outcomes establish the effectiveness of the sensor in accurately detecting and distinguishing various blood components, underscoring its immense potential for medical sensing applications. Furthermore, the favourable performance of the sensor in terms of confinement loss, chromatic dispersion, and nonlinear coefficient suggest potential applications in the field of optical communications. The versatility and high relative sensitivity of the PCF-based blood component sensor open up new possibilities for advancing both medical diagnostics and optical communication technologies.

References

[1] Nyachionjeka K, Tarus H and Langat K 2020 Design of a photonic crystal fiber for optical communications application *Sci. Afr.* **9** e00511
[2] Gangwar A and Sharma B 2012 Optical fiber: the new era of high speed communication (technology, advantages and future aspects) *Int. J. Eng. Res* **4** 19–23
[3] Nagatsuma T, Ducournau G and Renaud C C 2016 Advances in terahertz communications accelerated by photonics *Nat. Photonics* **10** 371–9

[4] Mittleman D M 2018 Twenty years of terahertz imaging [Invited] *Opt. Express* **26** 9417

[5] Azman M F, Mahdiraji G A, Wong W R, Aoni R A and Mahamd Adikan F R 2019 Design and fabrication of copper-filled photonic crystal fiber based polarization filters *Appl. Opt.* **58** 2068

[6] Sridarshini T, Geerthana S, Balaji V R, Thirumurugan A, Sitharthan R, Sivanantha Raja A *et al* 2023 Ultra-compact all-optical logical circuits for photonic integrated circuits *Laser Phys.* **33** 076207

[7] Geerthana S, Syedakbar S, Sridarshini T, Balaji V R, Sitharthan R and Shanmuga Sundar D 2022 2D-PhC based all optical AND, OR and EX-OR logic gates with high contrast ratio operating at C band *Laser Phys.* **32** 106201

[8] Sivaranjani R, Shanmuga Sundar D, Sridarshini T, Sitharthan R, Karthikeyan M, Sivanantha Raja A *et al* 2020 Photonic crystal based all-optical half adder: a brief analysis *Laser Phys.* **30** 116205

[9] Maidi A M, Kalam M A and Begum F 2023 Unsafe food additive sensing through octagonal-core photonic crystal fibre sensor *Phys. Scr.* **98** 065528

[10] Maidi A M, Yakasai I, Abas P E, Nauman M M, Apong R A, Kaijage S *et al* 2021 Design and simulation of photonic crystal fiber for liquid sensing *Photonics* **8** 16

[11] Ang C S, Maidi A M, Kaijage S and Begum F 2023 Highly sensitive biosensor based on a microstructured photonic crystal fibre for alcohol sensing *Results Opt.* **12** 100433

[12] Sridarshini T, Chidambaram P, Geerthana S, Balaji V R, Thirumurugan A, Sitharthan *et al* 2022 Current and future horizon of optics and photonics in environmental sustainability *Sustain. Comput. Inform. Syst.* **36** 100815

[13] Zeltner R, Pennetta R, Xie S and Russell P S J 2018 Flying particle microlaser and temperature sensor in hollow-core photonic crystal fiber *Opt. Lett.* **43** 1479

[14] Sun B, Huang Y, Liu S, Wang C, He J, Liao C *et al* 2015 Asymmetrical in-fiber Mach–Zehnder interferometer for curvature measurement *Opt. Express* **23** 14596

[15] Fu H Y, Khijwania S K, Tam H Y, Wai P K A and Lu C 2010 Polarization-maintaining photonic-crystal-fiber-based all-optical polarimetric torsion sensor *Appl. Opt.* **49** 5954

[16] Rajan G, Ramakrishnan M, Semenova Y, Domanski A, Boczkowska A, Wolinski T *et al* 2012 Analysis of vibration measurements in a composite material using an embedded PM-PCF polarimetric sensor and an FBG sensor *IEEE Sens. J.* **12** 1365–71

[17] Bock W J, Jiahua C, Eftimov T and Urbanczyk W 2005 A photonic crystal fiber sensor for pressure measurements *Conf. Rec.-IEEE Instrum. Meas. Technol. Conf.* **2** 1177–81

[18] Ahmed S, Mou J R, Mollah M A and Debnath N 2019 Hollow-core photonic crystal fiber sensor for refractive index sensing *2019 IEEE Int. Conf. on Telecommunications and Photonics (ICTP)* (Piscataway, NJ: IEEE) pp 1–4

[19] Mathews S, Farrell G and Semenova Y 2011 All-fiber polarimetric electric field sensing using liquid crystal infiltrated photonic crystal fibers *Sens. Actuators A Phys.* **167** 54–9

[20] Thakur H V, Nalawade S M, Gupta S, Kitture R and Kale S N 2011 Photonic crystal fiber injected with Fe_3O_4 nanofluid for magnetic field detection *Appl. Phys. Lett.* **99** 161101

[21] Abbaszadeh A, Makouei S and Meshgini S 2022 New hybrid photonic crystal fiber gas sensor with high sensitivity for ammonia gas detection *Can. J. Phys.* **100** 129–37

[22] Hu P, Dong X, Wong W C, Chen L H, Ni K and Chan C C 2015 Photonic crystal fiber interferometric pH sensor based on polyvinyl alcohol/polyacrylic acid hydrogel coating *Appl. Opt.* **54** 2647

[23] Abdullah-Al-Shafi M and Sen S 2020 Design and analysis of a chemical sensing octagonal photonic crystal fiber (O-PCF) based optical sensor with high relative sensitivity for terahertz (THz) regime *Sens. Bio-Sens. Res.* **29** 100372

[24] Singh S and Kaur V 2017 Photonic crystal fiber sensor based on sensing ring for different blood components: design and analysis *2017 9th Int. Conf. on Ubiquitous and Future Networks (ICUFN)* (Piscataway, NJ: IEEE) pp 399–403

[25] Kaur V and Singh S 2019 Design approach of solid-core photonic crystal fiber sensor with sensing ring for blood component detection *J. Nanophotonics* **13** 1

[26] Islam M T, Moctader M G, Ahmed K and Chowdhury S 2018 Benzene shape photonic crystal fiber based plasma sensor: design and analysis *Photonic Sens.* **8** 263–9

[27] Ahmed K, Ahmed F, Roy S, Paul B K, Aktar M N, Vigneswaran D *et al* 20191 Refractive index-based blood components sensing in terahertz spectrum *IEEE Sens. J.* **19** 3368–75

[28] Hossain M B and Podder E 2019 Design and investigation of PCF-based blood components sensor in terahertz regime *Appl. Phys.* A **125** 861

[29] Sharma P and Sharan P 2015 Design of photonic crystal based ring resonator for detection of different blood constituents *Opt. Commun.* **348** 19–23

[30] Jaszczak E, Ruman M, Narkowicz S, Namieśnik J and Polkowska 2017 Development of an analytical protocol for determination of cyanide in human biological samples based on application of ion chromatography with pulsed amperometric detection *J. Anal. Methods Chem.* **2017** 1–7

[31] Liu L, Wang X, Yang J and Bai Y 2017 Colorimetric sensing of selenocystine using gold nanoparticles *Anal. Biochem.* **535** 19–24

[32] Timofeyenko Y G, Rosentreter J J and Mayo S 2007 Piezoelectric quartz crystal micro-balance sensor for trace aqueous cyanide ion determination *Anal. Chem.* **79** 251–5

[33] Murphy L R, Yerolatsitis S, Birks T A and Stone J M 2022 Stack, seal, evacuate, draw: a method for drawing hollow-core fiber stacks under positive and negative pressure *Opt. Express* **30** 37303

[34] Ebendorff-Heidepriem H and Monro T M 2007 Extrusion of complex preforms for microstructured optical fibers *Opt. Express* **15** 15086

[35] Luo Y, Canning J, Zhang J and Peng G-D 2020 Toward optical fibre fabrication using 3D printing technology *Opt. Fiber Technol.* **58** 102299

[36] Camacho Rosales A L, Núñez Velázquez M A, Zhao X and Sahu J K 2020 Optical fibers fabricated from 3D printed silica preforms ed H Helvajian, B Gu and H Chen *Laser 3D Manufacturing VII* (Bellingham, WA: SPIE) p 29

[37] Ebendorff-Heidepriem H, Schuppich J, Dowler A, Lima-Marques L and Monro T M 2014 3D-printed extrusion dies: a versatile approach to optical material processing *Opt. Mater. Express* **4** 1494

[38] Bulbul A A-M, Jibon R H, Biswas S, Pasha S T and Sayeed M A 2021 Photonic crystal fiber-based blood components detection in THz regime: design and simulation *Sens. Int.* **2** 100081

[39] Habib M A, Anower M S, AlGhamdi A, Faragallah O S, Eid M M A and Rashed A N Z 2021 Efficient way for detection of alcohols using hollow core photonic crystal fiber sensor *Opt. Rev.* **28** 383–92

[40] Habib M A, Abdulrazak L F, Magam M, Jamal L and Qureshi K K 2022 Design of a highly sensitive photonic crystal fiber sensor for sulfuric acid detection *Micromachines* **13** 670

[41] Yakasai I, Abas P E, Kaijage S F, Caesarendra W and Begum F 2019 Proposal for a quad-elliptical photonic crystal fiber for terahertz wave guidance and sensing chemical warfare liquids *Photonics* **6** 78

[42] Yakasai I K, Abas P E, Ali S and Begum F 2019 Modelling and simulation of a porous core photonic crystal fibre for terahertz wave propagation *Opt. Quantum Electron.* **51** 122

[43] Maidi A M, Abas P E, Petra P I, Kaijage S, Zou N and Begum F 2021 Theoretical considerations of photonic crystal fiber with all uniform-sized air holes for liquid sensing *Photonics* **8** 249

[44] Rahaman M E, Jibon R H, Mondal H S, Hossain M B, Bulbul A A-M and Saha R 2021 Design and optimization of a PCF-based chemical sensor in THz regime *Sens. Bio-Sens. Res.* **32** 100422

[45] Hasan M M, Pandey T and Habib M A 2021 Highly sensitive hollow-core fiber for spectroscopic sensing applications *Sens. Bio-Sens. Res.* **34** 100456

[46] Yakasai I K, Abas P E, Suhaimi H and Begum F 2020 Low loss and highly birefringent photonic crystal fibre for terahertz applications *Optik* **206** 164321

[47] Begum F, Namihira Y, Razzak S M A, Kaijage S F, Hai N H, Miyagi K *et al* 2009 Flattened chromatic dispersion in square photonic crystal fibers with low confinement losses *Opt. Rev.* **16** 54–8

[48] Begum F, Namihira Y, Kinjo T and Kaijage S 2010 Supercontinuum generation in photonic crystal fibres at 1.06, 1.31, and 1.55 μm wavelengths *Electron. Lett.* **46** 1518

[49] Maidi A M, Shamsuddin N, Wong W-R, Kaijage S and Begum F 2022 Characteristics of ultrasensitive hexagonal-cored photonic crystal fiber for hazardous chemical sensing *Photonics* **9** 38

[50] Podder E, Jibon R H, Hossain M B, Al-Mamun Bulbul A, Biswas S and Kabir M A 2018 Alcohol sensing through photonic crystal fiber at different temperature *Opt. Photonics J.* **8** 309–16

[51] Bulbul A A-M, Imam F, Awal M A and Mahmud M A P 2020 A novel ultra-low loss rectangle-based porous-core PCF for efficient THz waveguidance: design and numerical analysis *Sensors* **20** 6500

[52] Hai N H, Namihira Y, Kaijage S F, Kinjo T, Begum F, Abdur Razzak S M *et al* 2008 A unique approach in ultra-flattened dispersion photonic crystal fibers containing elliptical air-holes *Opt. Rev.* **15** 91–6

[53] Begum F and Abas P E 2019 Near infrared supercontinuum generation in silica based photonic crystal fiber *Prog. Electromagn. Res.* C **89** 149–59

[54] Agbemabiese P A and Akowuah E K 2020 Numerical analysis of photonic crystal fibre with high birefringence and high nonlinearity *J. Opt. Commun.*

[55] Kaijage S F, Ouyang Z and Jin X 2013 Porous-core photonic crystal fiber for low loss terahertz wave guiding *IEEE Photonics Technol. Lett.* **25** 1454–7

[56] Miyagi K, Namihira Y, Razzak S M A, Kaijage S F and Begum F 2010 Measurements of mode field diameter and effective area of photonic crystal fibers by far-field scanning technique *Opt. Rev.* **17** 388–92

[57] Hossain M, Podder E, Adhikary A and Al-Mamun A 2018 Optimized hexagonal photonic crystal fibre sensor for glucose sensing *Adv. Res.* **13** 1–7

[58] Habib A, Anower S and Haque I 2020 Highly sensitive hollow core spiral fiber for chemical spectroscopic applications *Sens. Int.* **1** 100011

[59] Rana S, Saiful Islam M, Faisal M, Roy K C, Islam R and Kaijage S F 2016 Single-mode porous fiber for low-loss polarization maintaining terahertz transmission *Opt. Eng.* **55** 076114

Chapter 9

Photonic biosensors for healthcare applications

Seemesh Bhaskar, Kalathur Mohan Ganesh and Sai Sathish Ramamurthy

Recently, there has been escalating demand for the utility of photonic biosensors for point-of-care diagnostics and related healthcare applications. In this context, nano-materials have revolutionized the applicability and practicality of such biosensors on account of their unique optical energy confinement and augmentation capability, beyond the diffraction limit. Among the several novel photonic strategies developed since the onset of the 21st century for healthcare applications, surface plasmon-coupled emission (SPCE) technology has progressed drastically on account of its unique characteristics: sharp directionality, high polarization, low background noise, and high spectral resolution. The SPCE platform has presented unique physicochemical insights of photoplasmonic coupling of diverse nanosystems including intermetallics, metals, dielectrics, and low-dimensional (0D, 1D, 2D), graphene-π-plasmon & hybrid composites with novel architectural designs (spacer, cavity and extended cavity) for development of biosensors. In addition to the exploration of tunable first-, second-, third-, and fourth-generation plasmonic 'hotspots', in the classic work entitled 'Welcome to Nano 4.0', recently scientists have strategized novel nano-engineering approaches to develop alternative biosensing platforms such as ferroplasmon-on-mirror (FPoM), photonic crystal-coupled emission (PCCE), and Graphene oxide plasmon-coupled Soliton and Plasmon (GraSP) emission platform in an attempt to improve the sensitivity of related biosensor frameworks for monitoring human and environmental well-being. These pioneering works with next-gen photoplasmonic coupling attributes yielded unprecedented 1000-fold enhancements in sensitivity, consequently fostering cost-effective smartphone-based healthcare monitoring technologies amenable for resource-limited settings. While different unified principles of materials chemistry and opto-electronics are synergized to develop myriad photonic biosensor with tunable functionalities, this chapter focuses on the utility of disruptive engineering of nanohybrids in SPCE and PCCE platforms for achieving detection of ions, molecules, and analytes of interest at ultra-sensitive concentrations all the way from femtomolar to zeptomolar limits, thereby enabling early healthcare diagnostics. The platforms elaborated on in this chapter are user-friendly, robust, reliable, and cost-effective for biosensing, especially in low-and-middle income countries.

9.1 Introduction

The utility of biosensors has surpassed conventional areas of use such as medical diagnostics and is being greatly expedient in different fields including food and water quality monitoring, homeland security, pharmaceuticals, and forensics, to name a few [1–4]. In the domain of medical diagnostics, newer technologies are rapidly emerging to enhance the performance of existing approaches in terms of hastening the process of quality feedback to physicians about the health status of the patient. In conventional approaches, the polymerase chain reaction (PCR) in addition to the cell culturing and microscopy experiments are employed with or without the use of luminescent tags (depending on need) to detect the analytes of interest [3, 5, 6]. The label-dependent methodologies such as lateral-flow assay (LFA), which are generally employed in home-based pregnancy tests and rapid COVID-19 tests, are used worldwide for swift screening prior to determining hospitalization or appropriate quarantine regulations [7]. Although these are easy to handle and enable rapid detection in resource-limited settings, the results obtained are semiquantitative and suffer the drawbacks of sensitivity and selectivity [8–10]. The instruments that are otherwise used in hospital settings for analysis of a broad range of analytes with high sensitivity are cumbersome and demand personnel with adequate training and expertise to handle the complex exorbitant instruments [11]. In light of these observations, there is a growing need for the development of biosensor devices for early diagnosis especially in the cases of infectious and other inflammatory diseases in addition to cancer, autoimmune, neurological, respiratory, and cardiovascular disorders, where significant reduction in the time interval of evaluation, improved sensitivity, selectivity, and the cost-effectiveness of the device play critical roles. Different biomarkers such as antibodies, proteins, hormones, DNA, RNA, prions, amino acids, and other analytes associated with diverse disease conditions are being studied by different research groups across the world [12–16]. Interestingly, different types of ions (cations and anions) as well as other biomolecules that may enter the patient body via normal diet are being studied since some of these molecules drastically effect the health conditions of the existing patient suffering a particular disease [17–21]. It has been observed that the biomarker and bioanalytes concentrations are extremely low (typically in the range of sub-picomolar concentrations) at the starting stages of the occurrence of the disease, thereby demanding highly sensitive biosensors for active health monitoring.

To address the current challenges of the conventional biosensing technologies, recently photoplasmonic interfaces are being developed with novel functionalities and characteristics. The surface plasmon resonance (SPR)-based sensors are the first of its kind where the metal–dielectric interface are explored with bioanalytes of interest captured in the dielectric interface about the metallic thin film [8, 22–25]. The advantages of the SPR-based biosensors have been maximized using the

Nanoparticle-on-Mirror (NPoM) configurations in order to enhance the performance of the metal–dielectric interface, with the introduction of plasmonic cavity and associated hotspots [26–29]. Hotspots are regions of high electromagnetic (EM) field intensity occurring at the nano-junctions between two metallic nanoparticles (NPs) and/or at the sharp nano-curvatures of sharp-edged nano-objects such as nano-stars and nano-cubes [30, 31]. The signal output and the readout are significantly improved with the incorporation of plasmonic NPs that are tagged to the analytes of interest, on account of the drastic changes in the refractive index at the interface of interest [5, 32, 33]. The photoplasmonic coupling efficiency between the localized surface plasmon resonance (LSPR) of metallic NP and the metallic thin film results in high field confinement in the nano-cavity between them. The broad photonics research community has introduced other low-dimensional materials (0D, 1D, and 2D) such as graphene, MoS_2, and WS_2 to name a few, in such nanocavities to further augment the performance of the photoplasmonic biosensor [34–37]. Such explorations have aided new possibilities where novel functionalities at metal–dielectric interfaces with respect to EM field confinement, electric-magnetic dipole resonances, electron transport, optoelectronic characteristic, and tunable bandgaps have been explored. The π-plasmon-metal plasmon hybrid coupling and the stacking affinity reinforced in such configurations has been explored in the presence of magneto-plasmonic and ferroplasmonic NPs to enable hastened biosensing approaches using external magnets [38–41]. The changes in the surface and the bulk refractive indices are being monitored using the nanomaterials of metal–dielectric hybrid characteristics and the different sizes and shapes to effectively alter the evanescent field (from tens of nanometers to hundreds of nanometers) [42, 43].

Although numerous biosensing applications for healthcare monitoring have been detailed in different research articles and reviews, the recent developments in the domain of surface plasmon-coupled emission (SPCE) in this direction is seldom discussed comprehensively [44–46]. In the early 2000s, Lakowicz and co-workers developed an innovative biosensing framework termed SPCE, which is fundamentally a prism-coupling technique, where the emission from the radiating dipoles are coupled to the surface plasmon polaritons (SPPs) of metallic thin film in Kretchmann–Raether (KR) and reverse Kretchmann (RK) configuration [47–49]. The out-coupled emission from the prism side is highly directional and carries the polarization properties of the metallic thin film. Consequently, $\sim$10-fold enhancement in the fluorescence intensity has been recorded vis-à-vis conventional fluorescence spectrophotometer [50]. Further, citrate-capped plasmonic AgNPs were interfaced over the SPCE platform using the silane-based monolayers, which resulted in further enhancing the fluorescence signal intensity to 60-fold [50]. This augmentation is attributed to the hybrid coupling between the LSPR of the AgNPs and the propagating SPPs of the metal thin film. Further, this strategy was utilized for development of different biosensing frameworks where dissimilar interactions between the NPs and the radiating dipoles are investigated [51–56]. In addition to realizing novel physicochemical insights from the photoplasmonic perspective, such explorations assisted in the realization of novel fluorescence-based diagnostic technologies. While there are different applications of the SPCE platform, the latest advancements pertaining to substrate nano-engineering are discussed in detail in this chapter, in addition to capturing the key highlights of analogous platforms such as Ferroplasmon-on-Mirror (FPoM) [57] and photonic crystal-coupled emission (PCCE) [38]. Photonic crystal and associated interfaces are currently being utilized for transdisciplinary applications encompassing cancer detection [58], integrated photonic circuits [59–61], coupled with the environmental sustainable goals [62] and its next-generation systems applications [63]. The crucial advantages as well as the drawbacks of these technologies are discussed to present a comprehensive understanding to the broad audience of diagnostic technologies.

9.2 Biosensors: plasmonic and photonic platforms

Figures 9.1(a)–(c) present a Scopus analysis of the number of publications reported in the domain of biosensors, plasmonic biosensors, and photonic biosensors. At the outset, it is clearly observed that the number of publications in all the three classifications has exponentially increased in the past 20 years. Further, in order to give the overview, the number of publications in the broad domain of biosensors is presented. It is important to note that plasmonic and photonic biosensors are not the only two sub-categories of the broad domain of biosensors [3, 64, 65]. There are biosensors that are developed using electrochemical reactions and other domains of physics and chemistry that do not essentially involve the use of photonics and plasmonics. Hence, it is reasonable to observe comparatively lesser number of documents in the specific domains of 'plasmonic' (figure 9.1(b)) and 'photonic' (figure 9.1(c)) biosensors vis-à-vis 'biosensor' (figure 9.1(a)) [66, 67]. It is also important to mention that the categorization of plasmonic and photonic biosensors is typically blurred because of the strong inter-connections between these two domains of research. While all the plasmonic biosensors typically involve the use of metallic NPs in one form or other for the development of the final biosensor, the photonic biosensors need not necessarily incorporate the plasmonic NPs in their biosensor platforms. Hence, predominantly the plasmonic biosensors comprise an effective subset of the broad domain of photonic biosensors. Interestingly, from figures 9.1(b) and (c), this claim seems to be invalid, because the number of publications in the domain of photonic biosensors is apparently less than that of plasmonic biosensors. In the broad domain of photoplasmonics, the ability to

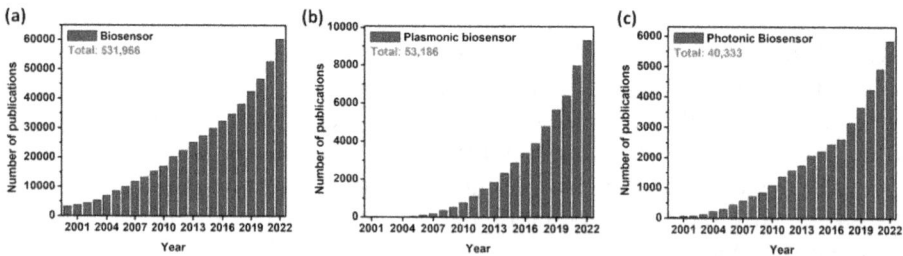

Figure 9.1. Analysis of the number of publications in the years 2000–22 in the domain of (a) biosensor, (b) plasmonic biosensor, and (c) photonic biosensor, as obtained from the Scopus dataset.

analyze the plasmonic properties of materials emerges with the intricate interactions between the light (photon) and appropriate optic elements. In this context, while it is advisable to utilize the terminology 'plasmonics' to describe the light–matter interactions occurring at nanoscale dimensions, a great deal of flexibility has been exercised by researchers in defining terminologies. For instance, there are certain studies where metallic nanomaterials are used for enhancing the sensitivity of the electrochemical-based biosensor. At such instances, it not uncommon to see researchers conveniently use the terminology of 'plasmonic biosensors', although in ideal conditions, the plasmonic resonance property of the nanomaterials are not utilized for development of biosensor (as light–matter interactions are not involved). However, this categorization is not completely biased because metallic nanomaterials are also addressed as plasmonic nanomaterials on account of their localized surface plasmon resonance attributes that significantly affect the functionality of the platform on which they are used. On the other hand, it can also be noted that the terminologies utilized for explaining the observations made in the photonic biosensors are more conveniently defined by their plasmonic interactions. Consequently, although the biosensor can be categorized as a photonic biosensor, scientists alternatively use the term 'plasmonic biosensor' to keep the description more focused on the subject of interest as well as on the audience/readers. Considering these observations, the increase in the number of publications for plasmonic biosensors is realistic as compared to that of photonic biosensors. More importantly, it is worth mentioning that the photoplasmonic-based biosensors have gained great interest in the past decade (2010–22), especially as compared to the progress in the previous decade (2000–10). It is self-evident from the number of publications that there are numerous photonic and plasmonic technologies that are being developed for the expansion of biosensors and it is practically not feasible to discuss the highlight of all such technologies. In this background, this chapter aims to discuss the latest advancements in the broad domain of plasmonic and photonic technologies for biosensor development where a few case studies are presented. In order to provide a comprehensive understanding of the photoplasmonic biosensors, the subsequent sections are devoted to discussing the details of burgeoning fields of surface plasmon-coupled emission (SPCE) and photonic crystal-coupled emission (PCCE). We have chosen these two domains of research in order bring out the novel insights pertaining to the plasmonic (SPCE) and photonic (PCCE) biosensors. The subsequent sections are hence being devoted to discussing the advancements made in the recent past in the domains of SPCE and PCCE technology where the key highlights and differences along with the advantages and disadvantages are discussed from the point of view of biosensors.

9.3 Overview of biosensing technologies and plasmonics for healthcare applications

In this section, we provide a broad overview of the different capabilities of the biosensing technologies as well as plasmonics technologies, prior to plunging into the details of the SPCE and PCCE platforms. On account of the ever-rising health complications worldwide, different biosensing technologies are being developed to achieve better performance of the sensing devices in addition to reducing the cost of the production [69]. Figure 9.2 succinctly captures the key capabilities of the different types of biosensors that are widely developed for healthcare applications [70]. By and large, the biosensors are utilized for monitoring waste management, agricultural studies, forensic analysis, food, air and water quality monitoring, and

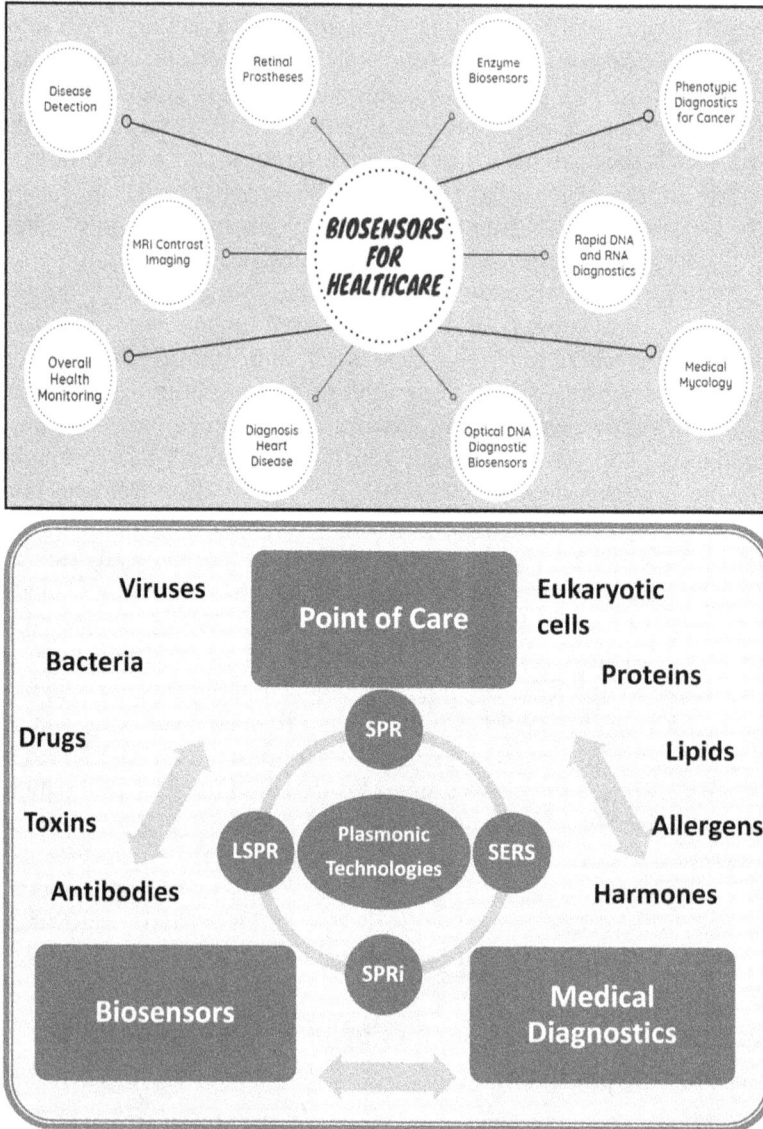

Figure 9.2. The applications of biosensors for different facets of healthcare (top). Reproduced from [11], CC BY 4.0. The utility of plasmonic technologies for different healthcare applications (bottom).

associated human healthcare applications [7, 64]. The human healthcare applications of biosensors include detection of different disease conditions such as retinal prostheses, MRIs, heart and other organ diagnosis, medical mycology, and health monitoring by analyzing the concentrations of RNA and DNA in body fluids to name a few [11]. In the past few decades, the need for improving the current disease diagnostic technologies from centralized hospital setup to mobile and/or decentralized healthcare platforms has been explored to cater to emergent health implications such as COVID-19. Moreover, home-based healthcare systems are gaining great interest on account of their relevance, especially when a pandemic requires social distancing and quarantine. A typical biosensor consists of components such as analyte of interest, bioreceptor, transducer, associated electronics, reader, and certain display systems [7, 11, 64]. While there are different types of biosensors explored for accomplishing societal healthcare, photoplasmonics technologies are extensively researched on account of their unique advantages. The lower panel of figure 9.2 presents an executive summary of the different types of applications arising from the plasmonic-based technologies on account of phenomena such as LSPR, SPR, and SERS, to name a few [7, 68]. While there are numerous reports highlighting the applications of these platforms for healthcare applications, the plasmonic resonances of metallic systems are investigated with luminescent moieties for different point-of-care diagnostics applications. As it is not feasible to comprehensively cover all the domains of plasmonic technologies, we have chosen to deliberate on the recently developed SPCE and PCCE technology in this chapter where a few specific examples are extensively described.

9.4 Overview of nanophotonics and plasmonics for healthcare applications: introducing SPCE and PCCE technology

In the broad domain of nanophotonics there are different methodologies being developed for exciting the surface plasmons at the metal–dielectric interface. To accomplish this, the energy and momentum conservation laws of the photon-metal combination need to be satisfied. In other words, the phase-matching conditions need to be satisfied where the momentum and energy of the incident photons couple with the momentum and energy of the plasmonic modes. This is achieved by utilizing certain optical coupling interfaces where elements such as prism, grating, as well as other waveguide generating structures such as photonic crystals and fiber optical couplers are employed. Fundamentally, these interfaces utilize the phenomenon of total internal reflection, refraction, diffraction, and evanescent wave coupling methods for achieving the conditions enabling surface plasmon excitation [68]. Some of these configurations are presented in the figure 9.3 to present a bird's eye view to the broad audience. Initially, the Otto and Kretschmann–Reather (KR) configurations were explored where the dielectric is interfaced between the metal and the prism in the former, and the metal is interfaced between the dielectric and the prism in the latter [10, 68]. A typical configuration of the KR is presented in figure 9.3(a) where the shift in the resonance coupling angle can be correlated to the quantitative binding of the desired analyte. That particular angle where the

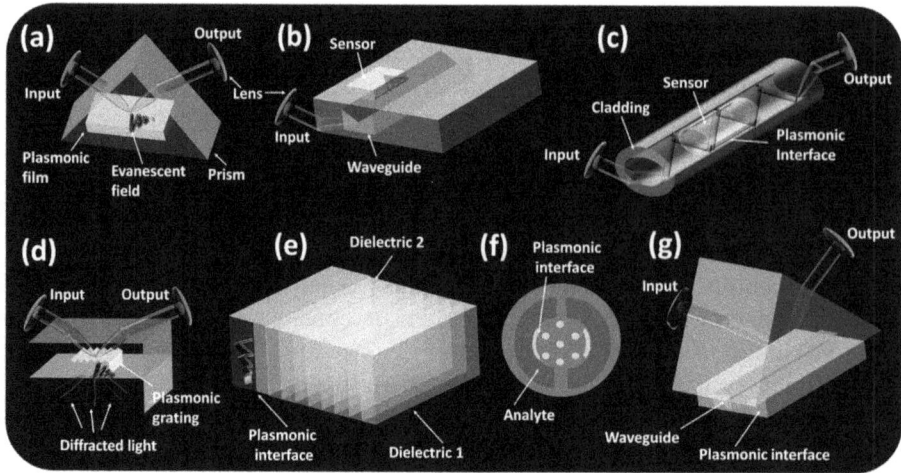

Figure 9.3. The generally adopted photonic configurations for generation and analysis of surface plasmon modes by effective coupling of incoming radiation: (a) Kretchmann–Raether (KR) using prism-coupling approach, (b) waveguide coupling approach, (c) fiber-optic coupling approach, (d) grating-based approach, (e) photonic crystal modes coupling to plasmonic modes, (f) honeycomb photonic crystal optical coupling, and (g) coupling of waveguide modes at MIM interfaces using a Kretchmann–Raether (KR) using prism-coupling approach.

incoming light transfers most of the energy to the plasmonic modes to generate surface plasmon polaritons or coupled oscillations is the resonance coupling angle.

A typical waveguide-based surface plasmon excitation and coupling is presented in figure 9.3(b). Here, the propagating waveguide modes assist in the excitation of the surface plasmons at the metal-sensing medium, where the phenomenon is generally studied in a narrow range of frequencies. The spectra at the output region of the metal-sensing medium can be evaluated using the shift induced in the resonance coupling wavelength. A typical example of such waveguide coupling architectures include fiber-optic-based coupling sensing devices that work with the principle of total internal reflection phenomena [68]. Such an example is presented in figure 9.3(c), where the light that is coupled into the waveguide propagates on account of the total internal reflection occurring at the inner low refractive index core and the outer high refractive index cladding [66, 67]. While the metal-sensing medium is integrated into such a framework, an effective biosensing platform can be developed. Yet another approach that is adopted for performing the biosensing is based on utilizing the metallic grating to couple the incoming radiation to the plasmonic modes sustained by the interface [68]. On account of the grating morphology the phase-matching is achieved in relationship to the order of the underlying grating structure, where the light that is transmitted or reflected is utilized for investigating the strength, frequency, and the polar angular phenomenon of the coupling. Figure 9.3(e) demonstrates a simple approach where the photonic crystal is interfaced with the metal thin film to excite the surface plasmons, so as to explore the possibility of developing a biosensor at the metal–analyte interface. Further, micro-structured photonic crystals are being studied where the holes within are filled with air to generate a periodic structure, and the channels are gold-coated so as to excite the surface plasmons [13, 49]. It is important to mention that the above discussed technologies are being synergized with one another to enhance the performance of the overall sensing device, and one such example is the development of metal–dielectric–metal interface where the established waveguide is in turn coupled to the prism and excited using the KR configuration. In summary, there are different such techniques that can be explored for the excitation of the surface plasmons, which in turn is utilized for developing a reliable sensing platform. There are different methodologies where such techniques are integrated with the luminescence-based approaches, such as fluorescence, phosphorescence, and chemiluminescence for enhancing the performance of the luminescence-based biosensors [65–67].

Fluorescence spectroscopy and the associated technologies have found widespread applications in biochemical sensing, immunoassays, and diagnostics, where scientists have reported single molecular limit of detection by integrating several high-end optics elements and the above discussed different photonic and plasmonic modes sustained by substrates with the fluorescence-based detection technologies [43, 48]. In the past few decades, the probes and devices associated with the conventional fluorescence-based methodologies are being synergized with the plasmonic materials, as a result of which interesting observations have been documented. The excitation and emission attributes of the luminescent molecules dramatically fluctuates in the presence of plasmonic nanomaterial in its vicinity. Depending on the real and imaginary part of the dielectric constant of the nanomaterial as well as the absorption and scattering cross-sections at different frequencies (of the electromagnetic spectrum), the emission intensities of the nearby radiating dipoles are modulated [30, 75]. While the charge density oscillations of the plasmonic NPs triggers a modulation in the radiative and non-radiative decay lifetime of the fluorescent molecules, such alterations assist experimentalists to overcome the traditional drawbacks of fluorescence spectroscopy in a new platform termed 'plasmon-enhanced fluorescence (PEF)' [42]. From this perspective, the metal–luminophore system has been identified as a plasmophore (plasmon + luminophore) that efficiently re-radiates (via quenching or dequencing) the emission characteristics of different types of luminescence probes (fluorescence, phosphorescence, and chemiluminescence and their rational combinations).

However, it has been noted that although significant modulation in the emission attributes can be realized using metal NPs, the long-standing challenges of low background noise, poor spectral resolution, and collection efficiency ($<1\%$) are present inevitably in the PEF-based biosensing platforms. In this context, the surface plasmon-coupled emission technology was developed by Lakowicz and co-workers, where the above-mentioned limitations were successfully overcome [10, 48, 76]. Further, the SPCE technology has been extensively utilized for the development of different diagnostic technologies on account of its high sensitivity and fast response time. In a typical SPCE experimentation, the laser source directly excites the luminescent moieties and the emitted photons in turn couple with the modes sustained by the metal–dielectric waveguide interface. A prism-coupling technique is utilized to extract the observable emission with sharp directionality, where the coupled emission carries the attributes of the fluorescent moieties as well the high polarization attribute of the waveguide [22]. Moreover, the metal–dielectric–metal interfaces were explored over the SPCE platform to realize beaming or steering emission attributes. While there are different types of nano-engineering achieved over the SPCE framework, in this chapter we focus on a few of such approaches by highlighting the utility of nano-engineering for enhancing the performance of biosensing platforms.

9.5 CryoSoret nano-engineering techniques for SPCE-based biosensing applications

While plasmonic and dielectric NPs are extensively explored for photoplasmonics-based biosensing applications, in the last decade nano-assemblies have gained significant interest for the development of advanced functional interfaces enabling better sensitivity of the analytes of interest [28, 77]. Although different nano-assembly approaches (co-assembly, hierarchical, and directed self-assembly) are being explored for biosensing applications, the self-assembly techniques, especially the ones based on stimuli-responsive driven self-assemblies, have been exponentially researched for the realization of monometallic and heterometallic nano-assemblies with integrated functionalities [30].

Although electric and magnetic field-based as well as high temperature-based self-assemblies are significantly explored for myriad applications, low-temperature-based self-assemblies are a rarity. Recently, cryosoret nano-engineering (CSNE) methodology has been reported for the generation of precise nano-assemblies using adiabatic-cooling methodology [30]. The schematic representation of the formation of the CSNE is presented in figure 9.4(a). In this technique, the homogenous NP solution is cooled to extremely low temperatures in order to achieve the soret effect where the NPs move from the hot end of the vial to the cold end through thermodiffusion. This is basically driven by the thermal gradient that is intrinsically generated while the NP solution is subjected to low temperatures. Consequently, the electrostatic repulsion in the NPs that are capped using the conventional stabilizing agents (citrate ions) is overcome by the thermal gradient resulting in the formation of precise nano-assemblies [30]. This methodology has been demonstrated to be applicable at different lower temperatures with a decrease in the resultant time needed for the formation of the nano-assemblies.

In order to obtain metal–dielectric hybrid structures, the CSNE methodology was undertaken by admixing the plasmonic AgNPs with different low refractive and high refractive index dielectric NPs such as silica and titania. The cryo-sorets obtained with HRI TiO_2 nano-rods presented ghd highest fluorescence enhancements when studied in the SPCE platform. This is attributed to the high field enhancement contributed by the rod-shared nano-architectures of titania that assists in the formation of lightning rod effect in addition to the localized Mie and delocalized Bragg plasmons sustained by the entire cryo-soret [30]. Multiple TEM images of the hybrid cryo-soret of AgNPs and the titania nanorods are presented in figures 9.4(b)–(e)

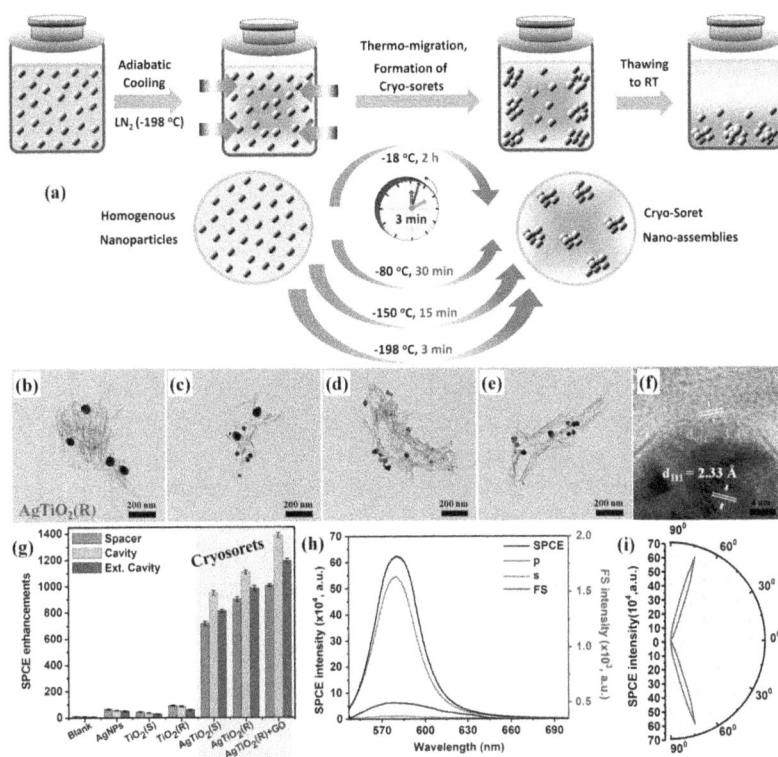

Figure 9.4. (a) Schematic representation of the formation of cryosorets (CSs) utilizing the adiabatic-cooling method and thermomigration of NPs under temperature gradient. The reduction in the synthesis time to obtain soret nano-assemblies at different cooling temperatures (-18 °C, -80 °C, -150 °C, and -196 °C) is also depicted. (b–e) Multiple TEM images of silver-titanium dioxide (rod), AgTiO$_2$(R) cryosorets. (f) HRTEM image of AgTiO$_2$(R) cryosorets. (g) SPCE enhancements obtained using AgNPs, TiO$_2$(spheres, S; rods, R), and AgTiO$_2$(S, R) in the spacer, cavity, and extended cavity nanointerfaces. (h) SPCE, FS, p, and s spectra for the AgTiO$_2$(R) CS sample presenting the highest enhancements in the cavity nanointerface in (g). (i) Corresponding angularity plot presenting sharp directional emission, validating the characteristic SPCE spectra. Reprinted with permission from [30]. Copyright © 2022, American Chemical Society.

and the HRTEM image presenting the characteristic lattice fringes of silver and titania is shown in figure 9.4(f).

The SPCE enhancements realized with different nanohybrids are shown in figure 9.4(g). Tunable enhancements were realized using spacer, cavity, and extended (ext.) cavity nano-interface [36, 78–80]. The spacer nano-interface is fabricated by spin coating the polymer embedded with the nanomaterials over the SPCE substrate (50 nm Ag thin film in this case) [81, 82]. Further, the radiating dipoles are spin coated after embedding in the polymer matrix. Such a configuration results in the sandwiching of the nanomaterials between the radiating dipoles and the metal thin film. Consequently, the nanomaterial essentially functions as a spacer nano-layer region. Further, in the cavity nanointerface, the nanomaterial and the radiating dipoles are admixed and spin coated as a single layer over the SPCE substrate. This yields an interface where the radiating dipoles are interfaced between the nanomaterials and the metal thin film, essentially in the regions of high electric field intensity in the infinitesimal nano-gaps, called nano-cavities [83, 84]. Further, in the extended cavity interface, this particular cavity region is extended to a particular size of interest by separately coating the radiating dipoles and the nanomaterials over the SPCE substrate. Depending on the nanomaterial, radiating dipole, and the configuration utilized the photoplasmonic coupling is modified and tailored fluorescence enhancements can be realized [17]. In an attempt to further maximize the SPCE enhancement the metallic thin film constituting the SPCE substrate was prefunctionalized using the graphene oxide (GO) in order to facilitate the judicious inter-plasmonic coupling between the cryo-sorets and the metallic thin film via the π-plasmons of the underlying GO nano-layer [30]. As a result of this in excess of 1300-fold SPCE enhancements were realized and the representative SPCE, free-space (FS), p and s-polarized spectra are presented in figure 9.4(h), along with the directionality of the emission captured in figure 9.4(i). Figures 9.5(a)–(c) present the summary of the CSNE protocol for the plasmonic AgNPs and AuNPs at different cooling temperatures. It is clearly observed that the number of NPs per assembly increases with increase in the cooling time interval. Further, lowering the cooling temperature from -18 °C to -80 °C to -150 °C to -196 °C results in the generation of cryo-sorets at lower time intervals. In other words, the time required to obtain similar cryo-sorets substantially decreased with decrese in the cooling temperature, thereby enabling the completion of CSNE to occur in less than 3 min at cryo-temperatures [30]. This observation is made in the case of Ag and AuNPs, and the understanding can be extrapolated to the other plasmonic and dielectric nanohybrids. The high fluorescence enhancement observed using the plasmonic Ag and dielectric titania nano-hybrid cryo-soret was utilized for the detection of environmentally hazardous organic dye, rhodamine B, and the results are presented in figure 9.5, along with the futuristic scope of the work.

From figures 9.5(d)–(g), we see that two linear sensing ranges were observed for detection of rhodamine B. The experiments were performed using the conventional Ocean optics spectrophotometer to determine the SPCE enhancements. Additionally, the out-coupled emission was captured using a smartphone camera and the results therefrom analyzed using a ColorGrab app to obtain the luminosity data [30]. The luminosity values are plotted against the SPCE enhancements to understand the direct correlation between them in figures 9.5(d) and (f), with the shade cards corresponding to the luminosity presented in figures 9.5(e) and (g), respectively. From this analysis, it may be noted that the smartphone-based detection can be performed with high reliability and reproducibility all the way down to single molecular limit of detection [30]. Figure 9.5(h) presents the schematic illustration of the different generations of plasmonic hotspots that has been explored to date and emphasizes the development of the fourth generation of the plasmonic hotspots using the cryo-soret study over the SPCE framework [30, 85]. An individual plasmonic NP assists in the concentration of the electric fields in the vicinity, as a result of which the EM field intensity increases, thereby facilitating enhanced light–matter interactions. Here, the local electric field intensity in the close vicinity of the

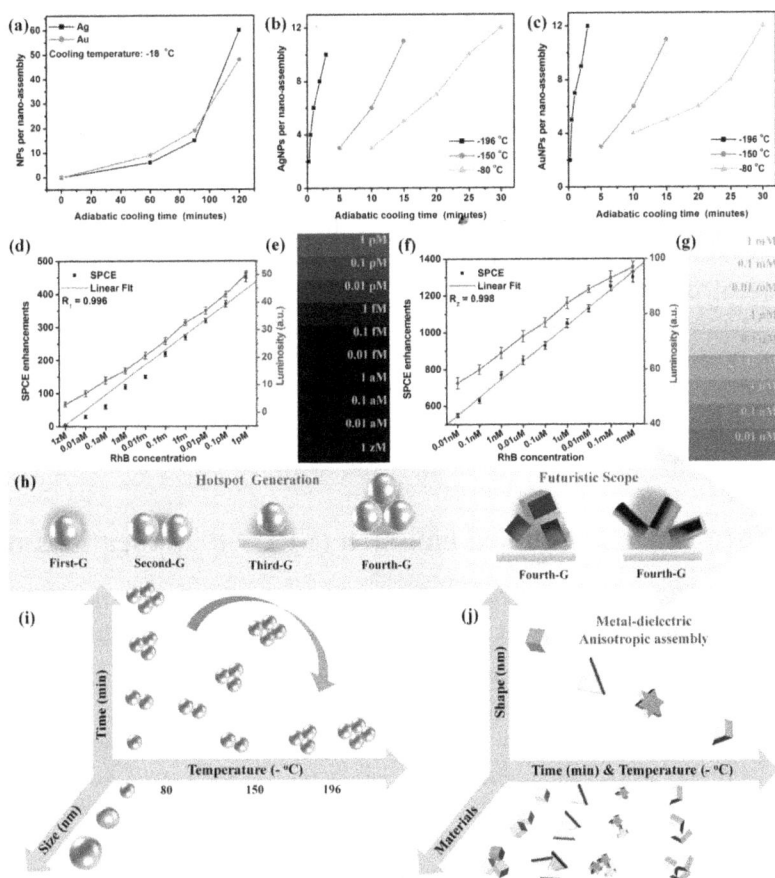

Figure 9.5. (a) Silver NPs (AgNPs) and gold NPs AuNPs obtained per soret assembly under −18 °C adiabatic-cooling conditions. (b) AgNPs obtained per soret assembly under −80 °C, −150 °C, and −196 °C adiabatic-cooling conditions. (c) AuNPs obtained per soret assembly under −80 °C, −150 °C, and −196 °C adiabatic-cooling conditions. (d, f) Sensing of RhB presenting a single molecular limit of detection (zeptomolar). The respective luminosity values are plotted in the right y-axis along with the corresponding shade cards shown in (e) and (g), respectively. (h) Development of different generations of hotspots in cryo-soret nanoengineering technology presenting opportunities for future explorations. (i, j) Snapshots of the possible permutations and combinations of cryo-soret-based nanoengineering that the current research work presents to the wide community of nanoscience and photoplasmonics. Reprinted with permission from [30]. Copyright © 2022, American Chemical Society.

nanomaterial is significantly enhanced especially when the scattering co-efficient of the overall extinction is dramatically maximized. Further, the research community has explored the dimer configuration with metal–metal, metal–dielectric and dielectric–dielectric NP configuration in order to comprehend the hotspot intensity, assisting in the formation of the second generation of plasmonic hotspots [30, 85]. While the metal–metal dimers yield intriguing observations and novel plasmonic effects, they suffer from intrinsic ohmic losses in metallic substrates. Consequently, recently, a variety of metal–dielectric nanohybrids are being evaluated, especially considering the dielectric nanomaterials of anisotropic shapes and high refractive index. Further, NPoM configurations support the formation of the intense plasmonic hotspots in the nano-gaps generated between the NPs and the thin film, thereby constituting the third generation of plasmonic hotspots. The cryo-sorets, on account of the nano-assembly architecture, support the formation of fourth generation of plasmonic hotspots while interfaced with the metallic thin film. In this scenario, the localized Mie plasmons and the delocalized Bragg plasmons of the cryo-sorets resonantly couple with the propagating plasmon polaritons of metallic thin films [85]. The executive summary of the results and the futuristic scope of the cryo-soret nanoengineering technology are captured in figures 9.5(h)–(j), where the numerous possibilities of exploring different metallic, dielectric, and low-dimensional substrates in assembly formation are depicted. The variations in the size and the shape of these nano-assemblies would hence render tunable plasmonic hotspots that can be further utilized for desired biosensing applications [30].

9.6 Surface plasmon-coupled emission (SPCE): applications in early diagnostics

9.6.1 Utility of chromaticity plot for tyrosine and spermidine sensing

Among the different methodologies that are being developed for the detection of analytes at extremely low concentrations, mobile phone-based detection techniques have gained great importance on account of several factors such as user-friendliness, cost-effectiveness, portability, and ease of mass production. Additionally, the smartphone-based rapid diagnostic tools are being extensively explored for the detection of different disease biomarkers, proteins, nucleic acids, and many more such biomarkers [39, 57, 73]. Artificial intelligence-based analytical tools are being incorporated to develop facile and rapid tools for further advancements in data analysis from disease diagnostics performed via smartphones. In light of these observations, this section presents the highlights of two sensing investigations performed using smartphone-based detection analysis. We have chosen metal–dielectric coupling and heterometallic coupling with the SPPs of the metal thin film as representative examples for the discussion of the utility of nanoengineering over the SPCE substrate as well smartphone-based detection systems. Fundamentally, there are two methods, namely (i) chromaticity and (ii) luminosity, using which the smartphone-based detection has been carried out in the past decade. While the chromaticity analysis comprises the analysis of the change in the color of the observed emission (in correlation with the SPCE enhancements), the luminosity analysis comprises the use of intensity (also called luminous value) of the emission (discussed with an example in the next section). The chromaticity analysis is hence preferred over the luminosity analysis, especially when there is an observable significant shift in the wavelength maxima of the observed emission from the sample under study. The chromaticity analysis has been utilized for evaluating the discernible changes in the pH, temperature, and formation of products in certain chemical reactions.

In the first approach metal–dielectric Au–SiO$_2$ nano-hybrids were interfaced with the SPCE platform in order to realize dequenched and augmented SPCE enhancements. Conventionally, the AuNPs quench the fluorescence signal from rhodamine moieties. Further, the high fluorescence enhancements were further utilized for the detection of spermidine molecule on account of its high biological relevance. This has been accomplished due to the strong interaction of spermidine molecule with the surface of the AuNPs [20]. The linear calibration plot presenting femtomolar sensing of spermidine is presented in figure 9.6(a) along with the spectra presented in figure 9.6(b). The out-coupled fluorescence was captured using the mobile phone-based detection system and analyzed using the chromaticity plot as presented in figure 9.6(c).

In the second approach, the heterometallic AgAu nanorods were explored over the SPCE substrate to gather dequenched and high SPCE intensity vis-à-vis quenched signal observed in the case of AuNPs. The high >1000-fold SPCE enhancements realized on account of the heterometallic coupling as well as the lightning rod effect were further utilized for the detection of tyrosine molecules at attomolar concentrations as presented in figures 9.6(d)–(f). Such high sensitivity could be achieved due to the high EM field intensity sustained by the AgAu nanorods. The sensing of biologically relevant tyrosine molecule was achieved by considering the simple interaction between the rhodamine B and the tyrosine, where the addition of the latter to the former results in the quenching of the fluorescence signal from tyrosine [86]. Further, the fluorescence signal intensity and the

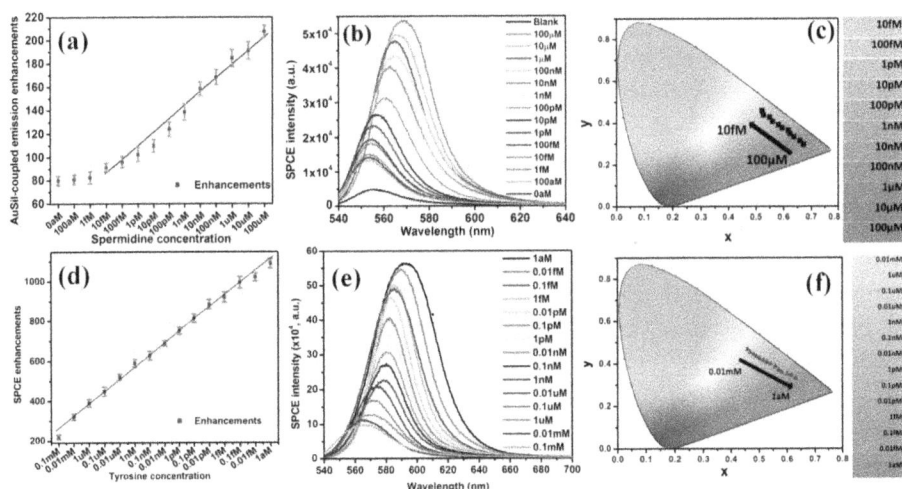

Figure 9.6. (a) SPCE enhancements for increasing concentrations of spermidine taken along with gold–silica (Au-SiO2) hybrids in ext. cavity interface. (b) SPCE spectra for corresponding emission enhancements observed in (a). (c) Commission Internationale de l'éclairage chromaticity plot and corresponding shade card indicating color change in emission pattern consistent with results in (b). (a)–(c) Reprinted with permission from [20]. Copyright © 2022, American Chemical Society. (d) SPCE spectra for decreasing concentrations of tyrosine in ext. cavity nanointerface. (e) SPCE enhancements for decreasing concentrations of tyrosine from 0.1 mM to 1 aM. (f) Chromaticity plot representing a consistent shift in color with change in tyrosine concentration. (d)–(f) Reprinted from [86], Copyright (2022), with permission from Elsevier.

chromaticity plots are shown in figures 9.6(b) and (c), respectively, for the sensing of tyrosine. Overall, it is observed that there is an excellent correlation between the smartphone-based detection system (with respect to chromaticity data) and the analysis performed using the conventional Ocean Optics detector (with respect to the SPCE enhancements). While this section deliberates on the utility of chromaticity data for analyzing the sensing performance of the device developed, the subsequent section is dedicated to the detection system based on luminosity analysis.

9.6.2 Utility of luminosity plot for the mercury ion sensing

While the chromaticity-based analysis of the smartphone-based detection technique was presented in the previous section, this section is focused on the luminosity-based analysis of smartphone-based detection technology. The luminosity analysis is predominantly carried out when there is a dramatic change in the intensity of the emission with a concomitant negligible change in the emission maxima of the emitted light. Although there are different approaches where the luminosity analysis has been performed, in this chapter we focus on the relevance of luminosity-based smartphone detection by considering a case study of mercury ion detection. While the previous section deliberates on detection of molecules of interest (such as tyrosine and spermidine), here an example of ion detection is discussed.

Among the several approaches that are reported in the literature for the synthesis of plasmonic, dielectric, and other semiconductor-based nanomaterials, the bio-inspired approaches stand out on account of their unique principles and relevance [52, 87]. Utilizing biomaterials for NP synthesis with the use of renewable resources assists in the realization of the twelve principles of green nanotechnology [21]. In this context, different approaches are developed for synthesis of nanomaterials using biodegradable and biocompatible materials as the only reducing and capping agent. In one of such approaches, Lycoat biopolymer was utilized for development of plasmonic AgNPs with fractal and cubic nano-morphologies [55]. Such nano-structures presented abundant EM field intensity while interfaced with plasmonic SPCE interface due to the generation of cavity plasmonic hotspots. The utility of nano-cubes synthesized using lycoat biopolymer assisted in accomplishing >900-fold SPCE enhancements, and was further utilized for mercury ion sensing as presented in figure 9.7. Figure 9.7(a) presents a schematic representation of the sensing strategy utilized for the sensing of Hg^{2+} ions. Here, it is observed that the AgNPs are initially synthesized by exposing a simple mixture of Ag^+ ions and lycoat polymer to UV irradiation. Further, the synthesized NPs were exposed to Hg^{2+} ion, which resulted in the release of Ag^+ ions into the solution with an effective dissolution of AgNPs. This presents an indirect method for sensing the Hg^{2+} ions where the LSPR of the AgNPs can be monitored using the UV–vis absorbance spectra. However, as observed in figure 9.7(b), we note that the sensing is dominant for higher concentrations and the negligible changes in the LSPR of AgNPs with low concentrations of Hg^{2+} ions cannot be detected. In this regard, to enhance the sensitivity of the platform, this was performed over the SPCE platform and the results are documented in figures 9.7(d) and (e). It is observed that the high

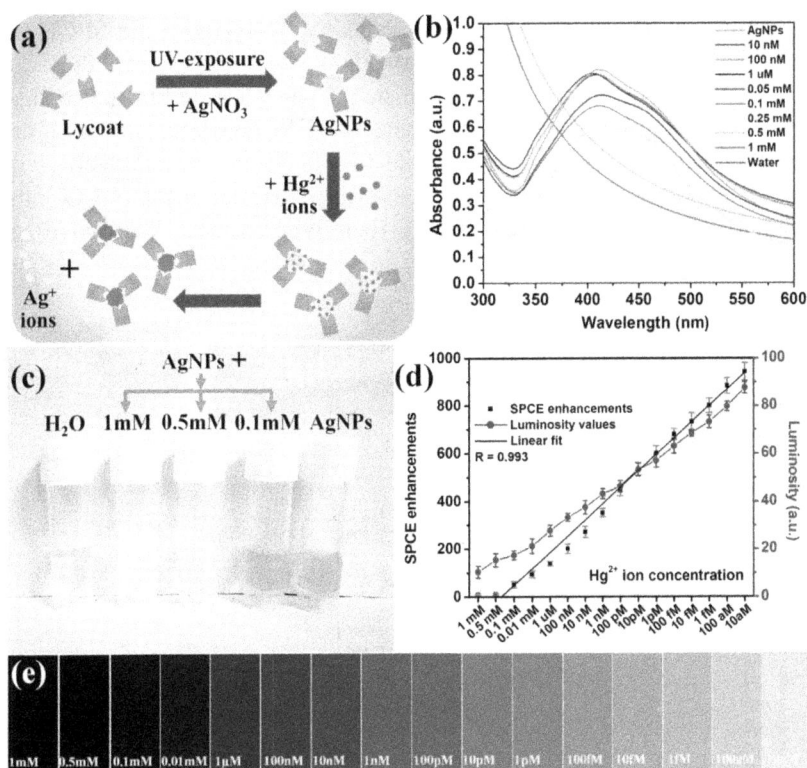

Figure 9.7. (a) Conceptual illustration of the steps involved in the disintegration of AgNPs to Ag$^+$ ions upon the addition of Hg^{2+} ions. (b) UV–vis absorbance spectra observed with different concentrations of Hg^{2+} ions addition. (c) Visual change in the color of the solution of AgNPs observed with addition of different concentrations of Hg^{2+} ions. (d) Correlation between the SPCE enhancements (gathered via spectrophotometer) and the luminosity data (gathered via smartphone). (e) Shade cards corresponding to different luminosity values shown in figure 9.8(d). Reprinted with permission from [55]. Copyright © 2021, American Chemical Society.

fluorescence enhancements realized with the lycoat-based AgNPs assisted in realization of better sensitivity. That is, the Hg^{2+} ions are detected in a long linear range as well as at ultra-low concentrations (attomolar limit of detection). Furthermore, the sensing is also validated using smartphone-based detection where the emission images were captured using the smartphone, analyzed using the color grab app to extract the luminosity values, and were correlated to the SPCE enhancements. An excellent correlation is observed between the SPCE enhancements and the luminosity values, thereby enabling low-cost and user-friendly smartphone-based analyte concentration monitoring.

9.7 Ferroplasmon-on-Mirror (FPoM): applications in early diagnostics

In order to explore the utility of the high refractive index (HRI) dielectric nanomaterials with a rod-shaped morphology, the rare-earth Nd$_2$O$_3$ nanorods were considered for the generation of plasmonic cavity hotspots in the SPCE platform [44]. The TEM image of the Nd_2O_3 nanorods is shown in figure 9.8(a). The rod-shaped morphology with a dielectric intrinsic property assisted in the realization of the quintessential lightning rod effect where the EM field intensity is significantly enhanced in the near field [44, 46]. While this facilitated the dielectric-enhanced fluorescence the immense forward light scattering attribute of the Nd_2O_3 nano-rods functioned as efficient 'Huygen sources', thereby presenting themselves as most

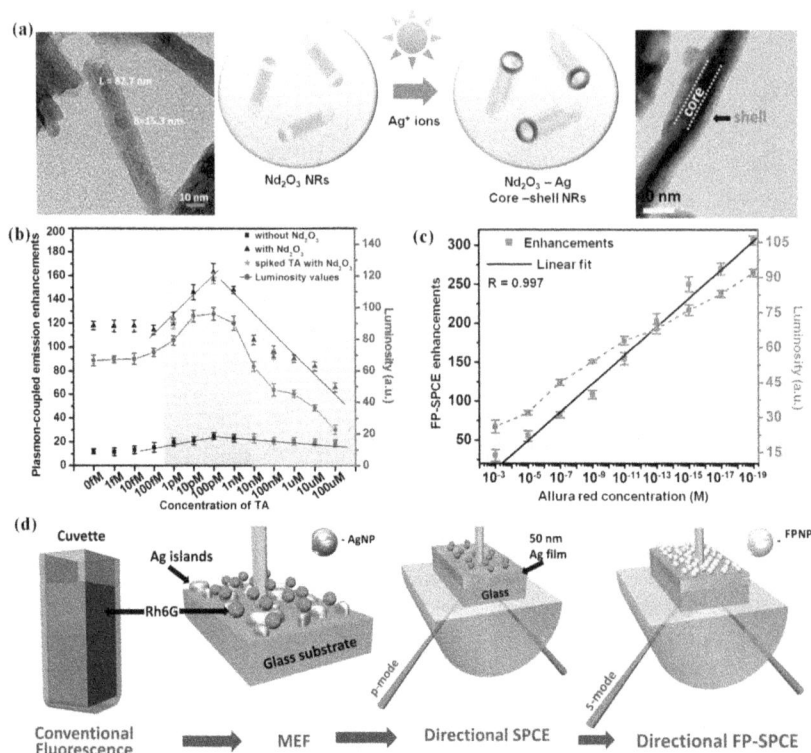

Figure 9.8. (a) Schematic representation of the formation of Nd_2O_3–Ag, core–shell nano-rods from Nd_2O_3 nano-rods, under the UV exposure. The representative TEM image is presented. (b) Left y-axis: modulation in plasmon-coupled emission enhancements with increasing tannic acid (TA) concentration (red stars indicate spiked samples). Right y-axis: luminosity plot presenting alteration in the emission intensity from Rh6G with increase in TA concentration. Bottom: gray scale shade card for different concentrations of TA. (a) and (b) Reprinted with permission from [44]. Copyright © 2019, American Chemical Society. (c) Attomolar sensing of Allura Red in ferroplasmon (FP)-SPCE platform using the cavity interface along with the luminosity plot. (d) Progress in Nanoparticle-on-mirror (NPoM) achieved with Ferroplasmon-on-Mirror (FPOM) to realize Ferroplamson-coupled SPCE substrate/platform. Here, we present such advancement with the use of high refractive index (HRI) and ferromagnetic Nd_2O_3–Ag nano-hybrids in conventional SPCE platform. The emission characteristic in conventional fluorescence and MEF is omnidirectional and only a part of it (<1%) is captured; the emission is highly directional (with >50% signal collection efficiency) and polarized in SPCE platform. Moreover, the polarization of the out-coupled emission can be tuned from p-polarized to s-polarized by changing the nano-interface from spacer to cavity for the same material. (c) and (d) Reprinted with permission from [57]. Copyright © 2021, American Chemical Society.

suitable candidates for SPCE platform. Different weight percentages of the Nd_2O_3 nano-rods were studied in different weight percentages of polymer matrix and explored in the SPCE platform to understand the photoplasmonic coupling efficiency. It was observed that the 0.01 wt percent Nd_2O_3 nano-rods had the highest fluorescence enhancements of the SPCE reporter molecule (RhB). This architecture was utilized for the detection of environmentally hazardous tannic acid in drinking water samples, by tapping the direct interaction of the Nd_2O_3 nano-rods with the tannic acid, and the results are presented in figure 9.8(b). Here, one can observe that the SPCE enhancements increased and decreased with the addition of Nd_2O_3 nano-rods to the tannic acid solution, with a limit of detection being 1 pM. Further details of the biosensor are presented in the related publication [44].

Further, a frugal and sustainable approach was adopted the synthesis of the metal–dielectric Ag–Nd_2O_3 nano-rods nano-hybrids and the resultant material portrayed intriguing characteristics. The Ag–Nd_2O_3 nano-rods nano-hybrids are fabricated using a user-friendly approach where the Nd_2O_3 nano-rods are admixed with Ag^+ ions and subjected to the high energy UV radiation, and the mechanism of the reaction can be explained as follows: Such an exposure to UV radiation triggers the photo-induced electron transfer from the Nd_2O_3 nano-rods to the adjacently located Ag^+ ions in the solution. Consequently, the monovalent Ag ions get reduced to the zero-valent-based Ag–Nd_2O_3 nano-rods and depending on the concentration of the metal ions and the time of UV exposure, the researcher can tune the observation of core–shell or decorated nano-architectures. In addition to the photo-induced electron transfer, the large oxide groups present over the Nd_2O_3 nano-rods assist in catalyzing the reaction on account of the formation of natural organic compounds via volatile organic species in the ambient conditions. With the different loading of Ag over the Nd_2O_3 nano-rods, the paramagnetic and diamagnetic nature of the Nd_2O_3 nano-rods and the AgNPs changed to ferromagnetic nature on account of the generation of the oxygen vacancies at the interface of interaction [57]. Further, such materials supporting ferromagnetic domains assist in the generation of metal–dielectric hybrid resonances as well as the electric and magnetic plasmonic hotspots in the nanocavities between the NPs and the metallic thin film. The extensive material characterization (surface area, microscopy, magnetic nature, bandgap calculations, and crystallographic properties) of the metal–dielectric nano-hybrids generated in this regard are presented for the different loading of silver in the related article [57]. The conceptual schematic as well the representative TEM image is presented in figure 9.8(a) where a simple mixture of Nd_2O_3 nano-rods and Ag^+ ions and the subsequent exposure to sunlight results in the generation of Ag–Nd_2O_3 nano-hybrid (with Nd_2O_3 nano-rod core and the Ag-shell). Interestingly, such materials where the dielectric nanomaterial is coated with plasmonic/metallic nanomaterial is expected to further enhance the refractive index of the bulk material, thereby enabling the formation of metasurfaces when studied over the metallic thin film [57]. An interesting observation that was made with the study of Ag–Nd_2O_3 nano-hybrid over the SPCE platform was the switch in the polarization of the emission attribute. Conventionally, the plasmon-coupled emission in the SPCE platform is highly directional and TM-polarized. With the use of

Ag–Nd$_2$O$_3$ nano-hybrid in the cavity nano-interface (where the radiating dipole is sandwiched between the NPs and the metallic thin film), the emission observed is completely TE-polarized. This switch in the polarization from TM to TE is attributed to the coupling of the radiating dipoles to the magnetic hotspots generated in the cavity nano-regime [57]. The high fluorescence enhancements realized using the Ag–Nd$_2$O$_3$ nano-hybrid in the SPCE platform was utilized for the detection of the food hazard allura red at attomolar concentrations as presented in figure 9.8(c). As the Ag–Nd$_2$O$_3$ nanohybrid presented strong ferromagnetic domains in addition to the stable LSPR resonance of the AgNPs present over the Nd$_2$O$_3$ nanorods, the resulting hybrid is termed as ferroplasmons rather than magnetoplasmons (which are nano-hybrids with magnetic property where the LSPR of the plasmonic counterpart might be concealed by the strong absorption of the magnetic NPs).

Both the sensing works of tannic acid (using Nd$_2$O$_3$ nanorods) and the allura red (using Ag–Nd$_2$O$_3$ nanohybrid) were performed using a mobile phone-based detection system. The luminosity values thus obtained are presented in the right y-axis of figures 9.8(b) and (c), respectively. The executive summary in the development of the SPCE technology using the ferroplasmons is given in figure 9.8(d). In brief, as the conventional fluorescence spectrophotometer reads the emission at 90°, the overall collection efficiency is lessened significantly. In order to enhance the fluorescence emission intensity, the metal NPs are explored along with the radiating dipoles over the glass slide, comprising metal-enhanced fluorescence (MEF). Furthermore, while the metal-radiating dipole combination is investigated over the SPCE platform the fluorescence enhancements increase as compared to that of the glass side, on account of the directional collection efficiency of the hybrid system. Essentially, it is to be noted that the addition of ferroplasmons over the SPCE platform results in the generation of FPoM configuration, which is analogous to the NPoM configuration that has been established for several years now. In this regard, the FPoM architectures are expected to revolutionize the SPCE portfolio, where the different types of magnetic NPs (superparamagnetic, antiferromagnetic, and other verities) can be explored with the novel material combinations for enhancing the performance of the biosensing devices. This is of paramount significance as such explorations can be coupled with the external magnetic field to drive the reaction in a short time interval, taking advantage of the intrinsic nature of the nano-hybrids, thereby significantly decreasing the time required for healthcare-based diagnostic applications.

9.8 Photonic crystal-coupled emission: applications in early diagnostics

Achieving high fluorescence enhancements is the need of the hour in order to attain enhanced sensitivity and selectivity of diagnostic tools. In this chapter, initially we discussed the efficiency of plasmonic metal nano-assemblies to outperform the pristine NPs in harvesting high fluorescence emission enhancements. While this has been demonstrated with the use of cryo-sorets, the rest of the chapter discusses heterometallic hybrid coupling as well as shape-edged plasmonic nano-structures resulting in augmented fluorescence enhancements. In the previous section, we presented the efficacy of HRI dielectric as well as metal–dielectric hybrid structures for realization of dramatically high enhancements in fluorescence signal intensity. This has been attributed to the less lossy nature of the dielectric structures to facilitate the reduction in Ohmic loss-induced quenching of fluorescence signal. However, it is important to note that the underlying substrate in all the cases discussed above relies on the SPCE substrate, which is essentially lossy from a plasmonics point of view. In order to overcome this drawback, in 2020 Bhaskar *et al* demonstrated the utility of alternating layers of quarter-wave plate thick high and low refractive index nanolayers to generate loss less dielectric-based fluorescence enhancements. The one-dimensional photonic crystal (1DPhC) was utilized to accomplish this goal and the conventional RK configuration was utilized for the realization of an effective platform termed photonic crystal-coupled emission (PCCE) that can outperform the SPCE platform. That is, while the SPCE platform assisted in the realization of 10-fold fluorescence enhancements, the PCCE platform presented 40-fold enhancements due to the loss less characteristic [22, 46].

Before discussing the biosensing performance of the PCCE platform, it is instructive to discuss the optical framework and the typical observations made using this platform. The experiments with the PCCE platform are performed in an optical framework similar to that of the SPCE platform, where the SPCE substrates is replaced with the 1DPhC. In a typical PCCE experiment, the laser excites the radiating dipoles from the samples side (also known as the free-space side), and the emission is captured from the curved surface of the prism. The blank fluorescence measurements on the PCCE substrate yielded four times higher enhancements in the fluorescence as compared to that obtained with the SPCE substrate. In order to further enhance the fluorescence intensity, different types of nanoengineering have been carried out over the PCCE substrate presenting interesting insights on the photoplasmonic coupling efficiency at the dielectric (1DPhC)-NP interface. While the plasmonic AgNPs yielded ~60-fold increase in the fluorescence intensity in the SPCE platform, ~200-fold augmented intensity was captured using the PCCE platform, on account of the loss Ohmic losses [88]. These high enhancements were utilized for the detection of aluminium ions at femtomolar limit of detection, and the details are presented in the related publication [88].

In an attempt to accomplish unprecedented fluorescence enhancements, from an extensive literature review it is important to judiciously integrate the novel functionalities of different sub-domains of the plasmonics field. By and large, metal-dependent plasmonics, dielectric-dependent plasmonics, graphene (and related 2D materials)-based plasmonics, and silicon plasmonics comprise the major four domains of the plasmonics portfolio [26]. While the strategic advancement from the SPCE to the PCCE platform has several advantages, the goal to attain significantly higher fluorescence enhancements demands the careful consideration of materials from different sub-domains of plasmonics, for carrying out the substrate engineering. Recently, three different material combinations have been explored in the PCCE platform, namely, (i) use of metal–dielectric nanohybrids, (ii) nano-assembly-dielectric hybrids, and (iii) graphene—anisotropic metal nanohybrids. In the first approach, silver-based metal nanoprisms were interfaced over the PCCE platform with and without HRI dielectric Nd_2O_3 nanorods. The high fluorescence enhancements obtained were utilized for the detection of biologically relevant cortisol molecule at single molecular limit of detection. Additionally, the close interaction between the sharp-edged silver nanoprisms with that of the iodide ions were incorporated to indirectly quantify the concentration of the iodide ion concentration. In the second approach, the soret nano-assemblies studied with HRI dielectric Nd_2O_3 nano-rods presented attomolar sensitivity of the SPCE reporter molecule (rhodamine). While these two approaches yielded interesting results enabling ultra-low concentration detection of desired analytes, the third approach utilized for realization of photoplasmonic hotspots is discussed in detail in this section.

Graphene and other low-dimensional materials have been attracting nano-scientists and physicists on account of their exceptional properties for applications in opto-electronics. For instance, the 0D carbon substrates have been utilized for light-emitting diode application while integrated with plasmonic AgNPs. Such AgCD nano-hybrids have found applications in other inter-disciplinary applications. Further, the plasmonic nature of graphene parent structure and analogues 2D materials such as WS_2 and MoS_2 are distinctive by themselves due to the existence of delocalizable π-electrons. Flatland photonics is an offshoot of such exploration with 2D materials and their functionalities has been explored in UV–vis as well as NIR and far-IR regions of the EM spectrum. Moreover, graphene oxide (GO) has been utilized for development of 1DPhC using alternating nanolayers of GO and PMMA to accomplish excellent biosensing platforms. In this background, on account of the unique attribute of GO to concentrate the EM radiation at nano-regimes, it has been evaluated over the PCCE platform. The π-plasmon of the GO assisted in the formation of hotspots over the PCCE platform thereby increasing the fluorescence intensity.

In order to fully exploit the delocalized plasmonic attribute of the GO monolayer of the PCCE substrate, it is pertinent to adopt that particular material that can promote extensive propagation of hybridized plasmons. Such an attempt would be futile if pristine plasmonic and/or dielectric NPs are investigated, because although GO renders extended interaction of π-plasmon with the underlying PCCE substrate, the plasmons in individual NPs are highly localized in space. Although sorets can yield delocalized plasmons to a certain extent, long-range hybrid coupling would not be feasible unless a plasmonic material with extremely high aspect ratio is adopted. In this regard, in order to further increase the signal intensity silver nano-wires (AgNWs) with high aspect ratio were interfaced on the PCCE platform where the GO is sandwiched between them. Such hybrid combination assists in the realization of extended propagation of plasmons with the use of AgNWs and GO.

The high >1300-fold PCCE enhancements obtained with the hybrid combination of AgNW and the GO were utilized for the detection of biologically relevant cholesterol and the results are presented in figure 9.9. The 1DPhC sustains surface electromagnetic waves called BSWs as well the photonic mode densities that are sustained within the 1DPhC termed internal optical modes (IOMs). While there are different reports highlighting the importance of BSWs for surface-based sensing applications, the utility of IOMs are seldom discussed from a sensing point of view.

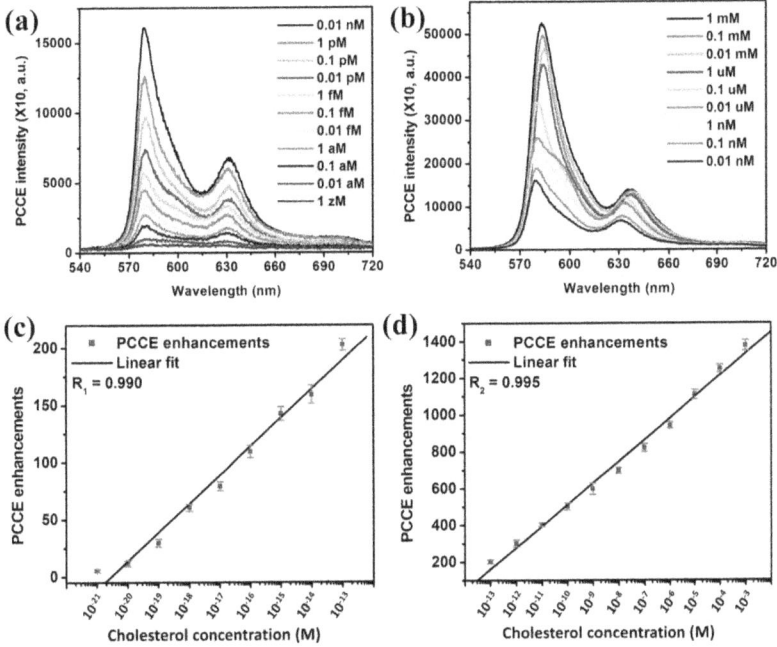

Figure 9.9. PCCE intensity spectra for different concentrations of cholesterol in the ranges from (a) 1 zM to 0.01 nM and (b) 0.01 nM to 1 mM. Linear calibration plots for sensing cholesterol in the ranges from (c) 1 zM to 0.1 pM and (d) 0.1 pM to 1 mM. Reprinted with permission from reference [38]. Copyright © 2023, American Chemical Society.

In this regard, the IOMs were studied for performing the sensing of cholesterol molecule at low concentrations. The ten bilayers of quarter-wave plate thickness 1DPhC made up of TiO_2 and SiO_2 nano-layers were utilized for carrying out the nano-engineering. The photoplasmonic coupling efficiency between the GO, AgNW, and the 1DPhC assisted in the realization of single-molecule detection of cholesterol. The rhodamine conjugated to cholesterol was synthesized using organic chemistry principles and the sensing performed by serially diluting the stock solution (1 mM) from 1 mM to 1 zM [38]. The decrease in the fluorescence intensity with decrease in the cholesterol concentration is shown in figures 9.9(a) and (b). The PCCE enhancements obtained as the ratio of the PCCE intensity counts to that of the FS intensity counts is presented in figures 9.9(c) and (d), respectively. A linear relationship observed with excellent limit of detection (0.01 aM) is attributed to the high EM field intensity at the nano-interface between the GO and AgNW where the cholesterol is sandwiched.

9.9 Futuristic scope and opportunities

Following a broad overview of photonic and plasmonic biosensing technologies, in this chapter we discussed in detail the latest advancements in SPCE and PCCE technology development. The effective nano-engineering protocols using cryo-soret technology, FPoM as well as with the judicious integration of metal–dielectric and heterometallic nanohybrids were discussed for enhancing the performance of plasmonic biosensors. These technologies present several insights for next-gen biosensing technology advancement due to their simple, user-friendly, rapid, and cost-effective modalities. The futuristic scope and opportunities therefrom can be categorized into two domains: (i) substate development and (ii) application-oriented device development. The first domain encompasses the exploration of different types of metal–dielectric nanohybrids for first-, second-, third-, and fourth- generation plasmonic hotspots. Further, the shape and the size of the nanomaterial explored can be tuned to obtain desired functionalities that are most suitable for the associated biosensing framework. Moreover, the chemistry and physical stability of the synthesized nanomaterial and/or nano-hybrid can be tailored to attain necessary properties. It is important to mention that the photoplasmonic interfaces are extensively studied using low-dimensional substrates as well as 2D materials such as graphene, WS_2, and MoS_2 to name a few. Such explorations over the SPCE and PCCE platforms along with inherently luminescent metal–dielectric nano-hybrids are expected to push the possibilities of realizing superior biosensing platforms. Further, the PCCE technology can be further developed using grating-based photonic crystals to extract beaming or steering emission characteristics that can significantly simplify the cost involved and the processing steps to perform ELISA experiments [13]. The incorporation of quantum dots and nano-diamonds in frameworks such as CSNE and FPoM would drastically modulate the lifetime and quantum yield of the luminescent moieties, and such explorations comprise the primary domain of the futuristic scope.

Furthermore, the second domain of the futuristic scope comprises the incorporation of the SPCE and PCCE technologies over flexible optoelectronic technologies, enabling the development of automated and autonomous platforms for real-time healthcare monitoring. Such explorations would assist in collecting the molecular profile information of patients from samples such as sweat, mucus, and urine so as to empower doctors and scientists in the development of precision and personalized medicine [7]. The realization of plasmonic-based wearable biosensors would find immediate application in screening large populations and advocating appropriate safety protocols, especially in scenarios pertaining to pandemics (such as COVID-19) and epidemics. Furthermore, the technologies discussed in this chapter can be integrated into private and societal healthcare monitoring infrastructures facilities such as transportation panels, universities, and office portfolios. Such real-time healthcare monitoring would generate awareness of regulating disease conditions as well as maintaining public health from biological threats [7]. The seamless incorporation of smartphone-based detection tools in such integrated healthcare facilities would essentially facilitate connection over the 'internet of things' engendering spatiotemporal health status dataset collection from inhabitants and migrants. Such explorations would necessitate the judicious combination of different artificial intelligence and neural network algorithms for analysis, correlation, storage, and interpretation of health status of the general public.

9.10 Conclusions

In order to achieve advanced healthcare technologies where the disease state can be continuously monitored at the bedside so as to obtain real-time feedback for physicians, in the recent past numerous approaches have been developed to enhance the sensitivity of the biosensing frameworks. While there are different methodologies adopted for the development of such biosensing frameworks, photonic and plasmonic biosensors have established a strong foothold on account of their easy operation, fast response time, and applicability for label-free biosensing. The first step that provides adequate liberty to the physician to tailor medication and/or quarantine regulations for a particular patient expected to be exposed to a disease condition is the sensitivity and selectivity of the biosensing framework. In this regard, the photoplasmonic frameworks are being refined using advanced nano-engineering techniques so as to realize higher sensitivity and augmented limit of detection of a probe molecule/a specific biomarker. In the past decade, new insights have been gathered with the advancement of surface plasmon-coupled emission (SPCE) technology [71, 72, 74], for not only detecting disease states, but also to analyze/monitor the structural changes in the biomarkers occurring during the course of disease incidence, progression, and termination. This chapter highlighted the recent advancements in the SPCE platform development for biosensing applications, where the utility of different nanoengineering strategies were succinctly presented. Following a broad overview of the relevance of photonic and plasmonic biosensing technologies, we presented the importance of cryo-soret nanoengineering for enhancing the performance of the SPCE framework. Further, the utility of smartphone-based diagnostic technologies were discussed by considering a few representative examples. The importance of chromaticity and luminosity analysis for obtaining an excellent correlation between the existing expensive methodologies vis-à-vis cost-effective mobile platforms was also discussed. Further, the importance and relevance of engineering dielectric nanomaterials over the SPCE platform were presented using a case study where high refractive index lanthanide oxide was studied with and without a plasmonic silver hybrid. The modulation in the physical, optical, and magnetic properties was discussed from the perspective of biosensing framework, where an effective platform termed 'ferroplamon-on-mirror' analogous to well-established 'nanoparticle-on-mirror' was discussed. Finally, this chapter also provided an elaborate discussion on improvising the shortcomings of the existing approaches in SPCE-based biosensing tools. In this context, the cogent progression from metal-dependent and lossy SPCE framework to metal-independent and non-lossy photonic crystal-coupled emission (PCCE) technology was deliberated. The single molecular limit of detection for analytes of interest was presented with rational combination of metal, dielectric, and graphene plasmonics. We believe this chapter will assist researchers working in the broad arena of photoplasmonics to plan and execute investigations catering to the development of devices with ultra-high sensitivity suitable for smartphone-based diagnostics enabling low-and-middle-income countries to use stable healthcare monitoring systems.

9.11 Exercises

1. What is the difference between the surface plasmon resonance (SPR) and Nanoparticle-on-Mirror (NPoM) configurations?
2. What is the meaning of hotspots in the context of nanoplasmonics?
3. Who innovated surface plasmon-coupled emission technology? State the relevance of the innovation from the perspective of fluorescence spectroscopy.
4. What are the differences between Kretchmann–Raether (KR) and reverse Kretchmann (RK) configuration?
5. List the advantages and disadvantages of SPCE technology.
6. State and explain the similarities and differences between FerroPlasmon-on-Mirror (FPoM) and Nanoparticle-on-Mirror (NPoM) configurations.
7. What are the fundamental differences between SPCE and PCCE biosensing platforms?
8. What are the different applications of SPR, SPCE, and PCCE biosensing frameworks?
9. What is the chromaticity approach utilized in the development of smartphone-based biosensors? Discuss with applications.
10. What is the luminosity approach utilized in the development of smartphone-based biosensors? Discuss with a few case studies.

References

[1] Agnihotri N, Chowdhury A D and De A 2015 Non-enzymatic electrochemical detection of cholesterol using β-cyclodextrin functionalized graphene *Biosens. Bioelectron.* **63** 212–7
[2] Bauch M, Toma K, Toma M, Zhang Q and Dostalek J 2014 Plasmon-enhanced fluorescence biosensors: a review *Plasmonics* **9** 781–99
[3] Borisov S M and Wolfbeis O S 2008 Optical biosensors *Chem. Rev.* **108** 423–61
[4] Li K, Wang J, Zhou W, Zeng S, Guo T and Wei L 2023 Lab-in-a-fiber biosensors *Microfluidic Biosensors* ed W C Mak and A H Pui Ho (New York: Academic) ch 3 pp 87–106
[5] Cunningham B T *et al* 2021 Photonic metamaterial surfaces for digital resolution biosensor microscopies using enhanced absorption, scattering, and emission *Proc. of the Integrated Sensors for Biological and Neural Sensing* (Bellingham, WA) (SPIE) vol 11663 pp 50–9
[6] Cunningham B, Lin B, Qiu J, Li P, Pepper J and Hugh B 2002 A plastic colorimetric resonant optical biosensor for multiparallel detection of label-free biochemical interactions *Sens. Actuators B Chem.* **85** 219–26
[7] Altug H, Oh S-H, Maier S A and Homola J 2022 Advances and applications of nanophotonic biosensors *Nat. Nanotechnol.* **17** 5–16
[8] Mao Z, Peng X, Zhou Y, Liu Y, Koh K and Chen H 2022 Review of interface modification based on 2D nanomaterials for surface plasmon resonance biosensors *ACS Photonics* **9** 3807–23
[9] Xiong Y *et al* 2022 Photonic crystal enhanced fluorescence emission and blinking suppression for single quantum dot digital resolution biosensing *Nat. Commun.* **13** 4647

[10] Dutta Choudhury S, Badugu R and Lakowicz J R 2015 Directing fluorescence with plasmonic and photonic structures *Acc. Chem. Res.* **48** 2171–80

[11] Haleem A, Javaid M, Singh R P, Suman R and Rab S 2021 Biosensors applications in medical field: a brief review *Sens. Int.* **2** 100100

[12] Wang X *et al* A target recycling amplification process for the digital detection of exosomal microRNAs through photonic resonator absorption microscopy *Angew. Chem. Int. Ed.* **62** e202217932

[13] Xiong Y, Shepherd S, Tibbs J, Bacon A, Liu W, Akin L D, Ayupova T, Bhaskar S and Cunningham B T 2023 Photonic crystal enhanced fluorescence: a review on design strategies and applications *Micromachines* **14** 668

[14] Xiong Y *et al* 2022 Microscopies enabled by photonic metamaterials *Sensors* **22** 1086

[15] Li N *et al* 2022 Label-free digital detection of intact virions by enhanced scattering microscopy *J. Am. Chem. Soc.* **144** 1498–502

[16] Liu L, Tibbs J, Li N, Bacon A, Shepherd S, Lee H, Chauhan N, Demirci U, Wang X and Cunningham B T 2023 A photonic resonator interferometric scattering microscope for label-free detection of nanometer-scale objects with digital precision in point-of-use environments *Biosens. Bioelectron.* **228** 115197

[17] Bhaskar S 2023 Biosensing technologies: a focus review on recent advancements in surface plasmon coupled emission *Micromachines* **14** 574

[18] Li L, Zhang Y, Zhou Y, Zheng W, Sun Y, Ma G and Zhao Y 2021 Optical fiber optofluidic bio-chemical sensors: a review *Laser Photonics Rev.* **15** 2000526

[19] Cao S-H, Cai W-P, Liu Q and Li Y-Q 2012 Surface plasmon–coupled emission: what can directional fluorescence bring to the analytical sciences? *Annu. Rev. Anal. Chem.* **5** 317–36

[20] Bhaskar S, Kowshik N C S S, Chandran S P and Ramamurthy S S 2020 Femtomolar detection of spermidine using Au decorated SiO_2 nanohybrid on plasmon-coupled extended cavity nanointerface: a smartphone-based fluorescence dequenching approach *Langmuir* **36** 2865–76

[21] Rai A, Bhaskar S, Reddy N and Ramamurthy S S 2021 Cellphone-aided attomolar zinc ion detection using silkworm protein-based nanointerface engineering in a plasmon-coupled dequenched emission platform *ACS Sustain. Chem. Eng.* **9** 14959–74

[22] Bhaskar S, Singh A K, Das P, Jana P, Kanvah S, Bhaktha B N S and Ramamurthy S S 2020 Superior resonant nanocavities engineering on the photonic crystal-coupled emission platform for the detection of femtomolar iodide and zeptomolar cortisol *ACS Appl. Mater. Interfaces* **12** 34323–36

[23] Rai A, Bhaskar S and Ramamurthy S S 2021 Plasmon-coupled directional emission from soluplus-mediated AgAu nanoparticles for attomolar sensing using a smartphone *ACS Appl. Nano Mater.* **4** 5940–53

[24] Tosi D, Shaimerdenova M, Sypabekova M and Ayupova T 2022 Minimalistic design and rapid-fabrication single-mode fiber biosensors: review and perspectives *Opt. Fiber Technol.* **72** 102968

[25] Shaimerdenova M, Ayupova T, Sypabekova M and Tosi D 2020 Fiber optic refractive index sensors based on a ball resonator and optical backscatter interrogation *Sensors* **20** 6199

[26] Jahani S and Jacob Z 2016 All-dielectric metamaterials *Nat. Nanotechnol.* **11** 23–36

[27] Khurgin J B 2015 How to deal with the loss in plasmonics and metamaterials *Nat. Nanotechnol.* **10** 2–6

[28] Bhaskar S, Das P, Moronshing M, Rai A, Subramaniam C, Bhaktha S B N and Ramamurthy S S 2021 Photoplasmonic assembly of dielectric-metal, Nd_2O_3-gold soret nanointerfaces for dequenching the luminophore emission *Nanophotonics* **10** 3417–31

[29] Bhaskar S, Visweswar Kambhampati N S, Ganesh K M, Mahesh Sharma P, Srinivasan V and Ramamurthy S S 2021 Metal-free, graphene oxide-based tunable soliton and plasmon engineering for biosensing applications *ACS Appl. Mater. Interfaces* **13** 17046–61

[30] Rai A, Bhaskar S, Ganesh K M and Ramamurthy S S 2022 Hottest hotspots from the coldest cold: welcome to nano 4.0 *ACS Appl. Nano Mater.* **5** 12245–64

[31] Bhaskar S and Ramamurthy S S 2021 High refractive index dielectric TiO_2 and graphene oxide as salient spacers for >300-fold enhancements *Proc. of the 2021 IEEE Int. Conf. on Nanoelectronics, Nanophotonics, Nanomaterials, Nanobioscience & Nanotechnology (5NANO)* pp 1–6

[32] Chauhan N *et al* 2022 Net-shaped DNA nanostructures designed for rapid/sensitive detection and potential inhibition of the SARS-CoV-2 virus *J. Am. Chem. Soc.* **145** 20214–8

[33] Ganesh N, Zhang W, Mathias P C, Chow E, Soares J A N T, Malyarchuk V, Smith A D and Cunningham B T 2007 Enhanced fluorescence emission from quantum dots on a photonic crystal surface *Nat. Nanotechnol.* **2** 515–20

[34] Arora D, Tan H R, Wu W-Y and Chan Y 2022 2D-oriented attachment of 1D colloidal semiconductor nanocrystals via an etchant *Nano Lett.* **22** 942–7

[35] Noori Y J, Cao Y, Roberts J, Woodhead C, Bernardo-Gavito R, Tovee P and Young R J 2016 Photonic crystals for enhanced light extraction from 2D materials *ACS Photonics* **3** 2515–20

[36] Rathnakumar S, Bhaskar S, Badiya P K, Sivaramakrishnan V, Srinivasan V and Ramamurthy S S 2023 Electrospun PVA nanofibers doped with titania nanoparticles in plasmon-coupled fluorescence studies: an eco-friendly and cost-effective transition from 2D nano thin films to 1D nanofibers *MRS Commun.* **13** 290–8

[37] Rai B, Sarma P V, Srinivasan V, Shaijumon M M and Ramamurthy S S 2021 Engineering of exciton–plasmon coupling using 2D-WS_2 nanosheets for 1000-fold fluorescence enhancement in surface plasmon-coupled emission platforms *Langmuir* **37** 1954–60

[38] Bhaskar S, Lis S S M, Kanvah S, Bhaktha BN S and Ramamurthy S S 2023 Single-molecule cholesterol sensing by integrating silver nanowire propagating plasmons and graphene oxide π-plasmons on a photonic crystal-coupled emission platform *ACS Appl. Opt. Mater.* **1** 159–72

[39] Bhaskar S, Jha P, Subramaniam C and Ramamurthy S S 2021 Multifunctional hybrid soret nanoarchitectures for mobile phone-based picomolar Cu^{2+} Ion sensing and dye degradation applications *Phys. E Low-Dimens. Syst. Nanostruct.* **132** 114764

[40] Bhaskar S, Rai A, Mohan G K and Ramamurthy S S 2022 Mobile phone camera-based detection of surface plasmon-coupled fluorescence from streptavidin magnetic nanoparticles and graphene oxide hybrid nanointerface *ECS Trans.* **107** 3223

[41] Rai A, Bhaskar S, Mohan G K and Ramamurthy S S 2022 Biocompatible gellucire® inspired bimetallic nanohybrids for augmented fluorescence emission based on graphene oxide interfacial plasmonic architectures *ECS Trans.* **107** 4527

[42] Lakowicz J R, Ray K, Chowdhury M, Szmacinski H, Fu Y, Zhang J and Nowaczyk K 2008 Plasmon-controlled fluorescence: a new paradigm in fluorescence spectroscopy *Analyst* **133** 1308–46

[43] Lakowicz J R (ed) 2006 *Principles of Fluorescence Spectroscopy* (Boston, MA: Springer)

[44] Bhaskar S and Ramamurthy S S 2019 Mobile phone-based picomolar detection of tannic acid on Nd_2O_3 nanorod–metal thin-film interfaces *ACS Appl. Nano Mater.* **2** 4613–25

[45] Bhaskar S and Ramamurthy S S 2021 Synergistic coupling of titanium carbonitride nanocubes and graphene oxide for 800-fold fluorescence enhancements on smartphone based surface plasmon-coupled emission platform *Mater. Lett.* **298** 130008

[46] Bhaskar S, Das P, Srinivasan V, Bhaktha S B N and Ramamurthy S S 2022 Plasmonic-silver sorets and dielectric-Nd_2O_3 nanorods for ultrasensitive photonic crystal-coupled emission *Mater. Res. Bull.* **145** 111558

[47] Lakowicz J R 2004 Radiative decay engineering 3. Surface plasmon-coupled directional emission *Anal. Biochem.* **324** 153–69

[48] Gryczynski I, Malicka J, Gryczynski Z and Lakowicz J R 2004 Radiative decay engineering 4. Experimental studies of surface plasmon-coupled directional emission *Anal. Biochem.* **324** 170–82

[49] Badugu R, Descrovi E and Lakowicz J R 2014 Radiative decay engineering 7: Tamm state-coupled emission using a hybrid plasmonic–photonic structure *Anal. Biochem.* **445** 1–13

[50] Chowdhury M H, Ray K, Geddes C D and Lakowicz J R 2008 Use of silver nanoparticles to enhance surface plasmon-coupled emission (SPCE) *Chem. Phys. Lett.* **452** 162–7

[51] Cao S-H, Cai W-P, Liu Q, Xie K-X, Weng Y-H, Huo S-X, Tian Z-Q and Li Y-Q 2014 Label-free aptasensor based on ultrathin-linker-mediated hot-spot assembly to induce strong directional fluorescence *J. Am. Chem. Soc.* **136** 6802–5

[52] Bhaskar S, Rai A, Ganesh K M, Reddy R, Reddy N and Ramamurthy S S 2022 Sericin-based bio-inspired nano-engineering of heterometallic AgAu nanocubes for attomolar mefenamic acid sensing in the mobile phone-based surface plasmon-coupled interface *Langmuir* **38** 12035–49

[53] Liu Q, Cao S-H, Cai W-P, Liu X-Q, Weng Y-H, Xie K-X, Huo S-X and Li Y-Q 2015 Surface plasmon coupled emission in micrometer-scale cells: a leap from interface to bulk targets *J. Phys. Chem.* B **119** 2921–7

[54] Xie K-X, Xu L-T, Zhai Y-Y, Wang Z-C, Chen M, Pan X-H, Cao S-H and Li Y-Q 2019 The synergistic enhancement of silver nanocubes and graphene oxide on surface plasmon-coupled emission *Talanta* **195** 752–6

[55] Rathnakumar S, Bhaskar S, Rai A, Saikumar D V V, Kambhampati N S V, Sivaramakrishnan V and Ramamurthy S S 2021 Plasmon-coupled silver nanoparticles for mobile phone-based attomolar sensing of mercury ions *ACS Appl. Nano Mater.* **4** 8066–80

[56] Xu L-T, Chen M, Weng Y-H, Xie K-X, Wang J, Cao S-H and Li Y-Q 2022 Label-free fluorescent nanofilm sensor based on surface plasmon coupled emission: *in situ* monitoring the growth of metal–organic frameworks *Anal. Chem.* **94** 6430–5

[57] Bhaskar S, Srinivasan V and Ramamurthy S S 2023 Nd_2O_3-Ag nanostructures for plasmonic biosensing, antimicrobial, and anticancer applications *ACS Appl. Nano Mater.* **6** 1129–45

[58] Balaji V R, Ibrar Jahan M A, Sridarshini T, Geerthana S, Thirumurugan A, Hegde G, Sitharthan R and Dhanabalan S S 2024 Machine learning enabled 2D photonic crystal biosensor for early cancer detection *Measurement* **224** 113858

[59] Sridarshini T, Geerthana S, Balaji V R, Thirumurugan A, Sitharthan R, Sivanantha Raja A and Dhanabalan S S 2023 Ultra-compact all-optical logical circuits for photonic integrated circuits *Laser Phys.* **33** 076207

[60] Geerthana S, Syedakbar S, Sridarshini T, Balaji V R, Sitharthan R and Sundar D S 2022 2D-PhC based all optical AND, OR and EX-OR logic gates with high contrast ratio operating at C band *Laser Phys.* **32** 106201

[61] Sivaranjani R, Sundar D S, Sridarshini T, Sitharthan R, Karthikeyan M, Raja A S and Carrasco M F 2020 Photonic crystal based all-optical half adder: a brief analysis *Laser Phys.* **30** 116205

[62] Sridarshini T, Chidambaram P, Geerthana S, Balaji V R, Thirumurugan A, Sitharthan , Madurakavi K and Dhanabalan S S 2022 Current and future horizon of optics and photonics in environmental sustainability *Sustain. Comput. Inform. Syst.* **36** 100815

[63] Rama Prabha K and Robinson S 2023 Two-dimensional photonic crystal-based filters review *Photonic Crystal and Its Applications for Next Generation Systems* (Singapore: Springer) pp 91–112

[64] Hamza M E, Othman M A and Swillam M A 2022 Plasmonic biosensors: review *Biology* **11** 621

[65] Inan H, Poyraz M, Inci F, Lifson M A, Baday M, Cunningham B T and Demirci U 2017 Photonic crystals: emerging biosensors and their promise for point-of-care applications *Chem. Soc. Rev.* **46** 366–88

[66] Bekmurzayeva A, Ashikbayeva Z, Myrkhiyeva Z, Nugmanova A, Shaimerdenova M, Ayupova T and Tosi D 2021 Label-free fiber-optic spherical tip biosensor to enable picomolar-level detection of CD44 protein *Sci. Rep.* **11** 19583

[67] Leitão C, Pereira S O, Marques C, Cennamo N, Zeni L, Shaimerdenova M, Ayupova T and Tosi D 2022 Cost-effective fiber optic solutions for biosensing *Biosensors* **12** 575

[68] Tokel O, Inci F and Demirci U 2014 Advances in plasmonic technologies for point of care applications *Chem. Rev.* **114** 5728–52

[69] Yu T and Wei Q 2018 Plasmonic molecular assays: recent advances and applications for mobile health *Nano Res.* **11** 5439–73

[70] Borghei Y-S, Hosseinkhani S and Ganjali M R 2022 'Plasmonic nanomaterials': an emerging avenue in biomedical and biomedical engineering opportunities *J. Adv. Res.* **39** 61–71

[71] Bhaskar S, Moronshing M, Srinivasan V, Badiya P K, Subramaniam C and Ramamurthy S S 2020 Silver soret nanoparticles for femtomolar sensing of glutathione in a surface plasmon-coupled emission platform *ACS Appl. Nano Mater.* **3** 4329–41

[72] Rai A, Bhaskar S, Ganesh K M and Ramamurthy S S 2021 Engineering of coherent plasmon resonances from silver soret colloids, graphene oxide and Nd_2O_3 nanohybrid architectures studied in mobile phone-based surface plasmon-coupled emission platform *Mater. Lett.* **304** 130632

[73] Rai A, Bhaskar S, Ganesh K M and Ramamurthy S S 2022 Gelucire®-mediated hetero-metallic AgAu nanohybrid engineering for femtomolar cysteine detection using smartphone-based plasmonics technology *Mater. Chem. Phys.* **279** 125747

[74] Geddes C D 2017 *Surface Plasmon Enhanced, Coupled and Controlled Fluorescence* (New York: Wiley) https://wiley.com/en-us/Surface+Plasmon+Enhanced,+Coupled+and+Controlled+Fluorescence-p-9781118027936

[75] Bhaskar S, Thacharakkal D, Ramamurthy S S and Subramaniam C 2023 Metal–dielectric interfacial engineering with mesoporous nano-carbon florets for 1000-fold fluorescence enhancements: smartphone-enabled visual detection of perindopril erbumine at a single-molecular level *ACS Sustain. Chem. Eng.* **11** 78–91

[76] Badugu R, Szmacinski H, Ray K, Descrovi E, Ricciardi S, Zhang D, Chen J, Huo Y and Lakowicz J R 2015 Metal–dielectric waveguides for high-efficiency coupled emission *ACS Photonics* **2** 810–5

[77] Moronshing M and Subramaniam C 2018 Room temperature, multiphasic detection of explosives, and volatile organic compounds using thermodiffusion driven soret colloids *ACS Sustain. Chem. Eng.* **6** 9470–9

[78] Sudha Maria Lis S, Bhaskar S, Dahiwadkar R, Kanvah S, Ramamurthy S S and Bhaktha B N S 2023 Plasmon-rich BCZT nanoparticles in the photonic crystal-coupled emission platform for cavity hotspot-driven attomolar sensing *ACS Appl. Nano Mater.* **6** 19312–26

[79] Bhaskar S and Sathish Ramamurthy S 2023 Performance enhancement of light emitting radiating dipoles (LERDs) using surface plasmon-coupled and photonic crystal-coupled emission platforms *Organic and Inorganic Light Emitting Diodes: Reliability Issues and Performance Enhancement* (Boca Raton, FL: CRC Press) pp 161–84

[80] Cheerala V S K, Ganesh K M, Bhaskar S, Ramamurthy S S and Neelakantan S C 2023 Smartphone-based attomolar cyanide ion sensing using Au-graphene oxide cryosoret nano-assembly and benzoxazolium-based fluorophore in a surface plasmon-coupled enhanced fluorescence interface *Langmuir* **39** 7939–57

[81] Thacharakkal D, Bhaskar S, Sharma T, Rajaraman G, Sathish Ramamurthy S and Subramaniam C 2023 Plasmonic synergism in tailored metal–carbon interfaces for real-time single molecular level sniffing of PFOS and PFOA *Chem. Eng. J.* **480** 148166

[82] Ganesh K M, Bhaskar S, Cheerala V S K, Battampara P, Reddy R, Neelakantan S C, Reddy N and Ramamurthy S S 2024 Review of gold nanoparticles in surface plasmon-coupled emission technology: effect of shape, hollow nanostructures, nano-assembly, metal–dielectric and heterometallic nanohybrids *Nanomaterials* **14** 111

[83] Ganesh K M, Rai A, Bhaskar S, Reddy N and Ramamurthy S S 2023 Plasmon-enhanced fluorescence from synergistic engineering of graphene oxide and sharp-edged silver nanorods mediated with castor protein for cellphone-based attomolar sensing *J. Lumin.* **260** 119835

[84] Ganesh K M, Rai A, Battampara P, Reddy R, Bhaskar S, Reddy N and Ramamurthy S S 2023 Optical coupling of bio-inspired mustard protein-based bimetallic nanohybrids with propagating surface plasmon polaritons for femtomolar nitrite ion sensing: cellphone-based portable detection device *Nano-Struct. Nano-Objects* **35** 101025

[85] Ding S-Y, Yi J, Li J-F, Ren B, Wu D-Y, Panneerselvam R and Tian Z-Q 2016 Nanostructure-based plasmon-enhanced Raman spectroscopy for surface analysis of materials *Nat. Rev. Mater.* **1** 1–16

[86] Rai A, Bhaskar S, Ganesh K M and Ramamurthy S S 2022 Cellphone-based attomolar tyrosine sensing based on Kollidon-mediated bimetallic nanorod in plasmon-coupled directional and polarized emission architecture *Mater. Chem. Phys.* **285** 126129

[87] Rai A, Bhaskar S, Battampara P, Reddy N and Sathish Ramamurthy S 2022 Integrated photo-plasmonic coupling of bioinspired sharp-edged silver nano-particles with nano-films in extended cavity functional interface for cellphone-aided femtomolar sensing *Mater. Lett.* **316** 132025

[88] Bhaskar S, Das P, Srinivasan V, Bhaktha B N S and Ramamurthy S S 2020 Bloch Surface waves and internal optical modes-driven photonic crystal-coupled emission platform for femtomolar detection of aluminum ions *J. Phys. Chem. C* **124** 7341–52

IOP Publishing

Advances in All-optical Communication

Shanmuga Sundar Dhanabalan, Arun Thirumurugan and Sridarshini Thirumaran

Chapter 10

Integrated photonic devices for cancer detection

M A Ibrar Jahan, V R Balaji, N R G Sreevani, D Sasikala, R G Jesuwanth Sugesh, C Jenila and Gopalkrishna Hegde

Cancer, alongside coronary heart disease and other major diseases, is a leading cause of death worldwide. If cancer cells are found at an early stage, there is a chance that a person's life can be saved. One of the important biological parameters that can be used to detect, identify, and segregate healthy and cancerous cells is their optical properties. Refractive index (RI)-based identification is one of the strategies utilized by numerous researchers, as it provides valuable information about cell anomalies. In recent years, advances in semiconductor nanofabrication technology, System-on-Chip (SoC), and miniaturized lab-on-a-chip components have allowed for major advancements in silicon photonic-based biosensors for the diagnosis of diseases. The integrated photonic devices utilizing evanescent waves offer benefits like high reliability, miniaturization, robustness, compactness, and high sensitivity. In particular, different silicon photonic structures based on the evanescent wave such as photonic crystal, waveguide Bragg grating-based devices, and SPR-based devices are available for the detection of cancerous cells. In this chapter, we introduce the various structures and then compare and contrast them in terms of their performance.

10.1 Introduction

The field of medical diagnostics has emerged as an important component of modern healthcare since it allows for the early identification and diagnosis of disease, as well as the enhancement of prompt and suitable care [1, 2], the protection of the safety of medical supplies like blood used for transfusions, and the reduction of expenses associated with healthcare [3, 4]. The vast majority of diagnostic systems have been developed to meet the requirements of clinical laboratories that are well-funded and operate in highly regulated environments [5]. However, these systems do not address the needs of the majority of patients and caretakers in developing countries, which typically have inadequate healthcare facilities and clinical laboratories. For example, the enzyme-linked immune sorbent assay (ELISA), which has been approved for

doi:10.1088/978-0-7503-5623-7ch10

use in biomarker detection for more than 40 years and is considered the gold-standard technology, is capable of obtaining an ultra-low detection limit. This method, on the other hand, is based on a label-based approach, which slows down the results, increases costs due to the use of specialized reagents, and necessitates complicated micro-evaluations utilizing huge automated analyzers [6]. Therefore, highly sensitive, rapid, and cost-effective methods of analysis are required for point-of-care (POC) diagnostic applications in both developing nations and developed countries, to improve access to healthcare technologies that are more cost-effective [7]. The ever increasing demand for reliable medical diagnostic technologies can be partially met by the creation of useful biosensors, which is one of the most promising new techniques. The global market for fiber-optic sensors is anticipated to reach a value of $4.9 billion by 2025, with an overall annual growth rate (CAGR) of 10.9% [8]. One of the primary motivating factors for the development of Si nanophotonics is its compatibility with more advanced complementary metal–oxide–semiconductor (CMOS) technology. This compatibility makes it possible to manufacture photonic-integrated circuits (PICs) in high volume at low cost. Optical measurements can be performed using a wide number of methods, including emission, absorption, fluorescence, refractometry, and polarimetry, all of which have been used effectively in the past. The detection of evanescent fields is the basic detection principle utilized by an extensive number of optical biosensors [9, 10].

In recent years, there has been a lot of focus placed on RI sensors that are based on lab-on-a-chip photonic structures. These sensors are becoming increasingly important since they are small in size, unaffected by electromagnetic interference, and can be easily integrated with other photonic components on a single chip [11–13].

Thus, extremely sensitive RI sensors that are based on photonic structures contribute to cost-effective, rapid, and precise determination of substances, making them appropriate for environmental monitoring, food quality control, and most importantly, medical diagnosis. Some examples of medical diagnosis include detecting bacterial infections, cancer, disorders, various physical and biological parameters for investigating and treating health conditions, and so on [14–17].

Cancer is currently one of the most widespread diseases in the world, and it has developed into a significant issue for the well-being of humans. Additionally, it is a type of virus that is known as an oncovirus. The impact of cancer's presence is determined by both the size of the virus and its divergence around the region of the cell. The smaller the virus, the greater the divergence. The rising incidence of cancer makes it critically important to develop the subsequent generation of cancer testing technology as quickly as possible to detect cancer cells at the earliest feasible stage. Early identification of cancer refers to the finding of tumors in the disease when it is still in its earliest stages of development. It is believed that early detection helps in the early recovery of patients. A comparison of existing technology and advance technology is given in figure 10.1.

Optical determination of a specific protein (Oestrogen Receptor [ER]) of breast cancer cells utilizing micro-structured fiber, detection of breast cancer utilizing light thermal effect on nanocomposite, and detection of cancer presence in body fluids such as urine, serum, blood, and tumour cells are some of the concepts that have been presented [18–21].

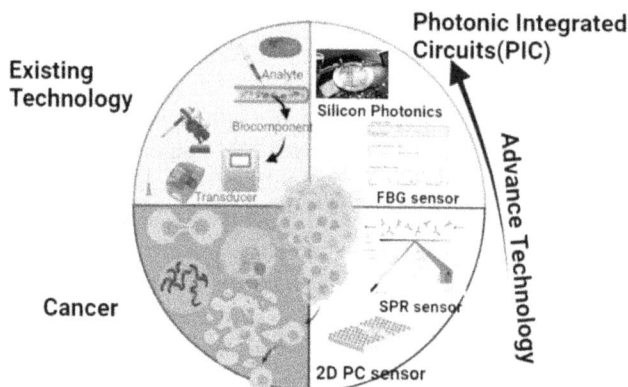

Figure 10.1. Comparison of existing and advance technologies.

Both SPR (surface plasmon resonance) and PIC have the advantage of being able to identify molecules in real time without the use of a label. In addition, they can detect numerous analytes inside a single sample by employing multiple sensor components concurrently, a process known as 'multiplexing'. In the case of SPR, this is accomplished by dividing the surface of the sensor into several different sensing locations. With the assistance of a digital image, a format known as a multi-array makes it possible to monitor the binding of hundreds of receptors to their targets at the same time. The intensity of the binding is represented by a scale of colours. When it comes to PICs, multiplexing is achieved by simply realizing numerous sensors on the same chip [22].

This chapter provides a comprehensive study of the development and performance of biosensors based on integrated photonic sensors such as waveguide Bragg grating, photonic crystal, and SPR for the detection of cancer cells. The chapter also includes a performance comparison of these sensors and their potential applications in the medical field for the diagnosis of other diseases.

10.2 Surface plasmon resonance (SPR)-based biosensors

SPR technology is one of the most promising tools since it is continuously improving and evolving to be ideal for user-friendly handheld devices. This makes it one of the most promising devices. The SPR approach has been proven to be one of the most versatile frameworks for the use of biosensors in some of the most pressing fields of medicine, biology, the environment, and food safety. With its label-free, real-time technique, SPR biosensors offer exceptional analytical performance in terms of high sensitivity, fast response, LOD, and reproducibility [23].

10.2.1 Principle of SPR

When a photon of incident light strikes the surface of a metal (gold layer), an effect known as SPR takes place as shown in figure 10.2.

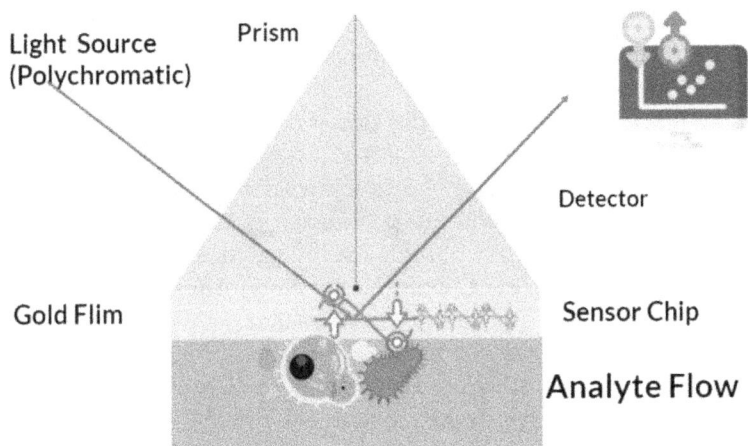

Figure 10.2. Basic principle of SPR.

A certain amount of the energy from light leaks through the coating of metal with the electrons on the metallic surface layer at a specific angle of incidence. This causes the electrons to accelerate due to excitation. These electron waves are now referred to as plasmons, and they travel in a direction that is parallel to the surface of the metal. An electric field is produced as a result of the plasmon oscillation, and its range is approximately 300 nm from the border between the sample solution and the metal surface. In a commercial SPR biosensor design, incoming light is utilized by employing a high-reflective index glass prism using the Kretschmann geometries of the attenuated total reflection (ATR) approach. Under the conditions of a light source with a constant wavelength and a metal thin surface, the determined SPR angle at which resonance occurs depends on the RI of the substance that is close to the metal surface. Plasmon cannot be created as a result when there is even a slight shift in the reflective index of the medium that is being sensed (e.g., due to the attachment of biomolecules). Measuring the variations in the reflected light that is collected by a detector is the method by which detection is carried out. In addition, the quantity of the surface concentration can be determined by observing the intensity of the reflected light or following the angle at which the resonance shifts. Both of these methods are described below. When developing SPR biosensors, probing molecules are initially immobilized on the surface of the sensor. When a mixture of target molecules flows in the vicinity of the sensor's surface, probe-target bonding via affinity interaction ensues. This, in turn, generates a change in the RI on the SPR sensing surface. Resonance or Response Units (RU) represent the change in the signal during the SPR experiment. One RU is equal to a shift in the critical angle of 10^{-4}. The RI change is given by

$$\Delta n_{d=(dn/dc)\mathrm{vol}\Delta\Gamma/h}$$

where Δn_d is the change in RI and h is the layer thickness; (dn/dc)vol is the RI increment as the volume concentration of the analyte increases; and Δ is the surface concentration of the bound target.

When the light couples into a propagating surface it allows for the real-time tracking of the change in the RI. In addition, the parameter associated with the RI can be used to identify and quantify target molecules when they are bound to a known probe that is immobilized on the surface of the sensor. In the SPR experiment, the limit of detection (LOD) is determined by a variety of factors. These factors include the molecular weight, optical property, and binding affinity of target-probe molecules, in addition to the surface area of the probing molecule. The SPR biosensing method seems to provide one of the most effective approaches for detecting the affinity binding of biomolecules and performing primary screening for druggable compounds. In the field of biomedical research, SPR-type sensors are being utilized more frequently to investigate a wide range of biological entities, including proteins, RNAs, DNAs, lipids, and carbohydrates as well as cells. However, a challenge that has been encountered with SPR technology is that its applicability for point-of-care (POC) diagnosis has not yet been resolved entirely. The potential interference of non-specific binds to the result signals is a limitation that is inherently present in SPR and is one of the method's drawbacks. In addition, the data on the dynamic rate and equilibrium values for biomolecular interactions that are provided by SPR biosensors are of little use in POC analysis. SPR technology is considered to be one of the most promising tools because it is continually improving and developing to be more suitable for user-friendly handheld devices. This is particularly relevant when considering the growing trend to employ biosensors in point-of-care testing (POCT), which is frequently carried out by using compact handheld devices. However, existing SPR devices remain cumbersome and expensive, which is a barrier that prevents the commercial use of the SPR technique. To make SPR an indispensable instrument for point-of-care (POC) diagnostics and routine clinical analysis, additional research will be required over the next several years to develop novel chip chemistry and antifouling tactics, as well as to combine these developments with amplification strategies and miniaturization [24].

In the literature, several studies are reported on the application of SPR technology for the identification of various types of cancer cells.

A graphene-coated multilayer-based SPR sensor for the detection of cancer in the early stage is presented. The RI component is utilized in the design and analysis of a BK7/TiO$_2$/Au/graphene-based SPR sensor. The design is presented for detecting cervical (HeLa), breast (MDA-MB 231 and MCF-7), blood (Jurkat), adrenal gland (PC12), and skin (basal) cancer cells. The sensor works on the Kretschmann configuration with a wavelength of 633 nm. As the structure and properties of multilayer arrangements have a substantial impact on the optical responses of plasmonic sensors, a comprehensive numerical analysis of the employed materials is performed. The graphene layer increases the diversity of biosensing applications and the biological detection capability of the biosensor by absorbing biomolecules and bonding with the abundant carbon-based rings in biomolecules. From the numerical analysis angular sensitivities of 278.57 deg/RIU for the breast (MDA-MB-231 and MCF-7), 292.86 deg/RIU for blood, 285.71 deg/RIU for the adrenal gland (PC12), 264.285 deg/RIU for cervical (HeLa), 245.83 deg/RIU for skin (basal) cancer types.

The reported detection limit is 0.263/deg, signal-to-noise ratio of 3.84, and a figure of merit of 48.02/RIU. [25].

Exosome lysates were analyzed using nanohole-based SPR for the detection of transmembrane (EpCAM and CD63) and intravesicular (AKT1) proteins. One benign cancer cell line and three ovarian cancer cell lines produced the exosomes. In addition to the atypical nanohole SPR, which happens to be more sensitive than conventional SPR, gold nanoparticles were employed to boost the analytical signal [26].

A customized fluidic SPR technology has been used to quantify the cancer exosomes within the total exosome population extracted from patient serum, which may provide insight into the disease's stage. Initially, the total exosome population was isolated using tetraspanin biomarkers (CD9, CD63), followed by the identification of exosomes containing the HER2 marker (breast cancer) through the construction of a sandwich with anti-HER2 [27].

Using arginine-glycine aspartic acid peptide as a linker, the breast cancer cell line MCF-7 was identified. MCF-7 breast cancer cells were captured by the interaction between human mucin-1 and a mucin-1-specific aptamer-modified gold surface. The SPR signal was amplified by the binding of NiO nanoparticles to the peptide's histidine tag [28].

On the surface of the metal, a one-of-a-kind two-dimensional heterostructure layer composed of titanium disilicide and a layer of black phosphorus was deposited. In the Kretschmann arrangement, a nanosheet made of titanium disilicide ($TiSi_2$) is positioned so that it is sandwiched between silver (Ag) and black phosphorus (BP). The performance of this biosensor is superior throughout a wide range of RI variances, which includes the distribution of biological cells in individual blood using the finite element method-based simulation technique; the sensitivity obtained was 195.4 deg/RIU, 167.6 deg/RIU, 212.4 deg/RIU, 168.4 deg/RIU, 212.4 deg/RIU, 186.6 deg/RIU, 218.6 deg/RIU, 195.4 deg/RIU, 203.6 deg/RIU, 202.6 deg/RIU 203.6 deg/RIU, and 202.6 deg/RIU for basal (skin cancer), basal (normal cell), HeLa (cervical cancer), MCF-7 (breast cancer), HeLa (normal cell), Jurkat (blood cancer), Jurkat (normal cell), PCI-2 (adrenal gland cancer), PCI-2 (normal cell), MDA-MB-231 (breast cancer), MDA-MB-231 (normal cell), and MCF-7 (normal cell) is achieved [29].

10.3 Grating-based biosensors

In the traditional sense, optical waveguides (WGs) are translucent geometries that have a RI difference and are used to direct optical rays through total internal reflection. Bragg grating (BG) is a regular waveguide structure with periodic changes in the refractive index. Because of these aberrations, a 1D photonic bandgap is produced, which reflects only a limited spectra of broadband signal that is sent through the WG. In a conceptual sense, BG WGs are analogous to the widely known fiber Bragg gratings (FBGs). Since the creation of grating structures, there has been a lot of interest in the field of optical sensing due to characteristics such as their compact size, low cost, high precision, real-time response, electromagnetic interference, and high sensitivity. Grating structures have also been the subject of a lot of research. It is possible to detect a wide number of parameters by utilizing devices that are based on gratings. Some of these qualities are strain, pressure, RI, and temperature. FBGs are used in a wide variety of applications today, including sensors for biological and medical devices, monitoring high temperature, structural health monitoring, biochemical sensing, and many more [30].

10.3.1 Bragg principle

The Bragg principle is based on the phenomenon of Bragg reflection, which occurs when a light wave interacts with a periodic structure, such as a grating. When a light wave encounters a grating with a periodic pattern of alternating regions of different RIs, it undergoes diffraction. Diffraction refers to the bending or spreading of light waves as they encounter an obstacle or pass through a narrow aperture.

In grating, the periodic structure causes the incident light wave to split into multiple diffracted waves that propagate in different directions. These diffracted waves interfere with each other, either constructively or destructively, depending on their relative phases. The Bragg principle focuses on the constructive interference that occurs between the diffracted waves in a grating. Specifically, it describes the condition for maximum constructive interference, resulting in a strong reflection of light at a specific wavelength [31].

The Bragg equation is used to calculate this resonant wavelength. It is given by:

$$\lambda_B = 2 \times \Lambda \times n_{\text{eff}}/m$$

where λ_B is the resonant wavelength; Λ is the period of the grating (the distance between adjacent RI perturbations); n_{eff} is the effective RI of the guided mode in the grating; and 'm' is the diffraction order.

According to the Bragg equation, the resonant wavelength is determined by the combination of the grating period, the effective RI, and the diffraction order. The diffraction order represents the number of times the wavelength fits within the grating period. It determines the angle at which the diffracted wave is emitted from the grating as shown in figure 10.3.

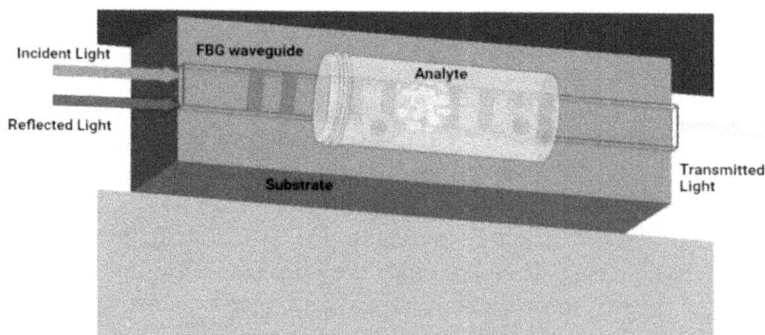

Figure 10.3. Basic working principle of FBG.

In the literature, many papers have been reported on grating-based biosensors for cancer detection. The biosensor proposed in the study utilizes a periodic grating-based on a multilayered structure to achieve high sensitivity for the detection of various biomarkers, including the Epidermal Growth Factor Receptor (EGFR), associated with cancer. The biosensor incorporates a periodic grating within a multilayered structure. This grating design generates a sharp resonance in the transmission spectrum, which allows for sensitive detection of cancer biomarkers. The Vroman Effect enables accurate detection of cancer biomarkers by enhancing the specificity and selectivity of the biosensor. The finite difference time domain (FDTD) method is employed for the detection process in the biosensor. Blood samples are analyzed for the presence of various cancer biomarkers by observing the resonance wavelength in the transmission spectrum of the biosensing platform. The changes in the resonance wavelength indicate the presence and concentration of the targeted biomarkers. The biosensor's performance is enhanced by optimizing the grating size and thickness of the various metal layers in the multilayered structure. The reported study highlights the high-quality performance of the biosensor, with a sensitivity of 26.91 and detection accuracy of 100.18 reported for the lung cancer biomarker Carcinoembryonic Antigen (CEA) at a RI of 1.33. These findings demonstrate the potential of the biosensor for the detection of cancer biomarkers and its usefulness in cancer diagnostics and research [32].

The biosensor utilizes a plasmonic tilted fiber Bragg grating (TFBG) sensor for the identification of breast cancer cells, specifically BT549 cells. The biosensor is fabricated using an 18° TFBG, which is a type of optical fiber with a tilted Bragg grating inscribed on its core. The TFBG is coated with a 50 nm-thick gold nanofilm, which enhances the plasmonic properties of the sensor. Additionally, a specific antibody against GPR30, a membrane receptor expressed in many breast tumors, is immobilized on the sensor surface. It has been shown to detect BT549 cells at concentrations as low as 5 cells/ml within a short timeframe of 20 min. This low detection limit is significant for early-stage cancer detection. The biosensor exhibits a linear response within the concentration range of 5–1000 cells/ml. This linearity indicates the sensor's suitability for practical applications, such as the detection of circulating tumor cells (CTCs) in real-world settings. The biosensor's performance has been validated through *in vitro* testing. The results of these tests demonstrate the biosensor's effectiveness in identifying breast cancer cells [33].

The study proposes a label-free and cost-effective biosensing platform for cancer detection using waveguide Bragg gratings. This platform leverages the distinct RI differences between cancer cells and normal cells, which leads to a unique reflection spectrum when the cells are deposited on the functionalized surface of the waveguide grating structure. To optimize the sensing performance of the grating structure, the researchers employed simulations combining different methods such as the effective index method, finite element method, and transfer matrix method. These simulation techniques allowed them to model the behavior of the grating structure and analyze its sensing capabilities. In particular, the sensitivity of the modeled grating structure was calculated to be 431 nm/RIU (nanometers per RI unit) using the transverse matrix method [34].

10.4 2D photonic crystal-based biosensors

A 2D photonic crystal (PhC) biosensor is a device that utilizes a two-dimensional array of periodically patterned structures to detect and analyze biological molecules or interactions. Photonic crystals are materials that exhibit a photonic bandgap, which means they have a range of frequencies or wavelengths of light that are forbidden to propagate through the crystal. The RI is periodically modulated, leading to the formation of photonic bandgap [35]. Figure 10.4 illustrates a 2D view of PhC sensing.

10.4.1 Basics of 2D PhC

A 2D PhC refers to a periodic structure that exhibits unique optical properties and is designed to manipulate the flow of light in two dimensions. It is constructed by creating a periodic arrangement of materials with alternating refractive indices or by introducing a pattern of air holes or defects into a high-index material [36–39].

Key characteristics and applications of 2D PhCs:
1. Formation of Bandgap
2. Light Localization
3. Waveguiding and Routing
4. Photonic Filters and Sensors
5. Nonlinear Optical Effects
6. Photonic Crystal Fibers

2D PhCs have attracted significant attention in research and development due to their potential for controlling and manipulating light at the nanoscale. They hold promise for advancing technologies in areas such as integrated photonics, optical communications, sensing devices, and quantum optics.

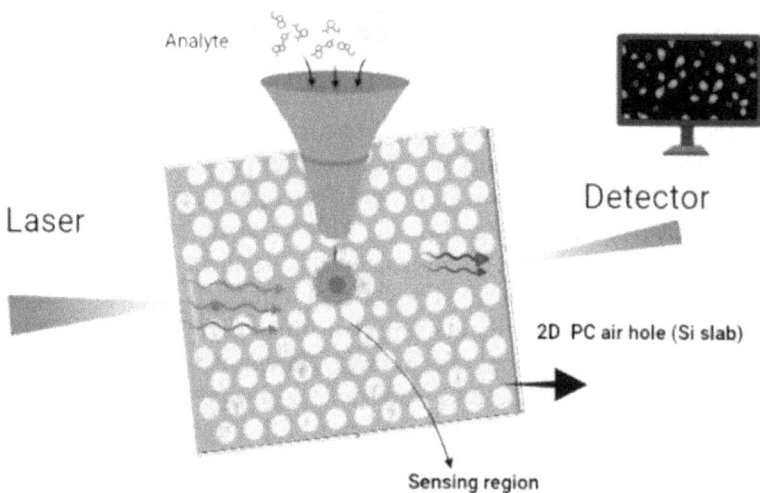

Figure 10.4. Basic working principle of 2D PhC biosensor.

10.4.2 Types of PhCs

2D PhCs are categorized based on their structural arrangements and the materials used [40]:

Square-lattice PhCs: In square-lattice PhCs, the periodic arrangement of dielectric or semiconductor materials forms a square-lattice structure. The dielectric or RI contrast between the materials creates a photonic bandgap.

Triangular-lattice PhCs: Triangular-lattice PhCs consist of a periodic arrangement of dielectric or semiconductor materials in a triangular lattice structure. The lattice points are equilateral triangles, and the RI contrast determines the photonic bandgap characteristics. These structures involve periodic arrangements of dielectric or semiconductor materials in the form of rods or slabs, respectively, to manipulate the propagation of light.

Rod-based 2D PhCs: Rod-based PhCs consist of a periodic array of cylindrical rods embedded in a background material. The rods and the surrounding material have different refractive indices, creating a RI contrast that leads to the formation of photonic bandgaps. The size, shape, and arrangement of the rods determine the bandgap properties and the ability to control light propagation. Rod-based PhCs are often used in applications such as waveguides, filters, and optical cavities. Figure 10.5 illustrates 3D and 2D views of rod-based 2D PhCs.

Slab-based 2D PhCs: Slab-based PhCs involve a periodic arrangement of planar slabs made of different materials. The slabs are typically placed in a layered structure, forming alternating layers with different refractive indices. The periodic variation in RI leads to the formation of photonic bandgaps. Slab-based PhCs can be used for creating waveguides, reflectors, and filters. They are particularly useful for manipulating light in planar waveguide structures. Figure 10.6 illustrates 2D and 3D views of slab-based 2D PhCs.

Both rod-based and slab-based 2D PhCs offer control over the flow of light and the formation of photonic bandgap. Their properties and applications vary based on the specific design, material choices, and fabrication techniques. These structures have been extensively studied for their potential in developing photonic devices with enhanced functionality, such as integrated circuits filter [41], sensors, routers [42], logic gates [43], and demultiplexers [44].

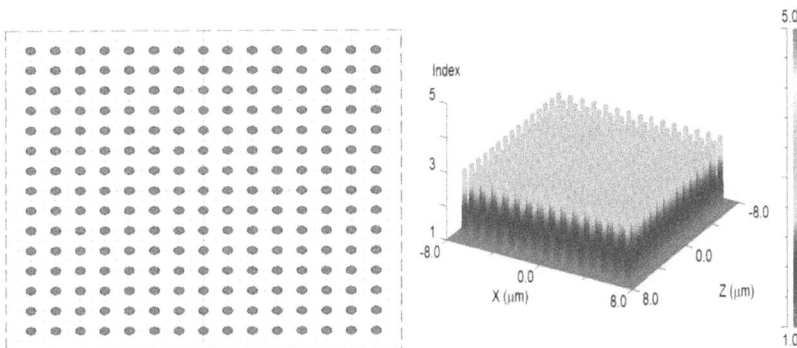

Figure 10.5. 3D and 2D views of rod-based 2D PhCs.

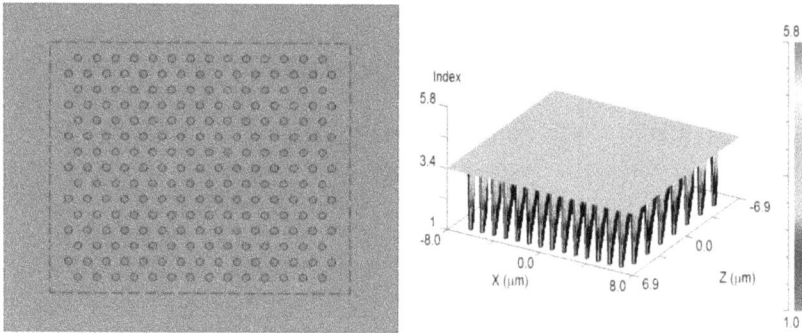

Figure 10.6. 2D and 3D views of slab-based 2D PhCs.

10.4.3 Numerical methods

Finite difference time domain (FDTD): FDTD is a widely used numerical method for simulating the electromagnetic behavior of PhCs. It discretizes the space and time domain using a finite difference grid and solves Maxwell's equations iteratively. FDTD is capable of capturing both the spatial and temporal behavior of electromagnetic fields and can provide insights into the transmission, reflection, and dispersion properties of PhCs [45].

Finite element method (FEM): FEM is a versatile numerical technique used for solving partial differential equations, including Maxwell's equations for PhCs. FEM divides the PhC structure into smaller subdomains (finite elements) and approximates the electromagnetic field within each element. By solving the equations for each element, the overall behavior of the PhC can be obtained. FEM is particularly useful for studying complex PhC geometries and material structures.

Plane wave expansion (PWE) method: PWE is a frequency-domain method used to analyze the photonic band structure of PhCs. It employs a plane wave basis set to expand the electromagnetic fields within the PhC structure and solves the eigenvalue problem to obtain the band structure and corresponding eigenmodes. PWE is computationally efficient and provides valuable information about the allowed and forbidden frequency ranges in the PhC.

Finite difference frequency domain (FDFD): FDFD is a frequency domain method that discretizes both space and frequency domains. It discretizes the PhC structure using a finite difference grid and solves Maxwell's equations at different frequencies to obtain the electromagnetic response of the PhC across a range of frequencies. FDFD is suitable for studying the spectral properties and dispersion characteristics of PhCs.

Transfer matrix method (TMM): TMM is a semi-analytical method that is particularly useful for analyzing layered PhC structures. It employs matrix operations to calculate the transmission and reflection properties of each layer and combines them to obtain the overall response of the PhC. TMM is efficient for analyzing multilayered PhCs and can provide insights into the behavior of guided modes and resonances.

10.4.4 Performance parameters of sensor

Resonant wavelength (λr) refers to the resonant wavelength of a photonic crystal, representing the wavelength at which the PhC structure exhibits strong resonance effects and significant light–matter interaction.

Full width at half maximum (FWHM) is a commonly used measure of spectral linewidth. It represents the width of a spectral feature at half of its maximum intensity.

Resonant shift (λ_s), also known as resonance shift, refers to the change in the resonant wavelength or frequency of a system or structure in response to various influences or perturbations. It describes the displacement of the resonance position from its original or reference value. The resonance shift, denoted as $\Delta\lambda$, is calculated as the difference between the shifted resonant wavelength (λs) and the original resonant wavelength ($\lambda 0$).

Refractive index (Δn) or 'delta N' denotes the variation in RI (Δn) in the medium. It observes the variation between the reference RI (n_0) and the changed RI (n).

Sensitivity (S) represents a characteristic of the sensor that measures its responsiveness or ability to detect and respond to changes in the measured quantity.

$$S = \Delta\lambda/\Delta n$$

$\Delta\lambda$ (delta lambda) represents the change or difference in the measured quantity that the sensor is designed to detect. This could be a change in wavelength, frequency, or any other parameter related to the quantity being measured.

Δn (delta N) represents the change or difference in the sensor's output.

Quality factor (Q) represents the quality factor of the sensor, λ represents the central wavelength of interest, and $\Delta\lambda$ represents the wavelength resolution or bandwidth of the sensor.

$$Q = \lambda r/\Delta\lambda$$

The quality factor (Q) of a resonant cavity or system in terms of the ratio of the energy stored in the cavity to the energy dissipated or lost in each cycle.

Resolution (R) represents the resolution (R) of a sensor in terms of the difference between the reference wavelength (λr) and the sensed or measured wavelength (λs).

$$R = \Delta\lambda = \lambda r - \lambda s$$

The detection limit (D) represents the smallest detectable change or variation in the parameter being measured. It is determined by dividing the wavelength resolution (R) by the sensitivity (S) of the sensor.

$$D = R/S$$

10.4.5 Schematic representation of 2D PhC-based biosensing

Light source: This component represents the light source used to illuminate the biosensor. It could be a laser or an LED that emits light of a specific wavelength.

Waveguide: The waveguide couples the light from the source, which guides the light towards the photonic crystal structure. The waveguide ensures efficient and controlled light propagation.

2D PhC: The PhC works as the core section in the sensor design, consisting of a 2D photonic crystal structure. It interacts with the target analyte or biomolecules in the sensing region, leading to changes in the optical properties of the structure.

Sensing region: This region is functionalized with capture molecules that specifically bind to the target analyte. When the target analyte binds to the capture molecules, it induces a variation in the RI or other optical properties in the structure.

Photodiode: A photodetector involves the absorption of photons by a photo-sensitive material, which generates a corresponding electrical current or voltage signal processing. The electrical signal from the photodetector is processed to extract relevant information. This may involve amplification, filtering, or digitization of the signal.

Data analysis: The processed signal is further analyzed to determine the presence, concentration, or specific interactions of the target analyte. This analysis may involve comparing the signal with calibration curves or employing algorithms for data interpretation. Figure 10.7 illustrates the schematic view of 2D PhC biosensing. In the literature, several studies have been reported on the application of 2D PhC technology for the identification of various types of cancer cells.

Early detection of cancer cells: This work reported the method for early detection of cancer cells using the PBG technique. The researchers conducted system-level cancer cell detection by using the dielectric constant of cells at optical frequencies as input. They compared the dielectric constant values of normal cells and cancerous cells and observed a precise frequency shift. Despite the small change in the input values, a micron-level frequency shift was observed. This indicates that a photonic crystal-based sensor can differentiate between normal and cancerous cells. This biosensor provides an efficient method for detecting cancer cells in the early stages. The sensor's resolution, which is in the range of 10^{-5}–10^{-2}, allows it to detect even minute changes, thereby increasing its sensitivity and accuracy. MEEP (MIT Electromagnetic equation Propagation) and MPB (MIT Photonic Bands) are simulation tools used in this design and optimize the properties of photonic crystal sensors [46].

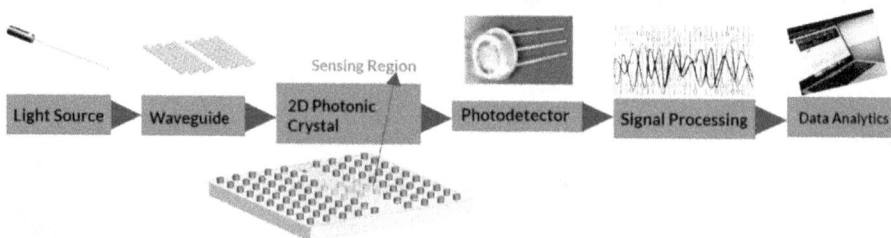

Figure 10.7. Schematic setup of PhC biosensor.

Detection of cervical cancer cells: A 2D photonic crystal-based biosensor for the diagnosis of cervical cancer cells. The biosensor design includes linear waveguides in the shape of 'L' and an inverted 'L' with sensing holes to enhance its sensing capabilities. The RI of normal cervical cells and cancer cells causes a corresponding shift in the intensity of the transmitted light. By measuring this shift in intensity, the biosensor can differentiate between normal cells and cervical cancer cells. The detection of cervical cancer cells relies on the change in RI, which leads to a shift in the transmission of light. The biosensor design incorporates 'L' and inverted 'L' shaped linear waveguides with sensing holes. Sensor design and analysis uses R-soft Tool, a software tool commonly used for the simulation and analysis of electromagnetic and photonic structures. The sensor reported a quality factor and sensitivity of 110 and 523, respectively [47].

RI-based cancerous cell detection: Sensing the nature of normal or cancerous infected cells using a 2D photonic crystal waveguide (PCW) [40]. The design consists of a 5×5 array of (Si) dielectric rods on a cubic lattice with a central defect. The background medium surrounding the rods and the defect is assumed to be air, which typically has an index of refraction close to 1. The PWE method numerically determines the electric field distribution and the peak reflected wavelength in a PCW. The simulation results show that for normal cells, the wavelengths drop in the color range of orange, and the infected cells fall in the range of yellow color. The sensor reported a high sensitivity of 2360.12 nm/RIU (RI unit), a low resolution of 1.78×10^{-6}, and a quality factor of 99.765. The report shows the accuracy of sensing infected and normal cells [48].

10.4.6 Inference of advance technologies

SPR, grating, and 2D PhC-based biosensors are all advanced technologies used for sensing and detecting biological substances.

Surface plasmon resonance (SPR): SPR is a widely used optical sensing technique that relies on the interaction between light and surface plasmons, which are collective oscillations of free electrons at the metal-dielectric interface. In SPR, a thin metal film (typically gold or silver) is coated onto a prism or a sensor chip. When polarized light is directed onto the metal-dielectric interface at a specific angle (known as the resonance angle), the surface plasmons are excited, leading to a dip in the reflected light intensity. By monitoring the changes in the resonance angle or the intensity of the reflected light, the presence and binding interactions of biomolecules can be detected and quantified.

Grating-based biosensors: Grating-based biosensors utilize the principle of light diffraction through a periodic structure, called grating, to enable sensitive detection of biochemical interactions. The grating consists of periodic patterns of lines or grooves, which can be fabricated on a sensor surface or integrated into waveguides. When light interacts with the grating, it is diffracted into different orders, and the angles and intensities of the diffracted light are sensitive to changes in the RI of the surrounding medium. By functionalizing the grating surface with biomolecules or capturing target analytes within the grating structure, the binding events can be detected by monitoring the shifts in the diffracted light patterns.

2D PhC-based biosensors: 2D photonic crystals are periodic dielectric structures that exhibit a photonic bandgap, which is a range of frequencies where certain wavelengths of light are prohibited from propagating. By introducing defects or modifying the photonic crystal structure, specific frequencies of light can be selectively transmitted or reflected. In biosensing applications, the introduction of biomolecules or analytes into the photonic crystal structure modifies the effective RI, leading to changes in the transmission or reflection spectrum. These changes can be monitored to detect the presence or concentration of target analytes.

Both grating-based and 2D PhC–based biosensors offer advantages such as label-free detection, high sensitivity, and potential for multiplexed analysis. They have found applications in diverse fields such as medical diagnostics, environmental monitoring, and food safety. Based on the requirements and needs of the application technology, it can be used to identify the nature of the analyte.

10.5 Conclusion

This chapter covered several types of biosensors based on integrated photonics and explained how a change in a sensor's parameter may either increase or decrease the effectiveness of the biosensor. In the case of cancer cells, in particular, it is possible to conclude that optical biosensors that make use of silicon photonics provide a detection method that is more accurate, faster, and economical if they are commercialized. On the other hand, the significant problem that was discussed in the previous part is still present, and a compelling study is being carried out all over the world. This chapter provides a comparison of the many types of biosensors as well as their respective operating principles. All of the articles presented have an operating wavelength in the range of 1500–1600 nm. This range is stated to be the most appropriate for creating sharp and distinct peaks for various cancer cells displaying other refractive indices compared to the behaviour of normal cells. However, this domain shows promise for accurate POC diagnosis, and if the challenges associated with practical implementation can be surmounted, it has the potential to open up new avenues for cancer cell detection and diagnosis in medicine in the years to come.

References

[1] Wulfkuhle J D, Liotta L A and Petricoin E F 2003 Proteomic applications for the early detection of cancer *Nat. Rev. Cancer* **3** 267–75

[2] McDermott U, Downing J R and Stratton M 2011 Genomics and the continuum of cancer care *New Engl. J. Med.* **364** 340–50

[3] Snyder E L, Stramer S L and Benjamin R J 2015 The safety of the blood supply—time to raise the bar *New Engl. J. Med.* **372** 1882–5

[4] Dhawan A P *et al* 2015 Current and future challenges in point-of-care technologies: a paradigm-shift in affordable global healthcare with personalized and preventive medicine *IEEE J. Transl. Eng. Health Med.* **3** 1–0

[5] Yager P, Domingo G J and Gerdes J 2008 Point-of-care diagnostics for global health *Annu. Rev. Biomed. Eng.* **10** 107–44

[6] Arlett J L, Myers E B and Roukes M L 2011 Comparative advantages of mechanical biosensors *Nat. Nanotechnol.* **6** 203–15

[7] Mascini M and Tombelli S 2008 Biosensors for biomarkers in medical diagnostics *Biomarkers* **13** 637–57

[8] 2021 Global fiber optic sensors market 2020–2025 *BCC Res.* https://researchandmarkets.com/reports/5321748/global-fiber-optic-sensors-market-2020-2025?utm_source=BW&utm_medium=PressRelease&utm_code=g6rqq5&utm_campaign=1577126+-+Global+Fiber+Optic+Sensors+Market+Report+2021%3a+10.9%25+CAGR+Forecast+Between+2020+and+2025%2c+with+Market+Forecast+to+Reach++%244.9+Billion+by+2025+&utm_exec=cari18prd

[9] Lechuga L M 2005 Optical biosensors *Compr. Anal. Chem.* **44** 209–50

[10] Dhanabalan S S, Thirumurugan A, Raju R, Kamaraj S-K and Thirumaran S 2023 *Photonic Crystal and Its Applications for Next Generation Systems* (Singapore: Springer) https://link.springer.com/book/10.1007/978-981-99-2548-3

[11] Thomas-Peter N *et al* 2011 Integrated photonic sensing *New J. Phys.* **13** 055024

[12] Luan E, Shoman H, Ratner D M, Cheung K C and Chrostowski L 2018 Silicon photonic biosensors using label-free detection *Sensors* **18** 3519

[13] Blanco F J, Agirregabiria M, Berganzo J, Mayora K, Elizalde J, Calle A, Domínguez C and Lechuga L M 2006 Microfluidic-optical integrated CMOS compatible devices for label-free biochemical sensing *J. Micromech. Microeng.* **16** 1006

[14] Properzi F, Logozzi M and Fais S 2013 Exosomes: the future of biomarkers in medicine *Biomarkers Med.* **7** 769–78

[15] Fuhrmann G, Neuer A L and Herrmann I K 2017 Extracellular vesicles–a promising avenue for the detection and treatment of infectious diseases? *Eur. J. Pharm. Biopharm.* **118** 56–61

[16] Park J *et al* 2021 An integrated magneto-electrochemical device for the rapid profiling of tumour extracellular vesicles from blood plasma *Nat. Biomed. Eng.* **5** 678–89

[17] Im H, Shao H, Park Y I, Peterson V M, Castro C M, Weissleder R and Lee H 2014 Label-free detection and molecular profiling of exosomes with a nano-plasmonic sensor *Nat. Biotechnol.* **32** 490–5

[18] Padmanabhan S, Shinoj V K, Murukeshan V M and Padmanabhan P 2010 Highly sensitive optical detection of specific protein in breast cancer cells using microstructured fiber in extremely low sample volume *J. Biomed. Opt.* **15** 017005

[19] Zhou J, Zheng Y, Liu J, Bing X, Hua J and Zhang H 2016 A paper-based detection method of cancer cells using the photo-thermal effect of nanocomposite *J. Pharm. Biomed. Anal.* **117** 333–7

[20] Bohunicky B and Mousa S A 2010 Biosensors: the new wave in cancer diagnosis *Nanotechnol. Sci. Appl.* **4** 1–10

[21] Ribaut C, Loyez M, Larrieu J C, Chevineau S, Lambert P, Remmelink M, Wattiez R and Caucheteur C 2017 Cancer biomarker sensing using packaged plasmonic optical fiber gratings: towards *in vivo* diagnosis *Biosens. Bioelectron.* **92** 449–56

[22] Yesudasu V, Pradhan H S and Pandya R J 2021 Recent progress in surface plasmon resonance based sensors: a comprehensive review *Heliyon* **7** e06321

[23] Steglich P, Lecci G and Mai A 2022 Surface plasmon resonance (SPR) spectroscopy and photonic integrated circuit (PIC) biosensors: a comparative review *Sensors* **22** 2901

[24] Homola J 2008 Surface plasmon resonance sensors for detection of chemical and biological species *Chem. Rev.* **108** 462–93

[25] Mostufa S, Akib T B, Rana M M and Islam M 2022 Highly sensitive TiO_2/Au/graphene layer-based surface plasmon resonance biosensor for cancer detection *Biosensors* **12** 603

[26] Park J, Im H, Hong S, Castro C M, Weissleder R and Lee H 2018 Analyses of intravesicular exosomal proteins using a nano-plasmonic system *ACS Photonics* **5** 487–94

[27] Reiner A T, Ferrer N G, Venugopalan P, Lai R C, Lim S K and Dostálek J 2017 Magnetic nanoparticle-enhanced surface plasmon resonance biosensor for extracellular vesicle analysis *Analyst* **142** 3913–21

[28] Mousavi M Z, Chen H Y, Hou H S, Chang C Y, Roffler S, Wei P K and Cheng J Y 2015 Label-free detection of rare cell in human blood using gold nano slit surface plasmon resonance *Biosensors* **5** 98–117

[29] Karki B, Uniyal A, Pal A and Srivastava V 2022 Advances in surface plasmon resonance-based biosensor technologies for cancer cell detection *Int. J. Opt.* **2022** 1–10

[30] Butt M A, Kazanskiy N L and Khonina S N 2022 Advances in waveguide Bragg grating structures, platforms, and applications: an up-to-date appraisal *Biosensors* **12** 497

[31] Bragg W L and Thomson J J 1914 Diffraction of short electromagnetic waves, etc. *Proc. of the Cambridge Philosophical Society: Mathematical and Physical Sciences* vol 17 (Cambridge: Cambridge Philosophical Society) p 43

[32] Teotia P K and Kaler R S 2018 1-D grating based SPR biosensor for the detection of lung cancer biomarkers using Vroman effect *Opt. Commun.* **406** 188–91

[33] Chen X, Xu P, Lin W, Jiang J, Qu H, Hu X, Sun J and Cui Y 2022 Label-free detection of breast cancer cells using a functionalized tilted fiber grating *Biomed. Opt. Express* **13** 2117–29

[34] Vishwaraj N P, Nataraj C T, Jagannath R P, Talabattula S and Prashanth G R 2023 Machine learning assisted cancer cell detection using strip waveguide Bragg gratings *Optik* **284** 170947

[35] Threm D, Nazirizadeh Y and Gerken M 2012 Photonic crystal biosensors towards on-chip integration *J. Biophotonics* **5** 601–16

[36] Sridarshini T, Dhanabalan S S, Balaji V R, Manjula A, Indira Gandhi S and Sivanantha Raja A 2023 Photonic crystal based routers for all optical communication networks *Modeling and Optimization of Optical Communication Networks* (Hoboken, NJ: Wiley) pp 137–62

[37] Geerthana S, Sridarshini T, Balaji V R *et al* 2023 Ultra compact 2D-PhC based sharp bend splitters for terahertz applications *Opt. Quantum Electron.* **55** 778

[38] Sridarshini T, Chidambaram P, Geerthana S, Balaji V R, Thirumurugan A, Madurakavi K and Dhanabalan S S 2022 Current and future horizon of optics and photonics in environmental sustainability *Sustain. Comput.: Inform. Syst.* **36** 100815

[39] Zhou W *et al* 2014 Progress in 2D photonic crystal Fano resonance photonics *Prog. Quantum Electron.* **38** 1–74

[40] Joannopoulos J D, Meade R D and Winn J N 1995 *Photonic Crystals: Molding the Flow of Light* (Princeton, NJ: Princeton University Press)

[41] Kavitha V, Balaji V R, Dhanabalan S S, Sridarshini T, Robinson S, Radhouene M, Hegde G and Sugesh R J 2023 Design and performance analysis of eight channel demultiplexer using 2D photonic crystal with trapezium cavity *J. Opt.* **25** 065102

[42] Geerthana S, Syedakbar S, Sridarshini T, Balaji V R, Sitharthan R and Sundar D S 2022 2D-PhC based all optical AND, OR and EX-OR logic gates with high contrast ratio operating at C band *Laser Phys.* **32** 106201

[43] Thirumaran S, Dhanabalan S S and Sannasi I G 2021 Design and analysis of photonic crystal ring resonator based 6 × 6 wavelength router for photonic integrated circuits *IET Optelectron.* **15** 40–7

[44] Yamunadevi R 2016 Characteristics analysis of metamaterial based optical fiber *Optik* **127** 9377–85

[45] Yang H Y 1996 Finite difference analysis of 2-D photonic crystals *IEEE Trans. Microwave Theory Tech.* **44** 2688–95

[46] Sharan P, Bharadwaj S M, Gudagunti F D and Deshmukh P 2014 Design and modelling of photonic sensor for cancer cell detection *2014 Int. Conf. on the IMpact of E-Technology on US (IMPETUS)* (Piscataway, NJ: IEEE) pp 20–4

[47] Sundhar A, Valli R, Robinson S, Abinayaa A and SivaBharathy C 2019 Two-dimensional photonic crystal based bio sensor for cancer cell detection *2019 IEEE Int. Conf. on System, Computation, Automation and Networking (ICSCAN)* (Piscataway, NJ: IEEE) pp 1–3

[48] Panda A and Devi P P 2020 Photonic crystal biosensor for refractive index based cancerous cell detection *Opt. Fiber Technol.* **54** 102123

IOP Publishing

Advances in All-optical Communication

Shanmuga Sundar Dhanabalan, Arun Thirumurugan and Sridarshini Thirumaran

Chapter 11

Absorbers as biosensors: leveraging absorption phenomena for enhanced biosensing

Madurakavi Karthikeyan, J Pradeep, M Harikrishnan, R Sitharthan, M Rajesh, Avinash Chandra and Rajkishor Kumar

The basic concepts and uses of absorbers in the field of biosensors are examined in this chapter. Due to its potential to improve sensitivity, selectivity, and detection limits in a variety of biological and chemical sensing applications, absorption-based biosensing has drawn a lot of attention. The refractive index of a substance or a medium can be measured via absorber-based refractive index sensing, which makes use of absorptive components or structures. The fundamental idea is to keep track of how the material absorbs light as a function of the refractive index of the environment. Depending on the refractive index of the surrounding medium, the absorptive material interacts with incident light and changes its absorption characteristics. Analysis is done on the absorption spectrum or the precise wavelengths at which absorption takes place. As the refractive index of the surrounding material varies, the absorption spectrum shifts or alters.

There are numerous industries that use absorber-based refractive index sensing, including biochemistry, environmental monitoring, material research, and optics. For label-free detection and monitoring of biological molecules, gases, and liquids based on changes in refractive index, it is especially helpful. The method is extremely sensitive and is adaptable for *in situ* and real-time sensing applications.

While absorber-based biosensors for refractive index detection provide intriguing features, they also have some limitations that may prevent their use in real-time applications. Sensitivity and dynamic range, interference and noise, specificity and selectivity, calibration and standardization, material and design constraints, integration and miniaturization, real-time monitoring difficulties, biocompatibility, and sample preparation are a few of the major restrictions.

This chapter explores the complex mechanisms that underlie absorption events to clarify their crucial function in biosensing. It covers the development, production, and incorporation of absorptive materials and structures into biosensor platforms,

doi:10.1088/978-0-7503-5623-7ch11

emphasizing the special qualities that make it possible to identify and measure biomolecules. The chapter also offers insights into recent developments, new patterns, and potential applications of absorber-based biosensors in an effort to spur additional investigation and development in this exciting area.

11.1 Introduction

In November 2002, the Chinese province of Guangdong was the epicentre of the SARS outbreak. Official reports of the first instances did not come in until February 2003. SARS-CoV, or SARS-associated coronavirus, is thought to have started in bats and spread to people, potentially via a civet cat or other intermediate host. Severe respiratory illnesses were the hallmark of the outbreak, which had a high fatality rate of almost 10%. Other portions of Asia, North America, Europe, and other regions were affected by the virus's spread. The outbreak was contained with the aid of efficient public health measures, such as case isolation, contact quarantine, and travel restrictions. In July 2003, the World Health Organization (WHO) proclaimed the global outbreak contained [1]. The first case of COVID-19 was discovered in Wuhan, China's Hubei province, in December 2019. The causative agent, SARS-CoV-2 (Severe Acute Respiratory Syndrome Coronavirus 2), is likewise thought to have started in bats and may have spread to people via an intermediary host, most likely a wild animal that was offered for sale at a Wuhan seafood market. The virus rapidly spread around the world, sparking a pandemic. When an infected individual coughs, sneezes, or talks, respiratory droplets are the main way that COVID-19 is transferred [2–5]. It can also spread by coming into contact with infected surfaces. The condition can cause severe respiratory distress and even death, as well as moderate or asymptomatic cases. Lockdowns, social distancing, mask wearing, and vaccine drives are just a few of the public health measures that have been put into place globally to stop the virus's spread. The global economy is severely impacted by a number of virus-related diseases in addition to the novel coronavirus. It goes without saying that early, accurate, and reliable virus detection can play a major role in containing the disease's spread and averting future pandemics such as COVID-19. Various diagnostic procedures are usually employed in conventional methods of sensing or detecting COVID-19 in order to determine the presence of the virus or its antibodies. Popular traditional techniques are given in figure 11.1. The principle and its drawbacks are given as follows:

In Testing with Polymerase Chain Reaction (PCR), the idea is that the viral RNA is amplified by PCR to identify its presence. The major drawbacks are

- Needs skilled workers and certain laboratory equipment.
- Results could come in a few days or a few hours.
- Results that are falsely negative might happen, particularly in the early or late stages of an infection.

The principle of Antigen testing is to examine patient samples for viral proteins. The major drawback involved are

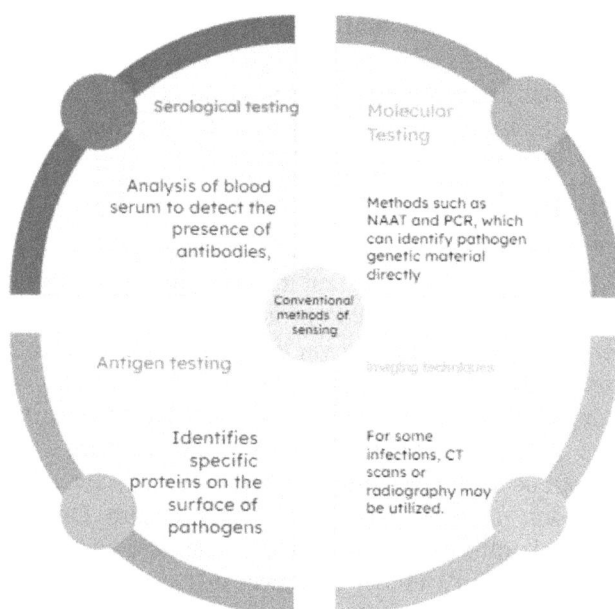

Figure 11.1. Various applications of biosensing.

- Less sensitive than PCR, which raises the possibility of false negative results.
- Depending on when the test is administered in relation to the beginning of symptoms, performance may change.

The main principle of Serological Tests is to recognize antibodies made in reaction to the infection. The major drawbacks involved are
- Since antibody development takes time, detecting antibodies is not ideal for early diagnosis
- Unable to distinguish between infections from the past and the present.
- May produce erroneous positive or negative results due to immune response variability or cross-reactivity.

Because it takes time for the body to develop antibodies, it is not appropriate for early diagnosis.

The principle behind CT scan is to utilize imaging methods to detect anomalies in the lungs linked to COVID-19. The main drawbacks are
- Not exclusive to COVID-19; comparable results can be obtained from other respiratory conditions.
- Exposure to radiation when using x-rays.

Considering the above disadvantages of the existing techniques, metamaterial-based biosensing is growing in popularity. Utilizing synthetic materials with distinct electromagnetic properties, metamaterial-based biosensing improves biological material identification and analysis. The development of electronic devices for medical application has gained popularity in recent years since it provides easy diagnosis [4, 6–9]. These implantable antennas are derived from the conventional antennas [10–12]. The dominance of this method is due the following features: Extremely Sensitive-It is possible to create metamaterials with a high degree of sensitivity to changes in the electromagnetic environment, which makes it possible to identify even the smallest changes in biological samples. This is very helpful for finding biomolecules at low concentrations.

Particulars-Enhancing the specificity of biosensors, metamaterials can be engineered to selectively interact with specific biological molecules. For the precise detection and diagnosis of specific biomarkers or pathogens, this specificity is essential.

Label-Free Detection-This is made possible by certain metamaterial-based biosensors, which do not require extra markers or labels on biological samples. This lowers the possibility of artifacts and streamlines the sensing process.

Monitoring in real time-Real-time biological process and reaction monitoring is possible using metamaterial-based biosensors. For dynamic studies of enzymatic processes, cellular activities, or variations in biomarker concentrations over time, this is crucial.

Ability to Multiplex-Multiple biomolecules can be detected simultaneously on a single sensing platform thanks to the engineering of metamaterials that facilitate multiplexing. This advantage is beneficial for thorough analysis and diagnosis.

Figure 11.2 lists the numerous applications for metamaterial-based sensing. In [13], a biosensor was created by forming a CTS_5 phase change material using a new alloy combination of $Ge_2Sb_2Te_5$. The sensor identifies the various haemoglobin and urine concentrations. It has been noticed that the suggested sensor sensitivity variation lies between 773 and 2667 nm RIU^{-1}. Glioma cell molecular categorization was accomplished in [14] by employing a label-free biosensing technique.

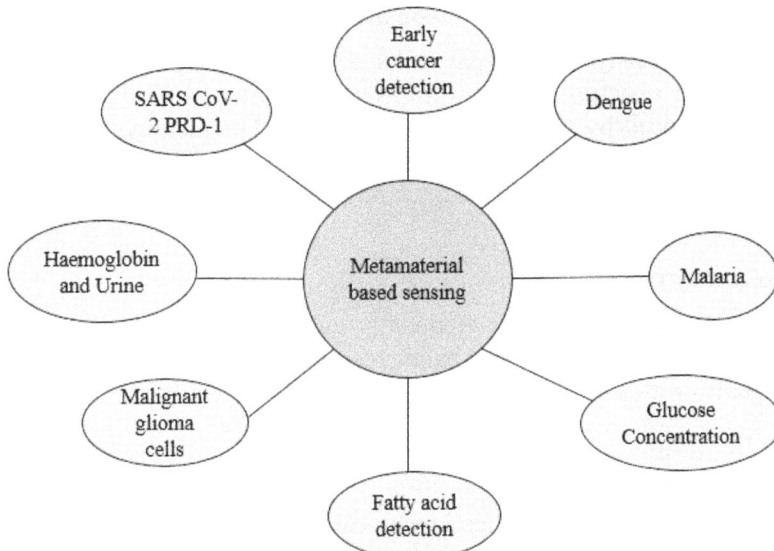

Figure 11.2. Typical uses of metamaterial-based sensors in biosensing.

At THz frequency, polarization- independent characteristics are likewise displayed by the suggested structure. The suggested metamaterial can differentiate between mid-type and mutant glioma cells without the need for an antibody, as the authors have measured. Fatty acid identification using terahertz time-domain spectroscopy and a metamaterial-based THz sensor were demonstrated in [15]. The results show how quickly fatty acid may be detected with a THz sensor.

A dual-band metamaterial absorber-based sensor has been created in [16] with a straightforward structure. Additionally, it has demonstrated a high level of sensitivity in the detection of glucose. It has been shown in [17] that COVID-19 SARAS-COVZ spike protein detection is possible by utilizing AuNRs, or gold nanorods. With the suggested design, a sensitivity of 111.11 deg. RIU^{-1} has been attained. It has been suggested in [18] to use hexagonal gold layers on a polyamide substrate to detect cancer early on. It has been shown that cancer cell sensing at $f = 3.15$ THz can achieve a sensitivity of 1649.8 GHz RIU^{-1}. For the purpose of early cancer detection, the PC12, MCF-7, basal cell, breast cell, and cervical cell have all been used. A metamaterial-based resonator has been used to detect dengue, as described in [19], and the suggested sensor was able to detect a change in blood permittivity.

11.2 Factors influencing sensing performance

A family of sensors known as metamaterial-based sensors uses specially designed materials with distinct electromagnetic properties to measure and identify a range of physical parameters. The efficiency and performance of metamaterial-based sensors are influenced by a number of factors.

11.2.1 Metamaterial design's composition and geometry

A significant part is played by the particular materials and how they are arranged in the metamaterial structure. The sensitivity and reaction of the sensor to specific stimuli are influenced by the selection of materials. The metamaterial's resonance frequencies and sensitivity to particular wavelengths or signals are determined by its geometry, which includes its size, shape, and elemental arrangement.

11.2.2 Frequency range

Metamaterial sensor designs are frequently customized for particular frequency ranges. Resonance frequencies and bandgaps, for example, must coincide with the desired signals or phenomena.

11.2.3 Electromagnetic properties

The metamaterial's capacity to interact with electromagnetic waves is determined by its electrical permittivity and magnetic permeability. The sensor's performance can be improved by adjusting these parameters.

11.2.4 Sensitivity and selectivity

The term 'sensitivity' describes a sensor's capacity to pick up on minute alterations in its surroundings or stimuli. The capacity to react to particular signals or parameters in a selective manner is known as selectivity. For metamaterial sensors to be accurate, both are essential.

11.2.5 Integration with other materials/devices

To improve their usefulness, metamaterial sensors are frequently combined with other materials or devices. The way the metamaterial interacts with the surrounding elements can affect how well the sensor works.

11.2.6 Fabrication techniques

The performance and structural accuracy of metamaterials can be affected by the fabrication techniques used to create them, such as 3D printing or lithography.

11.2.7 Environmental factors

Metamaterial-based sensors are susceptible to performance variations in temperature, humidity, and other environmental factors. It is imperative to take into account the behavior of these sensors in various operating scenarios.

11.2.8 Signal processing

The overall performance and reliability of the sensor can be greatly impacted by the signal processing methods used to retrieve data from its output.

11.2.9 Application-specific considerations

The sensor's design characteristics are influenced by its intended use. For uses like imaging, communication, or medical diagnostics, many sensors may be optimized.

11.2.10 Power usage

The sensor's power consumption may be crucial, depending on the application. Generally speaking, lower power consumption is preferable, particularly for remote- or battery-operated devices.

11.3 Materials employed for designing biosensor absorbers

Engineered materials with characteristics not present in naturally occurring materials are known as metamaterials. Their distinct electromagnetic characteristics allow for customization for certain uses, such as biosensors. It is possible to improve sensitivity, selectivity, and other performance metrics in metamaterial-based biosensors. The particular application and desired features determine which materials are best for these biosensors. The metamaterial absorber consists of three basic layers such as ground, substrate, and metasurface structure consisting of various

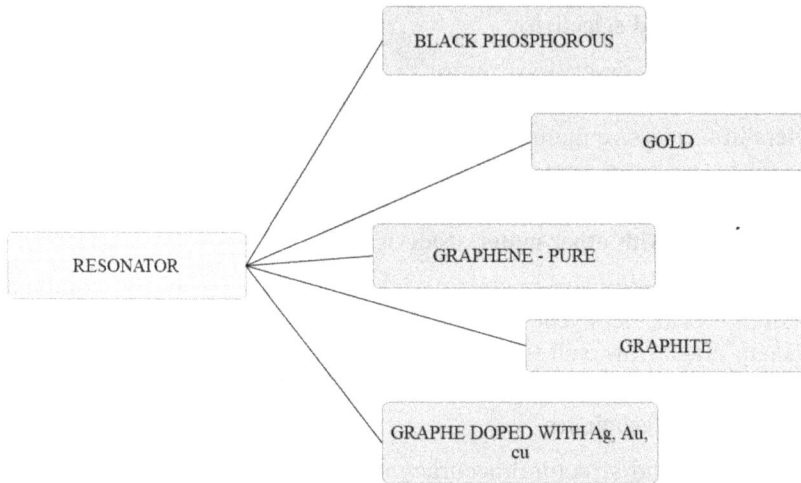

Figure 11.3. Popular resonators used in absorbers for biosensing applications.

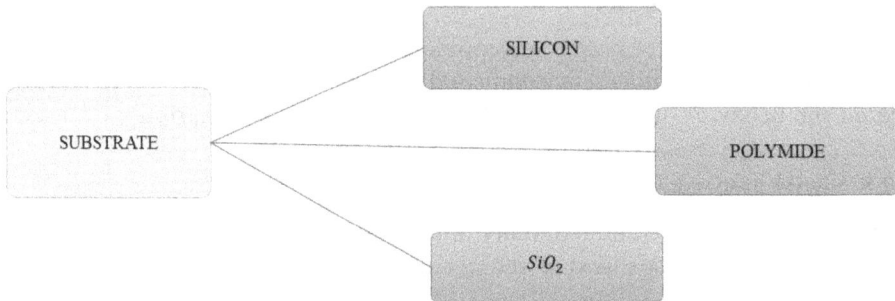

Figure 11.4. Popular substrate utilized to produce biosensors.

structure of metals. Figures 11.3 and 11.4 list the popular materials used for designing absorber-based biosensors. The substrate should have thermal stability.

11.3.1 Importance of metal layer and various metals used in metamaterial-based biosensor

Because of their superior conductivity and electromagnetic wave interaction, metals like gold, silver, and aluminium are frequently utilized in the building of meta-material constructions. To improve sensitivity and enable localized surface plasmon resonance, metamaterial structures can be enhanced by including nanoparticles composed of gold, silver, or other metals. Graphene is a novel substance that has attracted a lot of attention for a variety of uses, including possible use in metamaterials. Recently, several designs have been reported using graphene for biosensing applications. Graphene is a two-dimensional material consisting of a single carbon atom layered in a honeycomb configuration. It is the thinnest material in the universe and the strongest material ever measured. Graphene also possesses the highest mobility of all known materials and is electrochemically tunable in terms of conductivity [20]. Some of the other applications of graphene include tunable cloaks [21], modulators [22], nonlinear optic devices [23], etc.

In [24], metamaterial absorbers based on all-metal structures have been proposed to work in THz regime for various biosensing applications. A tri-layer structure with polymide as substrate, gold as ground plane, and InAs as a top layer for detecting various chemicals with a sensitivity of around 1109 GHz RIU^{-1} was proposed in [25]. In [26], vanadium dioxide was used to create the resonator. An Au-reflective layer was proposed for THz biodetection. Using indium tin oxide (ITO) and graphene, the authors in [27] proposed a THz-transparent absorber for molecular sensing

In [28], a THz sensor based on metamaterial absorber (MMA) has been proposed with five bands of resonance. They designed a ring-shaped gold resonator on a polymide substrate.

In [29], analysis of glucose absorption was caried out with transition metal-doped graphene. The transition atoms used were silver, copper, platinum, gold, and nickel. Through this design, an improved absorption and sensing improvement has been absorbed.

In [30], a dual-band absorber using black phosphorous was proposed. A silicon layer was used as substrate.

11.4 Popular designs of absorbers for biosensing

An adjustable dual-band absorber for glucose and malaria sensing was proposed and evaluated in [16]. The suggested absorber has a straightforward construction and a maximum sensitivity of 4.74 THz RIU^{-1}. Furthermore, the structure is non-symmetric and can demonstrate polarization insensitivity up to an 80-degree angle. The projected absorber's 3D perspective is displayed in figure 11.5. At the top is an array of gold metasurfaces. A series of 0.2-height and 0.4-width gold bars are arranged in an 2×2 array. Above the substrate and below the gold bars is a 0.35 nm-thick layer of graphene with chemical potential μ_c. Owing to the substrate's influence on the properties of graphene surface plasmons (GSPs), the graphene layer is produced on a 3 μm thick silicon dioxide (SiO2) substrate with an ε_r value of 2.25.

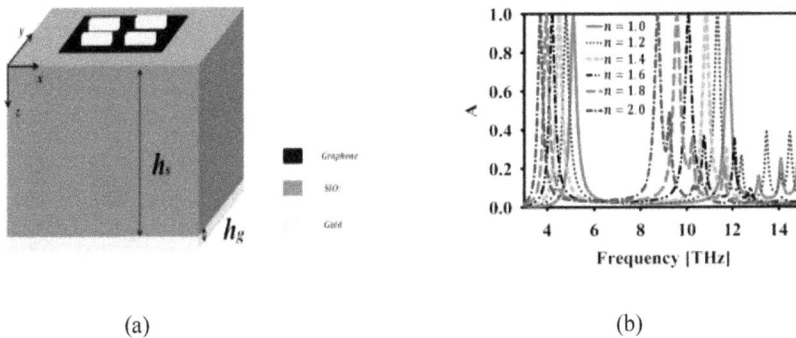

(a)   (b)

Figure 11.5. (a) Tunable dual-band absorber and its (b) absorbance for various refractive index medium. Reproduced from [16]. CC BY 4.0.

The authors in [31] describe a metamaterial (MM) absorber that provides dual-band resonance and perfect absorption by taking advantage of a graphene patch. At the THz spectrum, the structure provides an ultrathin geometry with a thickness approximately thirty times less than the wavelength in free space. In the lower and upper bands, the monolayer graphene patch offers absorption of over 99% and 50%, respectively. In order to form bipolar charged nodes in the graphene sheet and improve impedance matching in the upper band, a circular hole is placed into the sheet, trapping electric charge and increasing absorption to over 99%.

The proposed absorber's geometry is displayed in figure 11.6. This has a 2.5 μm high silicon dioxide (SiO$_2$) substrate with a relative permittivity of 2.25. At the top of the substrate, there is a graphene resonator with a square shape. The graphene resonator has a circular annular slit. The metallic sheet is positioned above the entire construction. Gold is the material utilized for the back reflector or metal plane in this arrangement.

The sensor described in [32] makes use of a metamaterial construction that is designed to demonstrate dual-band resonance in the THz frequency range. Changes in the biological composition can be detected with great sensitivity thanks to the metamaterial's ability to interact with the refractive index of the surrounding liquid. The ultrathin construction, polarization insensitivity, narrow Abs bandwidth, and high-quality factor of the suggested absorber are its key characteristics. Furthermore, the polarization angle of the input electromagnetic wave has no effect on the absorber.

A unique planar THz metamaterial sensor made of a corrugated metal stripe with three rectangular grooves pierced in it was presented by the authors in [33]. The sensor produces unusually strong resonance peaks in the transmission spectrum by taking advantage of the Fabry–Perot resonance of the spoof surface plasmons mode on the corrugated metal stripe, as demonstrated by numerical analysis. Because of these resonances' significant local field augmentation and high-quality factors, the sensor is extremely sensitive for applications requiring ultrasensitive sensing. Figure 11.7 shows the structure and transmission spectra of the metamaterial absorber proposed in [33].

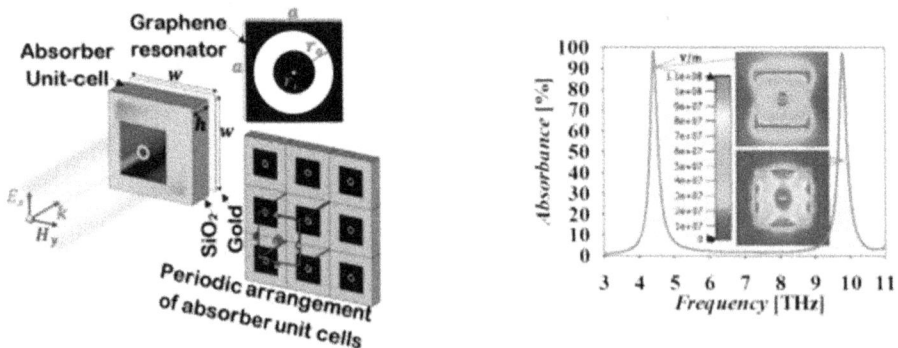

Figure 11.6. Dual band absorber and its absorption spectrum. Reproduced from [31]. CC BY 3.0.

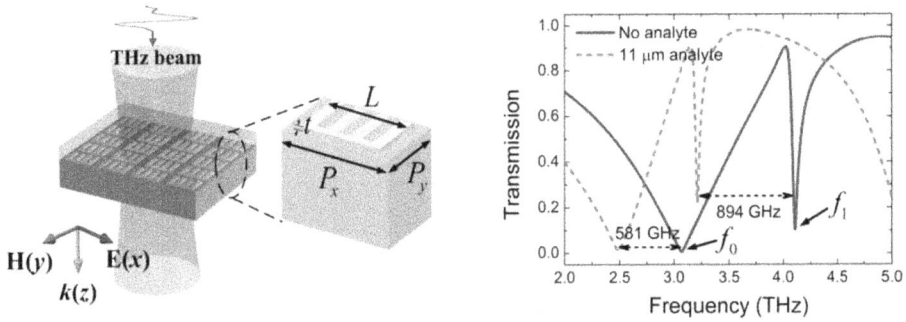

Figure 11.7. Metamaterial absorber and its transmission spectra. Reproduced with permission from [33].

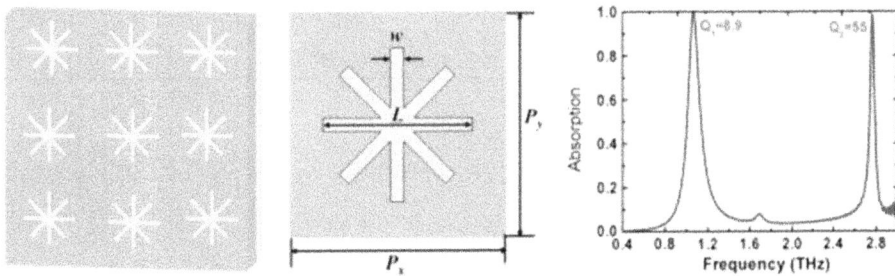

Figure 11.8. High sensitivity sensor design reported in [34].

The design in [34] is a flexible THz dual-band metamaterial absorber-based high-sensitivity sensing method. Two unique perfect absorption peaks can be seen in this absorber: a high-order resonance at 2.77 THz (f2) that results from interactions between neighboring unit cells, and a fundamental resonance at 1.08 THz (f1). Compared to f1, the quality factor and figure of merit are noticeably higher in f2. With changes in analyte thickness and refractive index, the method tracks shifts in resonance frequency. For f2, amplitude absorption and analyte thickness are shown to be linearly related, with a sensitivity of 31.2% refractive index units (RIU). The sensitivity for f1 is 13.7% RIU.

Furthermore, the absorption amplitude for $f1$ in aqueous solutions, such as ethanol-water mixes, falls linearly with an increase in dielectric constant. Figure 11.8 shows the design of the proposed absorber and its performance. Table 11.1 gives a comparison between the various parameters of recently proposed absorbers for biosensing.

11.5 Fabrication techniques

Graphene material has very strong sensing ability, even though absorbers for biosensing applications were made from a variety of materials. Researchers have investigated numerous potential methods for depositing graphene on diverse substrates. The fundamental steps of graphene deposition are covered in this section, along with a review of some earlier research. The most common method for creating

Table 11.1. Comparison of absorption spectra and sensitivity of recently proposed absorbers.

Reference	No. of bands	S (I Peak, II Peak) (THz RIU^{-1})	FOM (RIU^{-1})	f(THz)	Tunablility	Polarization insensitive up to the angle in °	Thickness in μm
[16]	2	2.08 4.72	6.7, 13.88	5.1, 11.7	Yes	80	3
[31]	2	14.08, 53	53.09, 41.1	4.4, 9.86	Yes	Insensitive	2.5
[32]	2	3.29, 3.28	470, 293	3.6, 4.6	Yes	Insensitive	2.25
[34]	0.147, 0.964	1.21, 17.28	1.08, 2.77	No	Not reported	—	10.4

graphene absorbers is chemical vapor deposition, or CVD. The following is a summary of the procedure.

11.5.1 Substrate preparation

To ensure that no impurities hinder the formation of graphene, a substrate—typically composed of silicon, nickel, or copper—is meticulously cleansed.

11.5.2 Catalyst deposition

A thin layer of a catalyst material—typically copper or nickel—is applied to the substrate. This catalyst provides carbon atom nucleation sites, which aids in the formation of graphene.

11.5.3 Graphene growth

Next, the substrate is put in a CVD chamber and heated to very high temperatures (usually between 1000 °C and 1100 °C) in a methane and hydrogen environment that is carefully regulated. Methane gas's carbon atoms break down on the catalyst's surface before recombining to create graphene. On the catalyst's surface, the carbon atoms self-organize into a single sheet of graphene.

11.5.4 Transfer process

The graphene film is usually moved onto a new substrate that is more appropriate for the intended purpose after graphene development. In order to carefully transfer the graphene layer onto the target substrate, this transfer procedure frequently entails etching away the original substrate using methods like polymer-assisted transfer or dry transfer methods.

11.5.5 Evaluation and optimization

In this stage, the graphene absorber is characterized to make sure that its perform-ance and quality fulfill the required requirements. To enhance the graphene layer's conductivity, homogeneity, and other characteristics, optimization procedures may be used.

With perfect control over thickness, conductivity, and other important features, this fabrication process produces high-quality graphene absorbers that are ideal for a variety of different uses beyond biosensing, including photodetectors, solar cells, and THz devices. The following recently published publications address the synthesis of graphene absorbers: A novel technique for large-scale pattern development of graphene sheets meant for usage as stretchable transparent electrodes is presented in the work reported in [35]. The main goal of the project is to provide a scalable production method for graphene films, which are needed for stretchable transparent electrodes—an essential component of flexible electronics and displays. To grow graphene films, the authors use a CVD technique. The controlled deposition of graphene in particular geometries is made possible by the introduction of a patterned copper substrate, which serves as a template for graphene growth. The technique solves a significant obstacle in the manufacture of graphene by allowing the large-scale development of graphene films. By precisely controlling the develop-ment process, CVD makes it easier to deposit graphene uniformly over wide surfaces. The resulting graphene films are highly mechanically stable and do not lose their conductivity even under severe stretching. The films can also be used as transparent electrodes in flexible electronic devices because they retain a high level of optical transparency. The creation of stretchy and flexible electronics, such as wearable electronics, flexible displays, and biomedical devices, is made possible by the scalable fabrication approach.

In [36], a novel technique for producing large-scale graphene films roll-to-roll with the express purpose of using them as transparent electrodes is proposed. The authors present a large-scale roll-to-roll production technique for creating graphene films. With this technique, lengthy rolls of graphene film can be produced by continually depositing graphene onto a flexible substrate, like polyethylene tereph-thalate (PET). By creating 30-inch graphene films, the research demonstrates the scalability of the roll-to-roll production process and reaches an important milestone. This capacity for large-scale production is essential for commercial applications because it makes it possible to fabricate transparent electrodes for mass-produced electronic devices at a reasonable cost. Overall, the work offers a scalable manufacturing technique that may completely transform the electronics sector and represents a major breakthrough in the manufacture of large-scale graphene films for transparent electrodes.

While graphene production technologies offer many advantages, there are still a number of obstacles that scientists and engineers are constantly trying to solve. Among these difficulties are the following:

Scalability: A variety of graphene production techniques, including CVD and mechanical exfoliation, have proven effective in yielding high-quality graphene in tiny quantities. It is still very difficult to scale up these procedures to generate graphene on an industrial scale while preserving cost-effectiveness, quality, and consistency.

Substrate compatibility: The electrical characteristics and adhesion strength of graphene can be greatly influenced by the substrate that is selected. It is still difficult to find appropriate substrates that work with large-scale production techniques and the intended applications.

Transfer techniques: Graphene is first generated on a metal substrate in a number of production techniques, including CVD, and then transferred onto a different substrate for device integration. Preserving graphene integrity during production requires the development of effective and dependable transfer methods that reduce contamination and flaws.

Uniformity and Quality Control: For many applications, such as composite materials and electronic devices, achieving uniformity in graphene films is essential. Maintaining constant and dependable performance in graphene layers requires careful control of the material's thickness, grain boundaries, and imperfections.

Cost: Graphene is still somewhat expensive as compared to other materials, despite great progress being made in lowering its production costs. The broad adoption of graphene-based devices depends on the development of efficient processes and cost-effective production techniques.

Environmental impact: The use of dangerous chemicals or excessive energy usage in several graphene production processes raises questions about how these procedures may affect the environment. The long-term viability of graphene technology depends on the development of more environmentally friendly and sustainable production techniques.

It will take multidisciplinary cooperation between researchers in materials science, chemistry, engineering, and other domains to tackle these problems. Resolving these issues and realizing the full potential of graphene-based technology will need ongoing developments in fabrication methods, materials synthesis, and process optimization.

11.6 Conclusion

This chapter offered a thorough summary of all the important factors that must be considered when designing and creating absorbers for biosensing applications. The substrate and resonator materials—which are crucial in defining the sensitivity and performance of biosensors—were the main topics of discussion.

The chapter explored the importance of choosing suitable substrate materials that provide desired attributes such stability, biocompatibility, and compatibility with production methods. As discussed, the selection of resonator materials has a substantial impact on the detection sensitivity and specificity of biosensors because these materials interact with biomolecules and analytes.

Recent developments in graphene production and its applications in biosensing were highlighted in particular. With its special qualities including high electrical conductivity, big surface area, and biocompatibility, graphene has a great deal of potential to improve biosensor performance. On the other hand, issues with cost-effectiveness, homogeneity, and scalability related to graphene manufacturing were also covered and must be addressed.

Additionally, the chapter covered a few newly released journal articles that investigate the use of graphene-based biosensors for a range of applications in food safety, environmental monitoring, and healthcare. These investigations demonstrate the encouraging developments and promise of graphene-based biosensors to transform the biosensing industry.

References

[1] *WHO Director-General's opening remarks at the media briefing on COVID-19—11 March 2020.* https://who.int/director-general/speeches/detail/who-director-general-s-opening-remarks-at-the-media-briefing-on-covid-19--11-march-2020 (accessed 31 December 2023)

[2] Sitharthan R *et al* 2022 Assessing nitrogen dioxide (NO_2) impact on health pre- and post-COVID-19 pandemic using IoT in India *Int. J. Pervasive Comput. Commun.* **18** 476–84

[3] Lu R *et al* 2020 Genomic characterisation and epidemiology of 2019 novel coronavirus: implications for virus origins and receptor binding *Lancet* **395** 565–74

[4] Dhanabalan S S *et al* 2022 Flexible compact system for wearable health monitoring applications *Comput. Electr. Eng.* **102** 108130

[5] Sridarshini T *et al* 2022 Current and future horizon of optics and photonics in environmental sustainability *Sustain. Comput. Inform. Syst.* **36** 100815

[6] Karthikeyan M *et al* 2023 Recent developments in flexible and printed reconfigurable antennas for medical and Internet of things applications *Adv. Flex. Print. Electron.* 9-1–9-25

[7] Kangeyan R and Karthikeyan M 2023 Implantable dual band semi-circular slotted patch with DGS antenna for biotelemetry applications *Microw. Opt. Technol. Lett.* **65** 225–30

[8] Kangeyan R and Karthikeyan M 2023 Miniaturized meander-line dual-band implantable antenna for biotelemetry applications *ETRI J.* **46** 413–420

[9] Kangeyan R and Karthikeyan M 2023 A novel wideband fractal-shaped MIMO antenna for brain and skin implantable biomedical applications *Int. J. Commun. Syst.* **36** e5509

[10] Karthikeyan M, Sitharthan R, Ali T, Pathan S, Anguera J and Shanmuga Sundar D 2022 Stacked T-shaped strips compact antenna for WLAN and WiMAX applications *Wirel. Pers. Commun.* **123** 1523–36

[11] Karthikeyan M, Sitharthan R, Ali T and Roy B 2020 Compact multiband CPW fed monopole antenna with square ring and T-shaped strips *Microw. Opt. Technol. Lett.* **62** 926–32

[12] Priyadharshini A S, Arvind C and Karthikeyan M 2023 Novel ENG metamaterial for gain enhancement of an off-set fed CPW concentric circle shaped patch antenna *Wirel. Pers. Commun.* **130** 2515–30

[13] Emaminejad H, Mir A and Farmani A 2021 Design and simulation of a novel tunable terahertz biosensor based on metamaterials for simultaneous monitoring of blood and urine components *Plasmonics* **16** 1537–48

[14] Zhang J *et al* 2021 Highly sensitive detection of malignant glioma cells using metamaterial-inspired THz biosensor based on electromagnetically induced transparency *Biosens. Bioelectron.* **185** 113241

[15] Tang M *et al* 2020 Rapid and label-free metamaterial-based biosensor for fatty acid detection with terahertz time-domain spectroscopy *Spectrochim. Acta A Mol. Biomol. Spectrosc.* **228** 117736

[16] Karthikeyan M, Jayabala P, Ramachandran S, Dhanabalan S S, Sivanesan T and Ponnusamy M 2022 Tunable optimal dual band metamaterial absorber for high sensitivity THz refractive index sensing *Nanomater* **12** 2693

[17] Behrouzi K and Lin L 2022 Gold nanoparticle based plasmonic sensing for the detection of SARS-CoV-2 nucleocapsid proteins *Biosens. Bioelectron.* **195** 113669

[18] Azab M Y, Hameed M F O, Nasr A M and Obayya S S A 2021 Highly sensitive metamaterial biosensor for cancer early detection *IEEE Sens. J.* **21** 7748–55

[19] Asghar Qureshi S, Zainal Abidin Z, A. Majid H, Ashyap A Y I and Hwang See C 2022 Double-Layered metamaterial resonator operating at millimetre wave for detection of dengue virus *AEU Int. J. Electron. Commun.* **146** 154134

[20] Zhang Y *et al* 2016 Broadband tunable graphene-based metamaterial absorber *Opt. Mater. Express* **6** 3036–44

[21] Farhat M *et al* 2013 A 3D tunable and multi-frequency graphene plasmonic cloak *Opt. Express* **21** 12592–603

[22] He X J *et al* 2014 Electrically tunable terahertz wave modulator based on complementary metamaterial and graphene *J. Appl. Phys.* **115** 17B903

[23] Nikolaenko A E *et al* 2012 Nonlinear graphene metamaterial *Appl. Phys. Lett.* **100** 181109

[24] Huang X, Ye W, Ran J, Zhou Z, Li R and Gao B 2023 Highly sensitive biosensor based on metamaterial absorber with an all-metal structure *IEEE Sens. J.* **23** 3573–80

[25] Niharika N and Singh S 2023 Highly sensitive tunable terahertz absorber for biosensing applications *Optik* **273** 170476

[26] Song C *et al* 2023 Dual-band/ultra-broadband switchable terahertz metamaterial absorber based on vanadium dioxide and graphene *Opt. Commun.* **530** 129027

[27] Xu W, Xie L and Ying Y 2023 Tunable transparent terahertz absorber for sensing and radiation warming *Carbon N. Y.* **214** 118376

[28] Jain P *et al* 2023 Machine learning assisted hepta band THz metamaterial absorber for biomedical applications *Sci. Rep.* **13** 1–12

[29] Devi P K and Singh K K 2023 A DFT studies on absorbing and sensing possibilities of glucose on graphene surface doped with Ag, Au, Cu, Ni & Pt atoms *Biosens. Bioelectron.* X **13** 100287

[30] Pan Y, Li Y, Chen F, Yang W and Wang B 2024 Black phosphorus–based metamaterial double-band anisotropic absorber for sensing application in terahertz frequency *Plasmonics* **19** 193–201

[31] Varshney G and Giri P 2021 Bipolar charge trapping for absorption enhancement in a graphene-based ultrathin dual-band terahertz biosensor *Nanoscale Adv.* **3** 5813–22

[32] Upender P and Kumar A 2023 Highly sensitive tunable dual-band THz refractive-based metamaterial sensor for biosensing applications *IEEE Trans. Plasma Sci.* **51** 3258–64

[33] Chen X and Fan W 2017 Ultrasensitive terahertz metamaterial sensor based on spoof surface plasmon *Sci. Rep.* **7** 1–8

[34] Yan X, Liang L-J, Ding X and Yao J-Q 2017 Solid analyte and aqueous solutions sensing based on a flexible terahertz dual-band metamaterial absorber *Opt. Eng.* **56** 027104

[35] Kim K S *et al* 2009 Large-scale pattern growth of graphene films for stretchable transparent electrodes *Nature* **457** 706–10

[36] Bae S *et al* 2010 Roll-to-roll production of 30-inch graphene films for transparent electrodes *Nat. Nanotechnol.* **5** 574–8

www.ingramcontent.com/pod-product-compliance
Lightning Source LLC
Chambersburg PA
CBHW080527220326
41599CB00032B/6230